ESSENTIALS OF HUMAN PHYSIOLOGY

ESSENTIALS OF HUMAN PHYSIOLOGY

UWE ACKERMANN, Ph.D

Professor of Physiology
University of Toronto
Toronto, Ontario, Canada

St. Louis Baltimore Boston Chicago London Philadelphia Sydney Toronto

Publisher: George Stamathis
Senior Managing Editor: Lynne Gery
Production Editor: Victoria Hoenigke

Printed in the United States of America

Mosby–Year Book, Inc.
11830 Westline Industrial Drive, St. Louis, MO 63146

Library of Congress Cataloging in Publication Data

Ackermann, Uwe.
Essentials of human physiology / Uwe Ackermann.
p. cm.
Includes index.
ISBN 1-55664-109-5
1. Human physiology—Outlines, syllabi, etc. I. Title.
[DNLM: 1. Physiology. QT 104 A182e]
QP41.A25 1992
DNLM/DLC
for Library of Congress

88-51913
CIP

92 93 94 95 96 CL/WA 9 8 7 6 5 4 3 2 1

PREFACE

As knowledge expands and textbooks get larger, our time to read them diminishes each year. Students most commonly express this dilemma by asking the question, "What am I really responsible for?"

Essentials of Human Physiology was written for those who need a framework for problem-solving or a quick reminder of the important points. I was thinking of senior students, stressed examination candidates, and curious practitioners in the life and health sciences.

Half of the information in each unit is presented in diagram form to allow ultra-quick review for some and to provide for others the substitute for a thousand words. The other half of the information is presented concisely, using bullets and short statements, so that even medical undergraduates should not need a highlighter pen to study from this book.

The content of *Essentials of Human Physiology* is similar to that of senior physiology texts that are now on the market, but differs from many by the inclusion of chapters on integrative physiology, fetal and perinatal physiology, the physiology of aging, and the pathophysiology of acid-base regulation.

Although physiology is, above all, an integrating science, I have attempted to write each chapter as a self-sufficient unit so that extensive cross-referencing is not necessary. When compromises needed to be made between scientific rigor and adherence to the essentials, I erred on the side of the harried student. Your teachers might not forgive me.

My students have helped me decide what must be included. Colleagues have corrected my errors. Dave Mazierski has put life into my sketches. Friends and family have endured my preoccupation. Walter Bailey of B. C. Decker has kept the faith.

Thank you.

Uwe Ackermann

CONTENTS

1

CELL PHYSIOLOGY

MEMBRANE TRANSPORT MECHANISMS

Cell Surface Membrane

This special barrier separates intracellular space from extracellular space, maintains their different compositions, and, thereby, maintains life.

Properties

1. Impermeable to proteins (except by special transport mechanisms)
2. Freely permeable to water
3. Highly permeable to K^+ (5×10^{-7} units)
4. Slightly permeable to Na^+ (5×10^{-9} units)
5. Slightly permeable to Cl^- (1×10^{-8} units)
6. Holds active transport mechanisms for many constituents

Functions

1. Separation of intracellular space from extracellular space
2. Regulation of transmembrane transport
3. Regulation of cell composition
4. Transmission of extracellular signals to intracellular space
5. Release of intracellular secretory materials to extracellular space
6. Recognition of other cells and initiation of appropriate response
7. Conduction of electrical signals to neighboring cells
8. Maintenance of physical shape of the cell

Passive Transport Mechanisms Across Cell Surface Membranes

Driving force

- The **difference in electrical potential** across the membrane (if the transported substance carries a charge) and/or
- The **difference in concentration (activity)** of transported substance **or in osmolality** on the two sides of the membrane*

Modulating influence

- The **permeability** (resistance, conductance) of the membrane

Active Transport Mechanisms Across Cell Surface Membranes

These **generally** (but do not necessarily) **work against an electrochemical gradient** and consume metabolic energy.

In the mammalian body there are mechanisms for active transport of many substances in specialized regions, but **the mechanism that transports Na^+ out of cells and K^+ into cells is ubiquitous.**

*Note that transport of water is commonly ascribed to differences in osmolality rather than to differences in the concentration of water. There is no conceptual difference between the two.

MEMBRANE TRANSPORT MECHANISMS

Differences Between Intracellular Space and Extracellular Space

Probably all constituents differ in concentration between intracellular and extracellular space. This skeleton shows the major components in the two spaces.

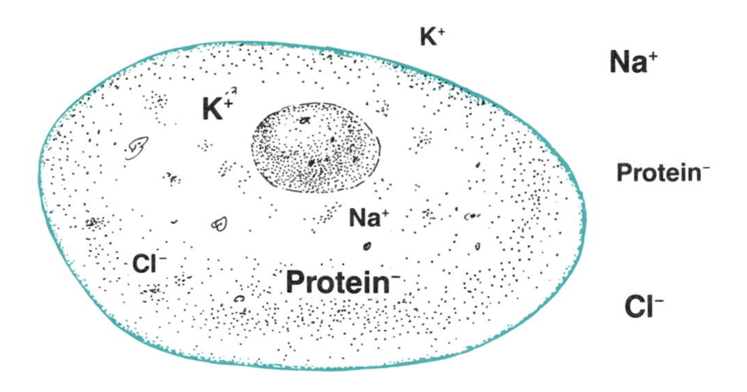

Structure of Cell Surface Membrane

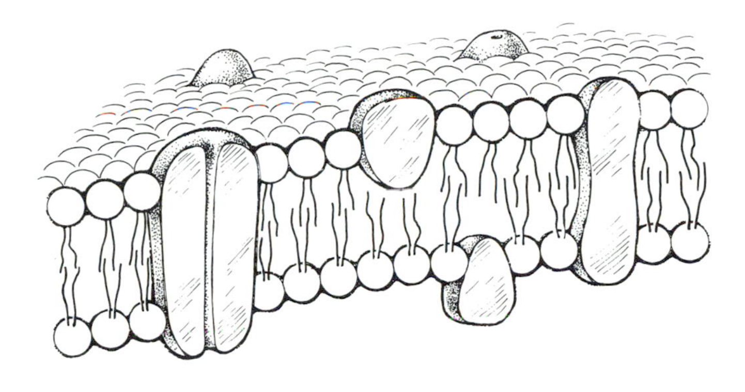

Passive Transport Mechanisms

- Passive transport mechanisms do not directly require metabolic energy, although such energy may have been used to create the driving force for the transport.
- The **amount** that is **transported** is determined by the product of **membrane permeability** and net **driving force.**

Active Transport Mechanisms

- Active transport mechanisms require the expenditure of metabolic energy.
- The **amount** that is **transported** is determined by **pump activity.**

MEMBRANE POTENTIALS

Balance of Forces Across Cell Surface Membranes

- Most ions are present at unequal concentrations across the cell surface membrane.
- The membrane is permeable to at least some of these ions and highly permeable to a few.

Why does each ion species not simply move from the region of high concentration to the region of low concentration until the concentration difference has been abolished?

- The movement of an ion down its concentration gradient results in the creation of an opposing gradient in electrical potential. As a result, **passive ion (net) transport stops when the force arising from the remaining concentration gradient is balanced by the opposing force arising from the gradient in electrical potential.**

Ion Equilibrium Potential

Definition

The **ion equilibrium potential** (E_{ion}), or the **Nernst potential,** of an ion is the electrical driving force that

- would be equal in magnitude to a concentration-gradient driving force;
- would prevent net passive transport of that ion species;
- would lead to a steady state with respect to the concentration gradient for that ion species.

Determination

E_{ion} can be measured directly only when there is but one ion species present. Therefore, E_{ion} is normally calculated from the existing concentrations of the ion species of interest and the valence of the ion.

Significance

If the calculated equilibrium potential for an ion is equal to the resting membrane potential (the voltage that can be measured across the surface membrane of the resting cell), **then it is likely that the steady-state distribution of the ion on the two sides of the cell membrane is determined by passive transport mechanisms only.**

If the calculated equilibrium potential for an ion across a cell surface membrane is different from the resting membrane potential of the cell, then active transport mechanisms are involved in maintaining the distribution of the ion across the surface membrane.

Resting Membrane Potential

Definition

The resting membrane potential of a cell is **the voltage that can be measured across the surface membrane of the resting cell.** It is not simply the algebraic sum of all ion equilibrium potentials, because that sum does not account for losses in potential resulting from the passage of each ion through the resistance of the membrane.

Determination

The resting membrane potential of a cell is usually determined by direct voltage measurement. However, it can be calculated with the help of the **Goldman-Hodgkin-Katz equation.**

MEMBRANE POTENTIALS

Balance of Forces Across Cell Surface Membranes

Concentration differences are maintained across the cell surface membrane even though it allows permeation of the major ions.

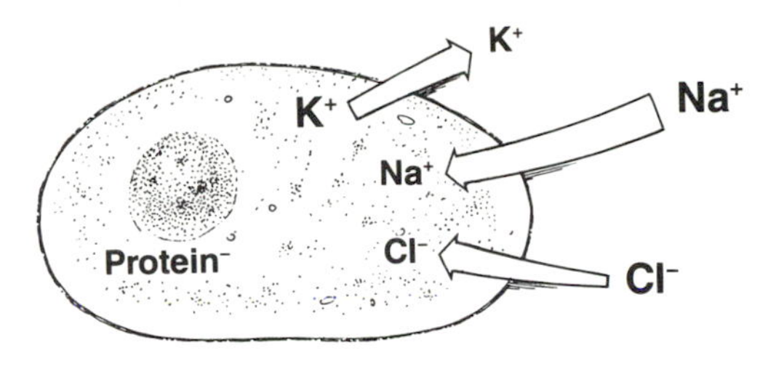

How?

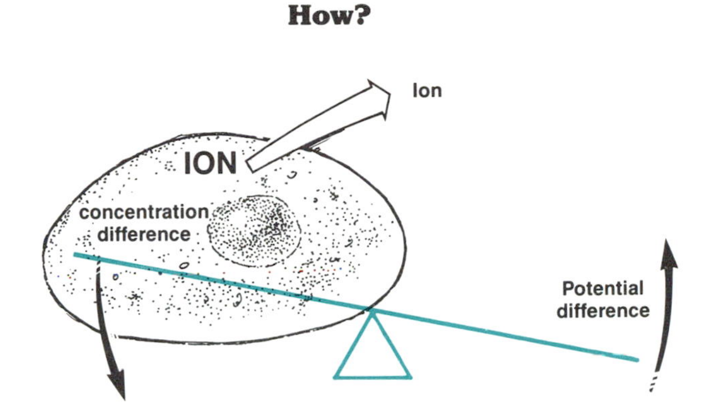

By balancing the force arising from a concentration gradient against the opposing force arising from the electrical gradient.

As an ion species moves across the membrane and its concentration gradient diminishes, the transmembrane charge gradient is enhanced.

At steady state, when there is no further net movement of ion,

$$E_{ion} = -\frac{61}{z}\log\frac{\text{Concentration inside}}{\text{Concentration outside}}$$

This is the **Nernst equation (z is the valence of the ion in question).**

Resting Membrane Potential

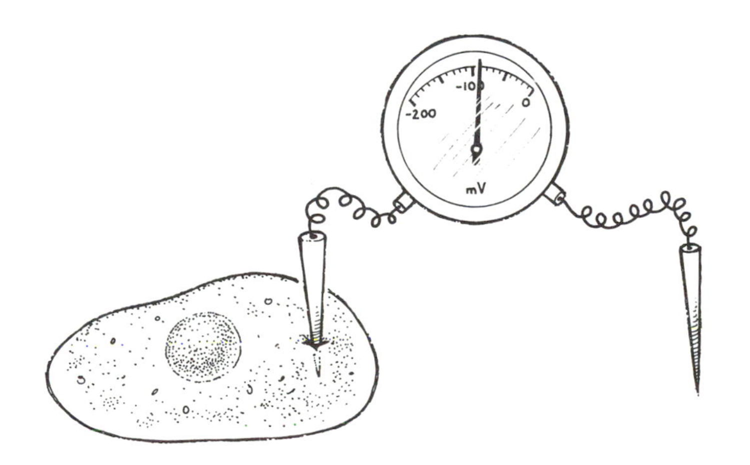

- Defined as the difference between intracellular potential and extracellular potential as measured with a voltmeter
- Normally about −90 mV inside vs. outside
- Can be calculated from the concentrations of all ions and the respective permeabilities of the cell membrane

$$E_m = 60\log\frac{P_K[K_o] + P_{Na}[Na_o] + P_{Cl}[Cl_i] + \cdots}{P_K[K_i] + P_{Na}[Na_i] + P_{Cl}[Cl_o] + \cdots}$$

Action Potential

Some cells (**excitable cells**) have membrane properties that allow their membrane potential to change temporarily from the stable resting level to a less negative (perhaps even a slightly positive) value.

Definition

An action potential is **a stimulus-triggered,** momentary **excursion in membrane potential from a threshold value to a peak value** (that is more positive than threshold) **and back to resting membrane potential.**

Sequence of events

Action potentials arise from instabilities in voltage-gated ion channels within the surface membrane. They are initiated only by **stimuli** that are strong enough to raise the membrane potential **to threshold.** At threshold the membrane permeability to Na^+ increases greatly and the following sequence is triggered:

1. **Na^+ ions enter the cell rapidly,** driven by the large concentration gradient between intracellular and extracellular space. **This produces the upstroke of the action potential.**
2. The rapid Na^+ influx would stop once the membrane potential reached E_{Na} (the Na^+ equilibrium potential; about +40 mV), but even before the membrane potential reaches this value, **open Na^+ channels spontaneously revert to the inactive state** and Na^+ permeability begins to return toward normal.
3. K^+ channels activate more slowly than Na^+ channels after threshold is reached. By the time a sufficient number of K^+ channels are open, Na^+ channels are already being inactivated. As a result, **K^+ permeability of the membrane reaches a peak when Na^+ permeability is decreasing.** At that point,
4. **K^+ ions leave the cell rapidly,** driven by the K^+ gradient. **This produces the downstroke of the action potential.** The outflow of K^+ ions stops when the membrane potential reaches the K^+ equilibrium potential (about −80 mV).

At the end of the action potential,

- **the membrane potential is back at its normal resting value;**
- the inside of the cell has **a slight excess of Na^+ ions and a slight deficit of K^+ ions.** (These ionic imbalances are so small that several hundred thousand action potentials could be generated before the cell would run low on K^+.)

The Na^+-K^+ pump restores ionic balances in the intervals between action potentials.

Action Potential

After an adequate stimulus, the membrane potential (E_m) of an excitable cell typically shows this behavior.

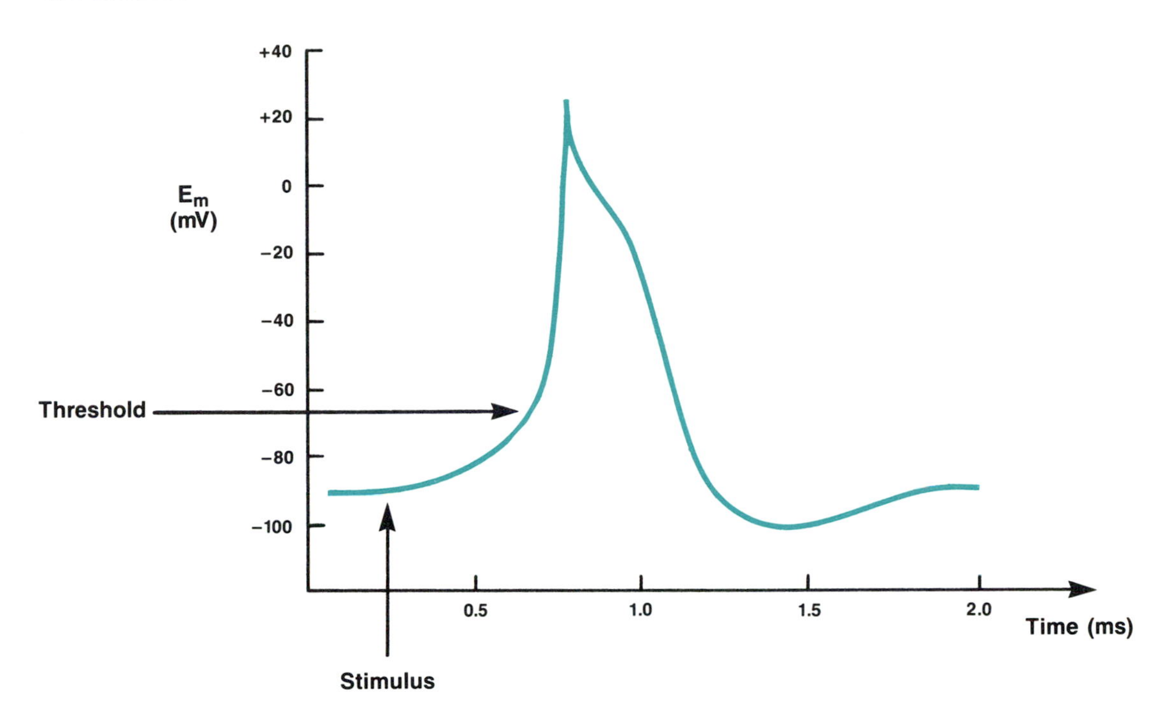

2

MUSCLE PHYSIOLOGY

SKELETAL MUSCLE

OVERVIEW

Muscle is **excitable, contractile tissue.** It is classified, on the basis of its microscopic appearance, as **striated** or **smooth** muscle. The difference in appearance is due to a difference in the physical arrangement of the contractile proteins.

SKELETAL MUSCLE ULTRASTRUCTURE

Muscle function depends on proximity of actin binding sites (abs) to heavy meromyosin (hmm) projections from the thick filaments.

- **Spontaneous coupling of heavy meromyosin to actin would occur if the actin binding sites were not blocked by tropomyosin.**

This blockade is removed only by a change in the conformation of the regulator protein, troponin.

SKELETAL MUSCLE

OVERVIEW

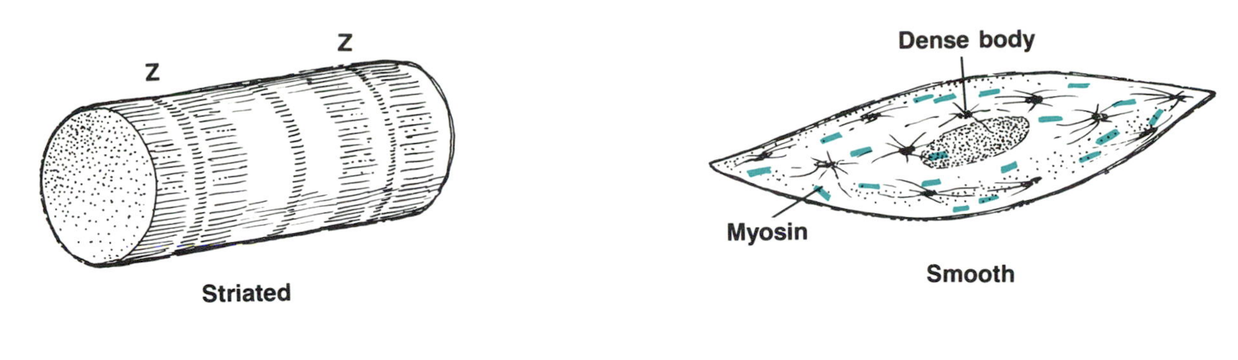

Striated		Smooth
Skeletal Muscle	**Cardiac Muscle**	
• Excitable • Contractile	• Excitable • Contractile • Automatic • Conductive	• Excitable • Contractile • Automatic (some) • Conductive

SKELETAL MUSCLE ULTRASTRUCTURE

The proteins **actin** and **myosin,** arranged in a repeating, interdigitating pattern, form the structural skeleton of **thin and thick filaments.**

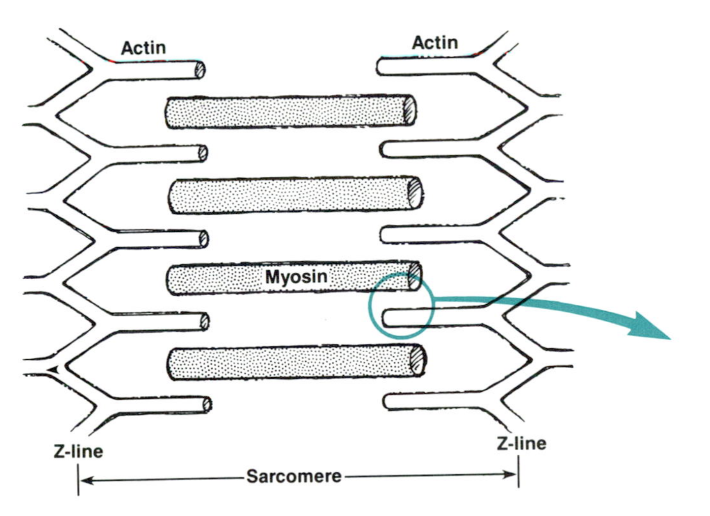

- In the region of overlap the **heavy meromyosin** (hmm) portion of myosin molecules **projects toward specific binding sites** (abs) on the actin filament.
- **Attachment of hmm to abs is prevented by tropomyosin.**
- Tropomyosin is held in place by **troponin,** a three-part molecule. A portion of it (troponin T) has **high affinity for tropomyosin;** another portion (troponin C) has **high affinity for Ca^{++};** and another portion (troponin I) is an **inhibitor of ATPase.**

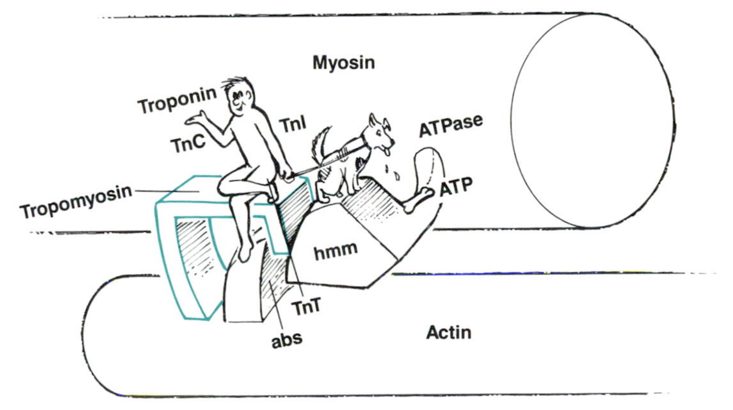

SLIDING FILAMENT THEORY OF MUSCLE FUNCTION

Removal of steric hindrance

When the intracellular concentration of Ca^{++} exceeds 10^{-6} M, sufficient Ca^{++} ions are available to interact with troponin C. This causes a change in the conformation of the troponin-tropomyosin complex, and

- the actin binding site (abs) is exposed to heavy meromyosin (hmm);
- abs and hmm couple and form a **cross bridge** between myosin and actin;
- **ATPase is no longer inhibited.**

Rotation of cross bridges

Hydrolysis of ATP releases energy that is required for **cross bridge rotation.**

◆ The extent of the motion obtained with one cross bridge rotation is only 50 to 100 Å. A sarcomere shortens by 1,000 to 3,000 Å during contraction. To achieve this degree of shortening, **repeated release and reattachment of cross bridges** are necessary.

Release of cross bridges

Cross bridges release when a new molecule of ATP is brought to the region of the abs-hmm complex.

- The released hmm then attaches to an adjacent abs that has been exposed by the action of a Ca^{++} ion.

◆ **If ATP is not available, the cross bridges do not release, the muscle remains contracted, and rigor mortis sets in.**

EXCITATION-CONTRACTION COUPLING

Excitation arises from muscle action potentials that spread over the sarcolemma and penetrate to the interior of the fibers along the membranes of the T-tubules.

◆ The depolarization spreads from the T-tubule to the terminal cisternae of the longitudinal sarcoplasmic reticulum (lsr) and causes the release of a burst of Ca^{++} from the lsr. This burst is sufficient to raise intracellular [Ca^{++}] well above the minimum needed for contraction.

Removal of steric hindrance

Interaction of Ca^{++} with troponin C removes the tropomyosin "shield" from the actin binding site and **allows a cross bridge to form.**

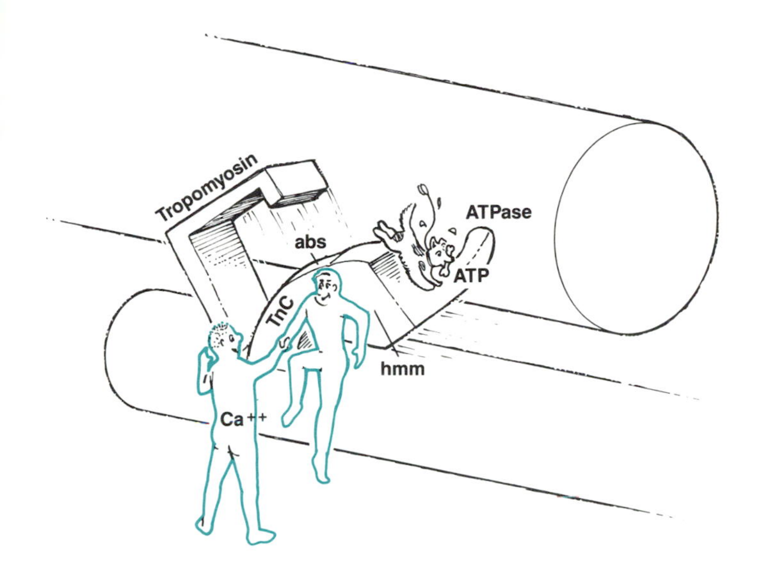

Rotation of cross bridges

- **ATP is hydrolyzed** (1), yielding **energy** (2).
- Energy is used to **rotate the myosin cross bridge** (3) and, thus, produce **motion** of the actin filament relative to the myosin filament (4).

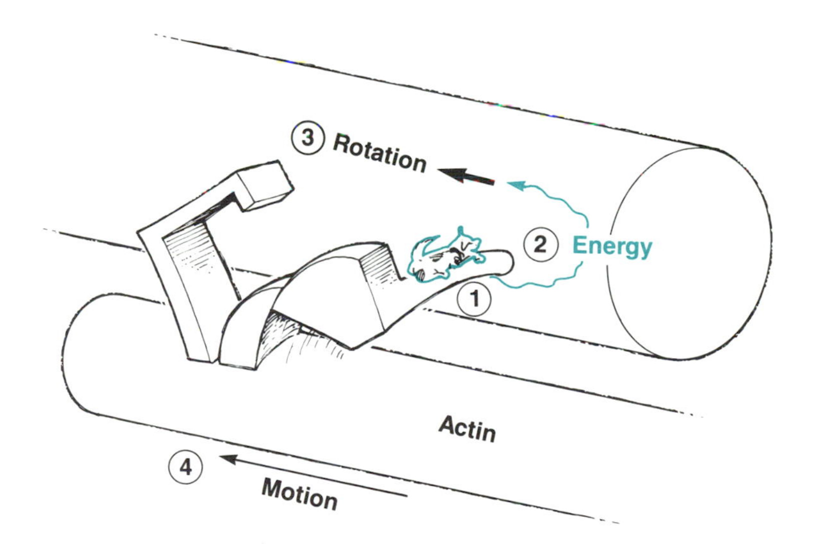

ORGANIZATION OF SKELETAL MUSCLE

Motor Units

The fibers in a whole skeletal muscle are arranged in functional groupings called **motor units.**

- **A motor unit consists of one motor nerve plus all the muscle fibers innervated by it.**
  - Motor units differ in size (many fibers or few fibers) and in type of muscle fiber **(fast, slow, or intermediate).**
  - Size is anatomically determined. **Fiber type is determined by motor unit function.**

Types of Skeletal Muscle Fibers

- **All fibers in a motor unit are of the same type, and the type is determined by the nature of the action potential activity in the motor nerve.**
  - Nerve activity that is chronically in a **phasic** manner and at a **high frequency** (more than 40 per second) leads to synthesis of **fast myosin** and the formation of **few mitochondria.**
  - Nerve activity that is chronically **tonic** and at a **low frequency** favors synthesis of **slow myosin** and the formation of **many mitochondria.**

Fast fibers

These are characterized by **fast myosin, few mitochondria, and predominance of the glycolytic pathway in metabolism** (little ATP, but produced very quickly).

Slow fibers

These are characterized by **slow myosin, many mitochondria, and predominance of the oxidative pathway in metabolism** (six times more ATP, but produced more slowly).

- **Conversion of fibers to the other type does not occur under physiologic conditions, but is possible in experimental settings.**

Types of Skeletal Muscle

Although all of the fibers in a given motor unit are of the same type, **any given region of muscle will show considerable anatomic intermixing of fibers from different motor units.** As a result,

- most muscles contain both fiber types as well as an intermediate type;
- **the proportion of fibers is determined by the nature of the muscle activity.**

Type I (red muscle)

This type contains mostly slow fibers. (Muscles controlling posture are red muscle.)

Type II (white muscle)

This type contains mostly fast fibers. (Muscles controlling eye movement are white muscle.)

ORGANIZATION OF SKELETAL MUSCLE

Motor Unit

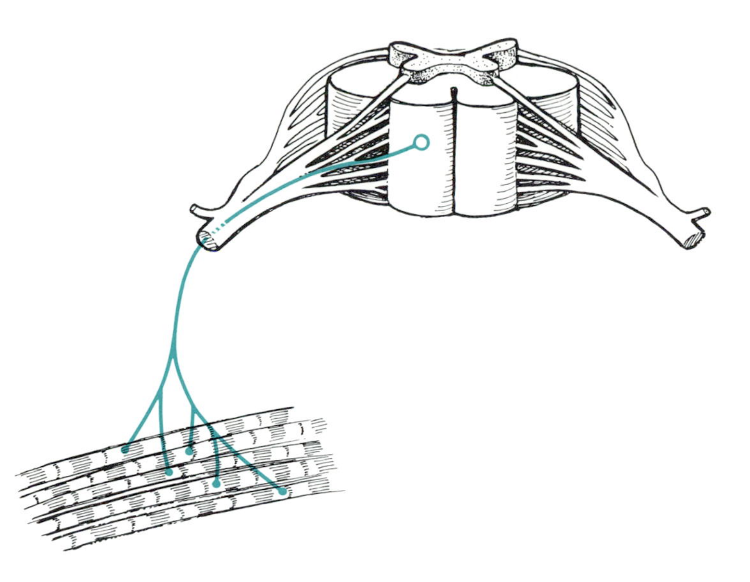

Types of Skeletal Muscle Fibers

Fast Fibers

- large motor nerve, therefore fast conduction
- glycolytic pathway:

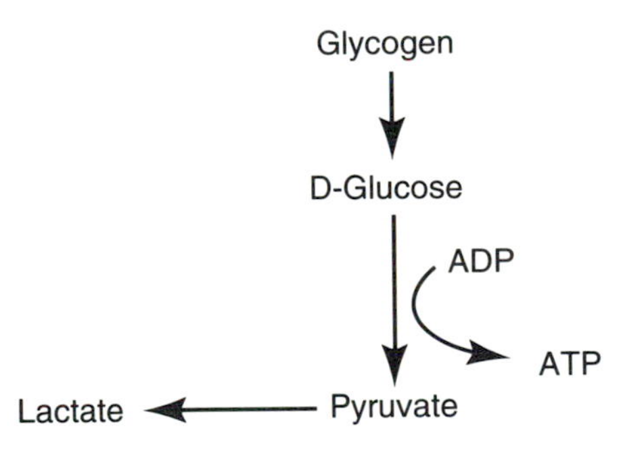

Slow Fibers

- small motor nerve, therefore slow conduction
- oxicative pathway:

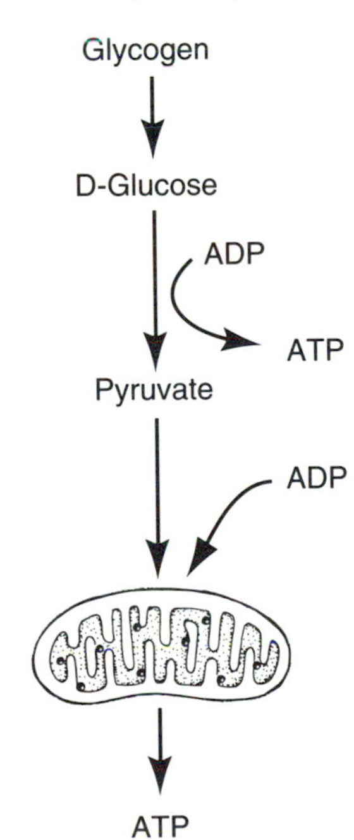

REGULATION OF SKELETAL MUSCLE CONTRACTION

Nerve Input to Motor Units

- Skeletal muscle **contraction** occurs **only in response to action potentials in motor nerves.**
- When the motor nerve fires, then **all of the fibers in that motor unit contract synchronously.**
  - An extensive **neuronal network ensures (1) integration of activity** with neighboring motor units and **(2) grading of activity** relative to the desired force.

Grading of Contractile Force

Force of contraction in a muscle is changed by two mechanisms:

1. Changing the number of active motor units **(recruitment)**
2. Changing the frequency of action potentials in motor nerves **(temporal summation)**

In addition, in any one active fiber, maximal contractile force is affected by **initial stretch** (preload, or length-tension relationship) and by the **biochemical environment** (contractility, or degree of activation).

Recruitment

Small motor units are activated first because their neurons are small and, therefore, reach the critical number of total ion flux before it is reached in larger neurons.

- The interplay between local muscle stretch receptors and input from higher central nervous centers determines whether the generated force is sufficient for the task or if activity is required from additional motor units.

Temporal summation

Skeletal muscle has a short cycle of electrical activity ($\approx$ 10 to 80 ms) compared to its cycle of mechanical activity ($\approx$ 160 to 250 ms). As a result,

- **it is possible to stimulate a motor unit before the force generated by the preceding stimulus has returned to zero.**

If the rate of stimulation is so high that there is no time for any relaxation between neighboring action potentials, then the muscle is said to be in a state of **tetanization** and is developing the **maximum force possible under the existing biochemical conditions.**

REGULATION OF SKELETAL MUSCLE CONTRACTION

Nerve Input to Motor Units

The frequency of motor unit action potentials is determined by a variety of inputs:

- From the central nervous system (1)
- From muscle receptor afferents synapsing in the spinal cord with the motor neuron (2)
- From receptor afferents of other muscles (3) and from other neurons in the spinal cord (4)

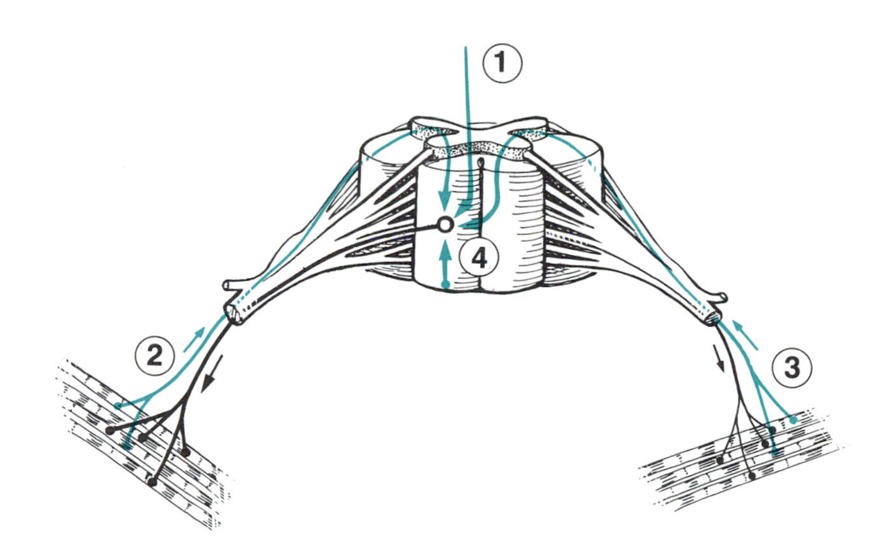

Grading of Contractile Force

Recruitment

The **force** generated by a muscle during contraction **is directly proportional to the number of contracting motor units.**

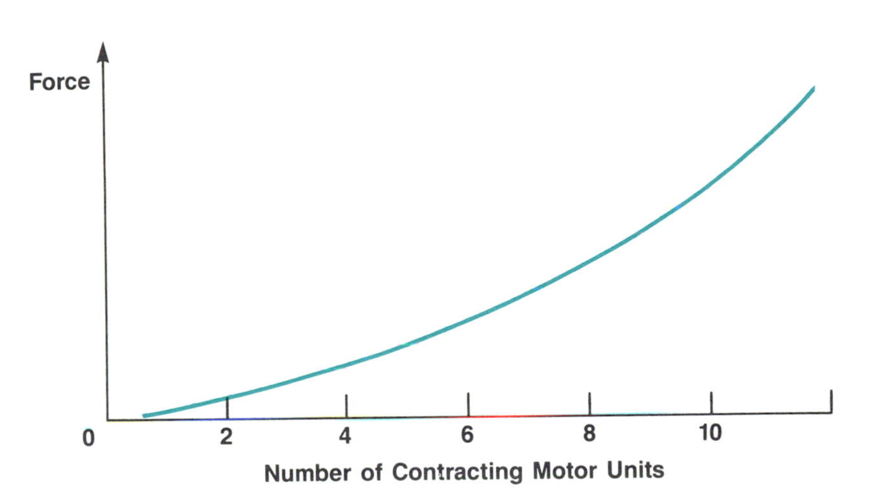

Temporal summation

- A single stimulus delivered to a muscle will yield a single muscle twitch of a certain peak tension.
- **If a subsequent contraction is initiated before the muscle has returned to its resting state, then the force developed by this contraction is added to the force remaining from the preceding contraction** and
- with continuing stimulation, a steady state will be reached at a higher average force than that observed with a single twitch.

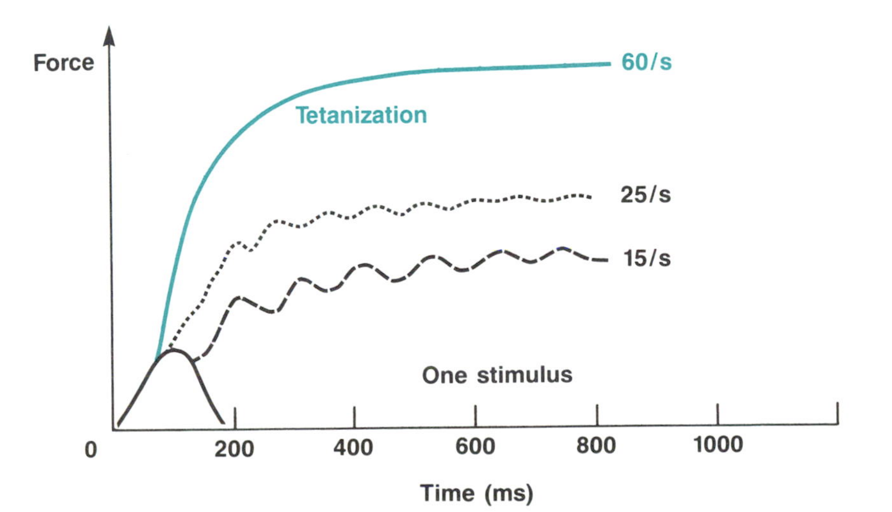

ASSESSMENT OF SKELETAL MUSCLE FUNCTION

The purpose of muscle is to develop force and to change length.

Although it is easy to determine maximum muscle strength, it is difficult to assess the relationships among biochemical changes and their mechanical correlates. Thus, **special techniques are needed to determine such indices of function as magnitude and rate of force development or velocity of shortening.**

Isolated Muscle

In experiments on isolated muscle it is possible to separate the dynamics of **tension development** from the dynamics of **length changes.**

Isometric function

When muscle is fixed rigidly at both ends so that it cannot shorten, then stimulation will be followed by the **maximum tension development that is possible under the existing preload, contractility, and rate of stimulation.**

- The pattern of change in tension with time after a single stimulus is called a **muscle twitch.** Its total duration is about 10 ms in a fast fiber or 100 ms in a slow fiber.

Isotonic function

When muscle is suspended in such a way that during contraction it will move a weight, W, then it is possible to study **shortening capabilities while contracting at a constant tension.**

- A plot of tension vs. time gives limited information in this setting.
- Plots of force vs. velocity of shortening give more information.
- **The performance of muscle can be characterized by v_{max}, the maximum velocity of shortening at zero load.**

ASSESSMENT OF SKELETAL MUSCLE FUNCTION

Isolated Muscle

Isometric function

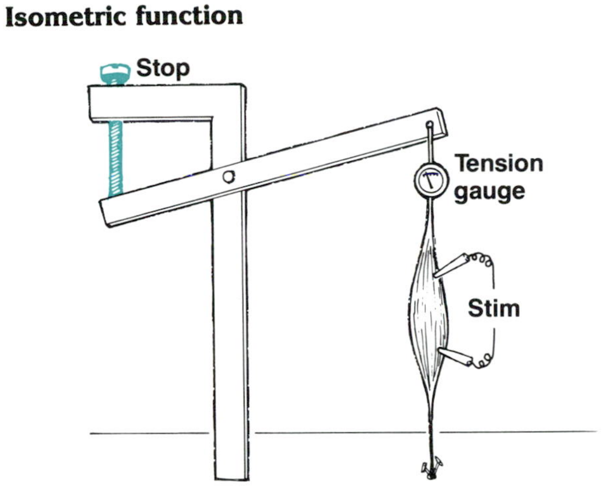

Isotonic function

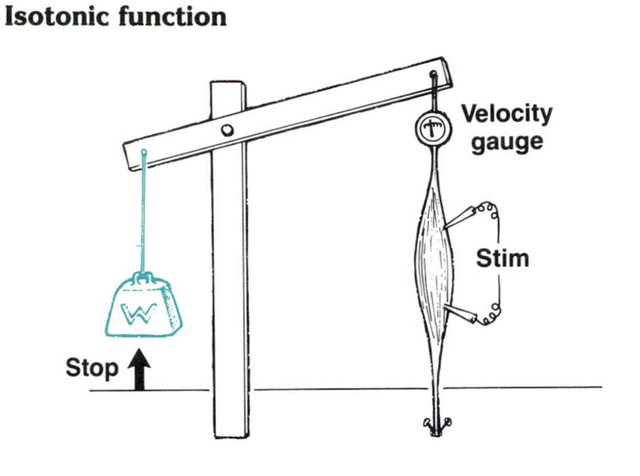

Tension-time plot (muscle twitch)

The performance of a muscle under different conditions of preload, contractility, or rate of stimulation can be characterized by several measurable features of the muscle twitch:

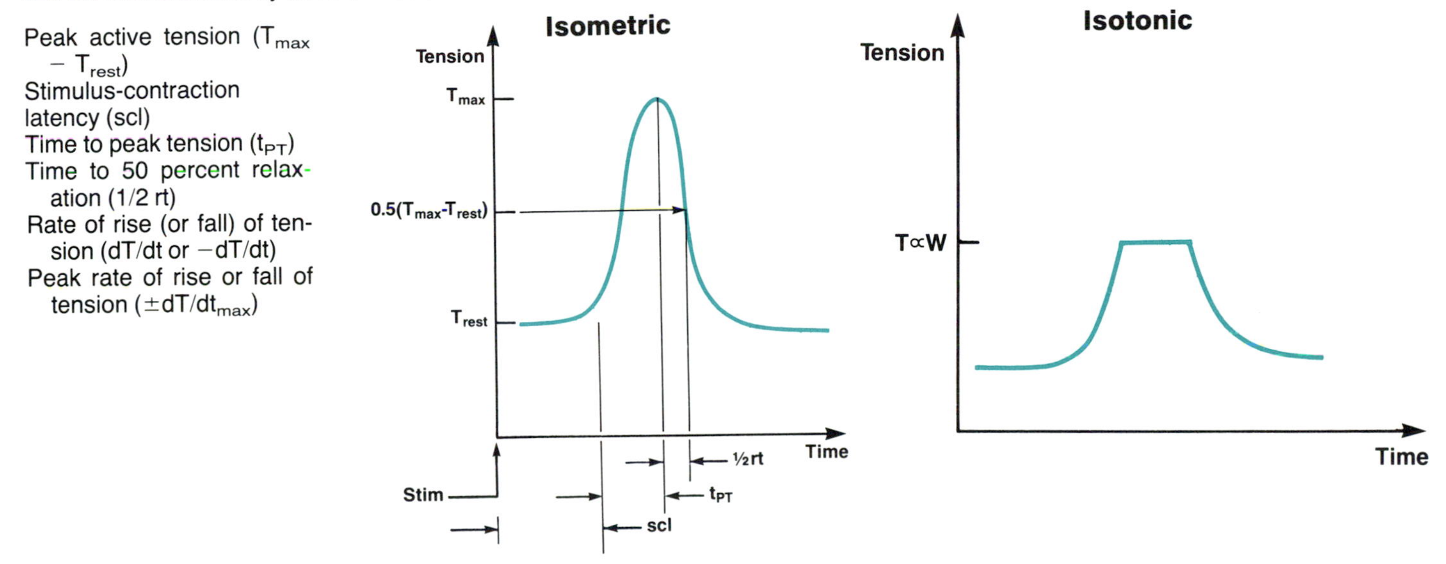

Force-velocity plot

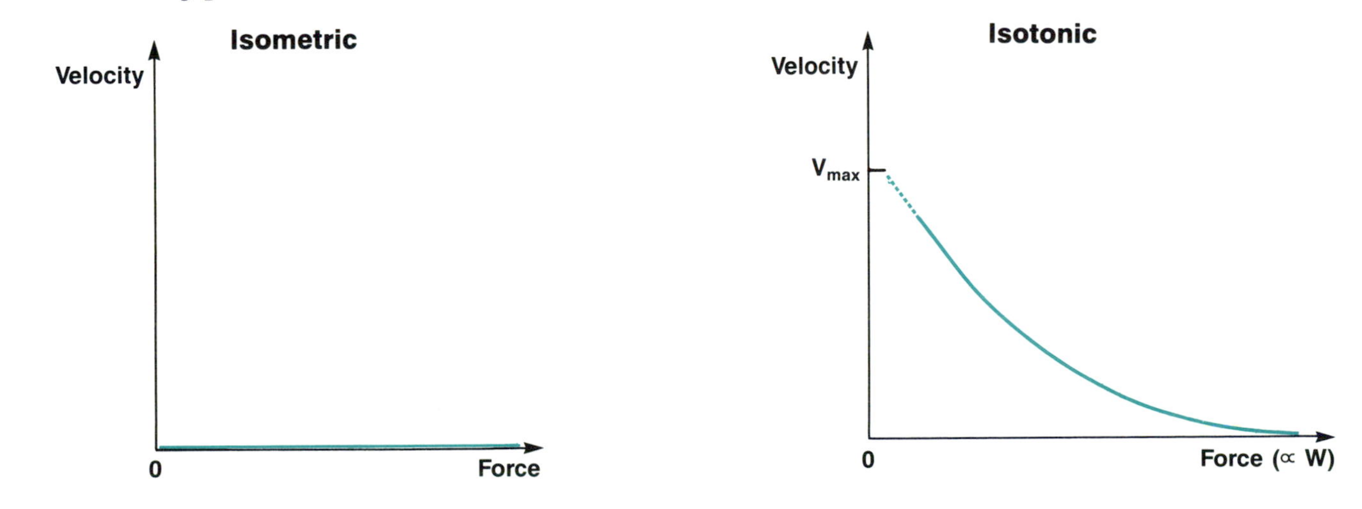

ASSESSMENT OF SKELETAL MUSCLE FUNCTION—CONT'D.

Whole Muscle In Situ

Qualitative information about muscle function can be obtained by recording electrical activity with a pair of electrodes placed on the surface of the muscle. The record thus obtained consists of motor unit action potential trains and is called an **electromyogram** (emg).

- **Changes in the amplitude** of emg deflections **are directly related to changes in firing frequency of active motor units** in the muscle.
- **Changes in the time interval between neighboring pulses** in the emg **are inversely related to changes in the number of active motor units** in the muscle.

These relationships were determined empirically and cannot easily be derived from basic principles.

ASSESSMENT OF SKELETAL MUSCLE FUNCTION—CONT'D.

Whole Muscle In Situ

Electromyography

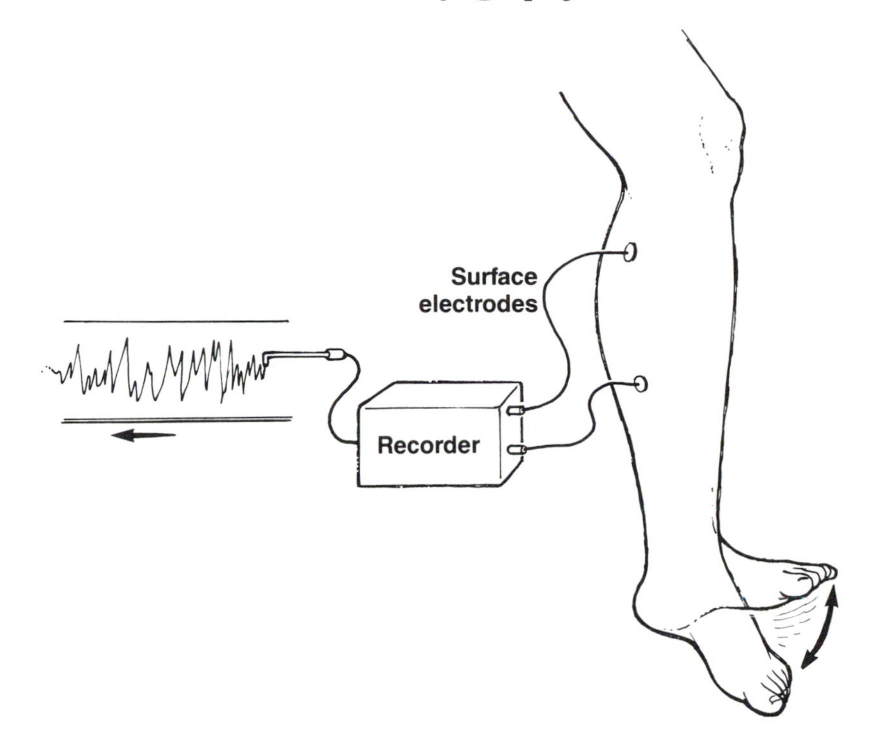

EFFECTS OF EXERCISE AND TRAINING

Most athletic endeavors involve three aspects: coordination, strength, and endurance. Training improves all three, or it can accentuate only certain aspects.

Coordination

With training, exercise-specific coordination improves in the central nervous system and leads to

- **better coordination of action potentials** in all involved muscles and
- **greater inhibition of antagonist muscles.**

The result is a **reduction of "wasteful" muscle activity** and an **improvement in the efficiency** of exercise performance.

Strength

Pure strength training consists of brief, maximal efforts involving all motor units in the affected muscle.

- High tissue pressure associated with the intense effort tends to compress blood vessels in the muscle and, thereby, reduce oxygen supply. As a result,
- **strength training favors fast, glycolytic motor units.**

The prominent **adaptive response** of muscle to such training is **hypertrophy.** This is especially noticeable with isometric exercises.

Endurance

Endurance training consists of sustained, submaximal efforts. Such activity leads to anatomic changes such as

- **mild hypertrophy,**
- **increased capillarization, and**
- **moderate thickening of connective tissue sheaths.**

- **The major effect of endurance training on muscle is to increase the muscle's capacity for metabolic activity. This is seen as an increase in the number of mitochondria as well as increased stores of oxidative enzymes.**

Training and Constancy of Muscle Fiber Types

Despite the frequent use of motor units, **conversion of slow units to fast units does not occur.** The explanation may lie in the changes of the action potentials with training:

- their average duration increases gradually with training, but
- they show a progressive diminution in their average frequency.

In experiments in which conversion was accomplished, it occurred after maintained elevation of nerve action potential frequency.

EFFECTS OF EXERCISE AND TRAINING

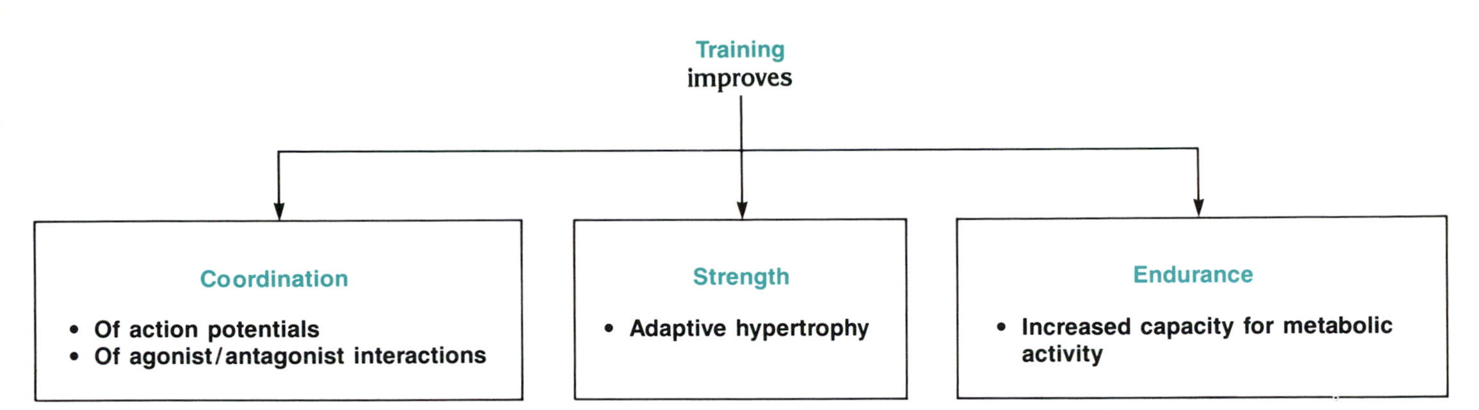

SMOOTH MUSCLE

COMPARISON AND CONTRAST WITH SKELETAL MUSCLE

Smooth muscle differs from skeletal muscle in ultrastructure, in control of function, and in endurance, but not in its mechanism of function (which is the sliding filament mode of actin-myosin interaction).

SMOOTH MUSCLE ULTRASTRUCTURE

- Actin filaments are arranged in groups, each of which shows **several actin filaments attached, like spokes, to a dense body.**
- The dense bodies are either attached to one another by structural proteins or attached to the cell membrane.
- Myosin is interspersed among actin filaments in such a way that each myosin filament is surrounded by a rosette of about 15 actin filaments.

EXCITATION-CONTRACTION COUPLING

Sources of Ca^{++}

Ca^{++} is required, just as it is in skeletal muscle, but

- **extracellular sources of Ca^{++} are more important in smooth muscle** because of the sparse longitudinal sarcoplasmic reticulum.

Extracellular sources

Entry of extracellular Ca^{++} can occur by one of two mechanisms:

1. Smooth muscle cells that are capable of generating action potentials (i.e., unitary smooth muscle) do so by permitting the influx of Ca^{++} through **voltage-sensitive Ca^{++} channels** once the membrane potential exceeds threshold (during both the upstroke and the "shoulder" phase of the action potential).
2. **Transmitter substances** such as acetylcholine bind with a specific membrane receptor and **cause Ca^{++} channels to open.** (The resulting influx of Ca^{++} can elicit a contraction and can also change the membrane potential, but this change may be too small to generate an action potential.) Thus, **action potentials are not necessary for smooth muscle contraction to occur,** provided that intracellular [Ca^{++}] is high enough.

Intracellular sources

Hormones may bind to specific membrane receptors, activating an intracellular enzyme that stimulates the formation of a **second messenger** such as cAMP or inositol triphosphate (IP_3). The **second messenger,** in turn, **can promote Ca^{++} release** from sarcoplasmic reticulum (if it is present) or from other intracellular organelles.

COMPARISON AND CONTRAST WITH SKELETAL MUSCLE

	Smooth Muscle	Skeletal Muscle
Cell size	2–100 μm	60 μm to several cm
Actin, myosin, tropomyosin	Yes	Yes
Troponin	No	Yes
Ultrastructure	Dense bodies and rosettes	Myofibrils and sarcomeres
T tubules	No	Yes
Longitudinal sarcoplasmic reticulum	Absent or poorly developed	Well developed
Innervation	Autonomic	Somatic
Receptors	Acetylcholine, norepinephrine, and others	Acetylcholine only
Receptor location	Throughout	At motor endplate only
Automaticity	Yes, in unitary smooth muscle	No
Response to passive stretch	Active tension in unitary smooth muscle	None
Upstroke of action potential	Ca^{++} influx	Na^{+} influx
Duration of action potential	30–300 ms	About 10 ms
Maximum duration of maintained tension	Years	Minutes

EXCITATION-CONTRACTION COUPLING

Entry of Ca^{++}

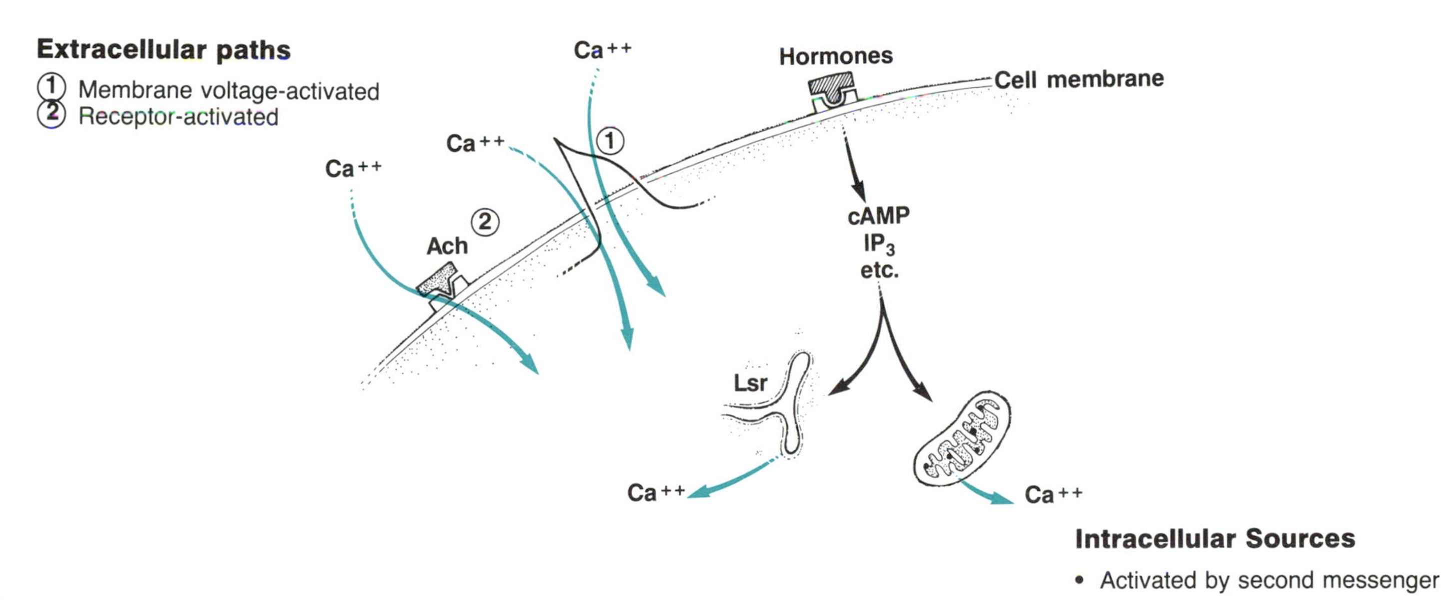

Effect of Ca^{++}

The ATPase inhibition that is exercised by troponin in skeletal muscle is exercised by **calmodulin** in smooth muscle.

- **Calmodulin binds Ca^{++} and, thereby, releases the inhibition on ATPase.** The subsequent steps toward a contraction are presumed to be identical to those described earlier for skeletal muscle.

Effect of Ca^{++}

Calmodulin inhibits ATPase.

Calmodulin
Myosin
Tropomyosin
ATPase
ATP
hmm
abs
Actin

- Calmodulin binds Ca^{++}.
- Coupling of Ca^{++} to calmodulin permits ATPase to act.

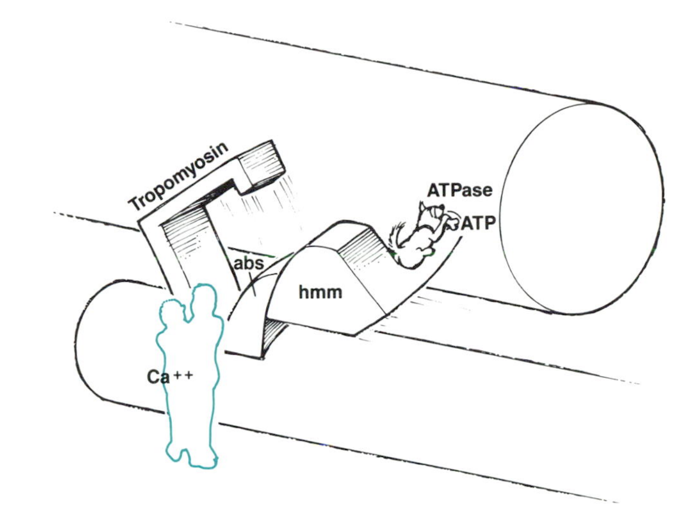

SMOOTH MUSCLE CONTRACTION

Types of Smooth Muscle

Unitary

Cells are arranged in **sheets** with multiple **gap junctions** for cell-to-cell ion transport.

Multiunit

Cells are arranged as **discrete muscle fibers,** each fiber **innervated by an individual nerve.** Electrical conductance between neighboring fibers is poor.

Factors Influencing Contraction

Spontaneous electrical activity

Spontaneous ***slow wave* fluctuations** of membrane potential **are a feature of unitary smooth muscle.** When these slow wave potentials reach **threshold,** then one or more spontaneous **action potentials** are generated, spread over neighboring cells, and cause a **contraction.**

Physical stretch

Stretching of unitary smooth muscle cells **makes the average membrane potential less negative** and, thereby, **increases excitability.**

- The resulting increase in the number of spontaneous action potentials leads to increased spontaneous muscle contraction.
- This automatic constrictor response of some smooth muscle to increased stretch is called the **myogenic reflex.**

Influence of nerves

Nerves **modulate activity in unitary smooth muscle, but initiate activity in multiunit smooth muscle.**

- Nerve influence is exerted by **transmitter substances** that are released from the nerve endings and by specific smooth-muscle-cell **membrane receptors** with which the transmitters interact.
- The dominant smooth muscle neurotransmitters are **acetylcholine** and **norepinephrine.**

Neurotransmitters, action potentials, and contraction

When acetylcholine and norepinephrine bind with their respective receptors,

- they **change membrane permeability to Ca^{++};**
- they also **change membrane potential;** but
- they **do not necessarily generate a muscle action potential.** (Whether or not a muscle action potential is generated depends only on whether the number of Ca^{++} ions needed for contraction is great enough to have moved the muscle cell membrane to its threshold for an action potential.)
- The final mechanical effect of each is determined by the type of receptor with which it interacts:
  - In some smooth muscle, acetylcholine is excitatory and norepinephrine is inhibitory.
  - In other smooth muscle, their effects are reversed.

Influence of chemicals

Local tissue factors such as P_{O_2}, P_{CO_2}, $[H^+]$, and blood-borne hormones **influence smooth muscle activity.**

- The mechanisms of action of tissue factors in this regard are not yet known.

SMOOTH MUSCLE CONTRACTION

Types of Smooth Muscle

Unitary

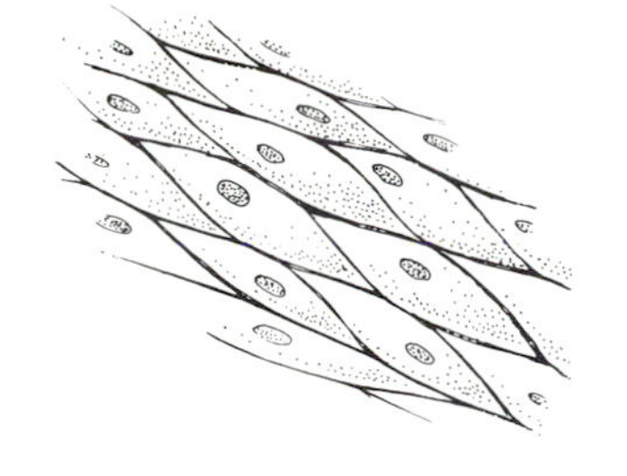

Multiunit

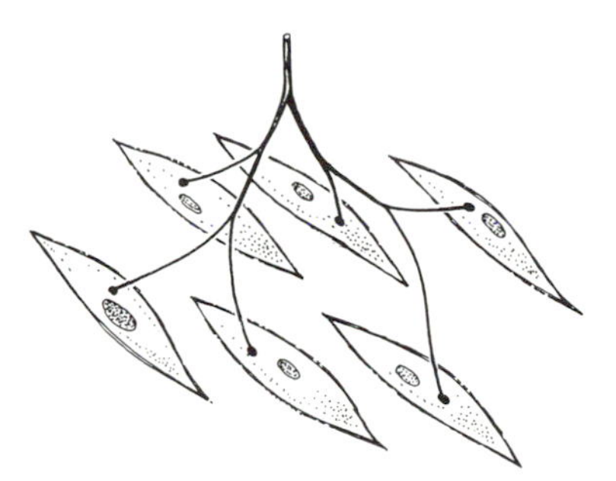

Factors Influencing Contraction

Spontaneous electrical activity

Unitary smooth muscle shows spontaneous periodic fluctuations of membrane potential **(slow waves).**

Occasionally the slow wave potentials reach threshold (about −35 mV) and then one or more action potentials are generated.

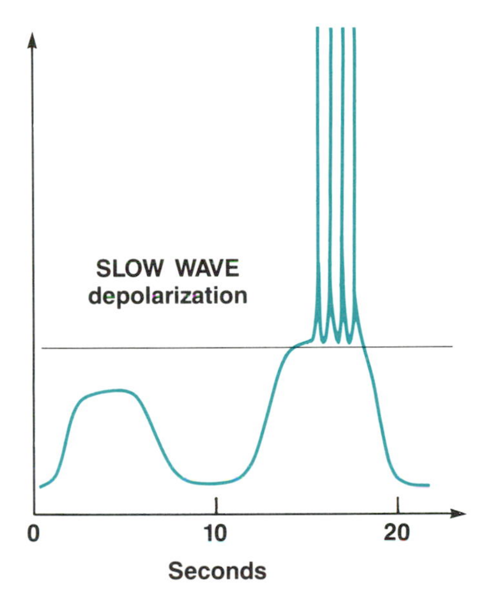

Influence of nerves

The ultimate effect depends on **receptor type.** Two patterns predominate:

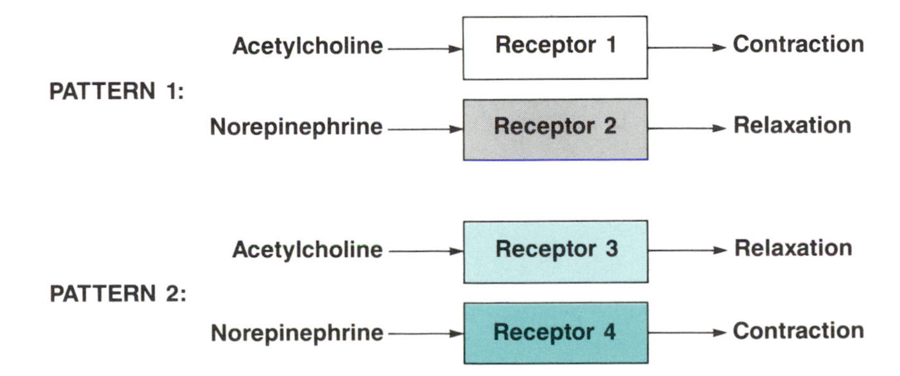

3 BLOOD

BLOOD AND PLASMA

OVERVIEW

Blood, a suspension of cells in plasma, is **a portion of the extracellular fluid volume.** It is confined to special transport channels, the **blood vessels,** and it has three major functions:

1. Transport
2. Communication
3. Organism preservation

These are performed by the **formed elements** (erythrocytes, leukocytes, and platelets) and by the **plasma.**

Formed Elements

Erythrocytes (red blood cells)

These cells are the primary oxygen transport system of the body.

Leukocytes (white blood cells)

These cells are broadly classified on the basis of histological appearance or affinity for certain dyes.

Granulocytes

- Neutrophils are the primary **defense against bacterial infection** and the primary **mediator of inflammatory responses.**
- Eosinophils } participate in allergic reactions.
- Basophils }

Monocytes

- Modulate **immune responses**
- **Scavenge** dead cells or denatured protein

Lymphocytes

- Mediators of **immune responses**

Platelets

The major function of these cells is the formation of hemostatic plugs after blood vessel injury.

Plasma

The fluid component of blood consists of **water** (90 percent), **minerals** (3 percent), and **proteins** (7 percent).

The major minerals are Na^+, Cl^-, and HCO_3^-. Na^+ and Cl^- provide osmotic equilibrium across the cell membrane; HCO_3^- is a major regulator of the H^+ concentration in body fluids.

Plasma proteins

The plasma protein constituents are **albumin** (4.5 g/dL), several **globulins** (2.5 g/dL), and **fibrinogen** (0.3 g/dL). Most are synthesized by the liver. They have **five major functions:**

1. They serve as **carriers** for hormones, trace metals, or drugs.
2. They serve as **proteolytic agents** in the cleavage of various hormonal and enzymatic precursors.
3. They serve as **inhibitors** for a variety of interstitial and plasma proteases.
4. They provide the **colloid osmotic pressure** in plasma.
5. They provide the **humoral immunity** portion of the immune system.

BLOOD AND PLASMA

Functions of Blood

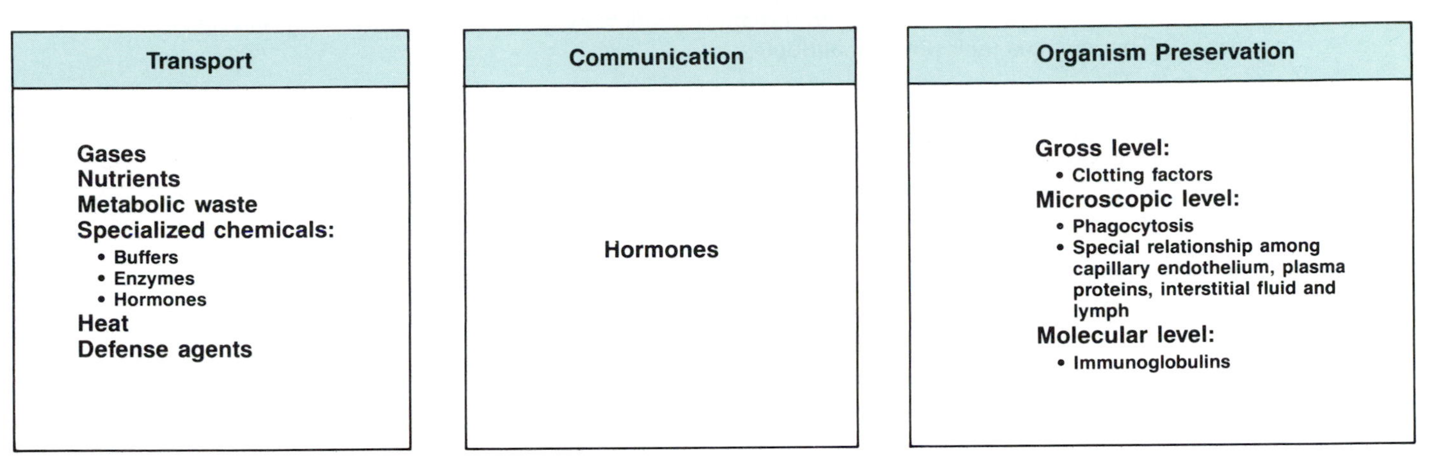

Transport
Gases
Nutrients
Metabolic waste
Specialized chemicals:
• Buffers
• Enzymes
• Hormones
Heat
Defense agents
Communication
Hormones
Organism Preservation
Gross level:
• Clotting factors
Microscopic level:
• Phagocytosis
• Special relationship among capillary endothelium, plasma proteins, interstitial fluid and lymph
Molecular level:
• Immunoglobulins

Composition of Blood

	% by volume	Differential Count (% of leukocytes)
Erythrocytes	42–48	
Leukocytes	<1	
Granulocytes		
Neutrophils		60–70
Eosinophils		2–4
Basophils		0.5–1
Monocytes		3–8
Lymphocytes		20–25
Platelets	<1	
Plasma	52–58	

Plasma Proteins

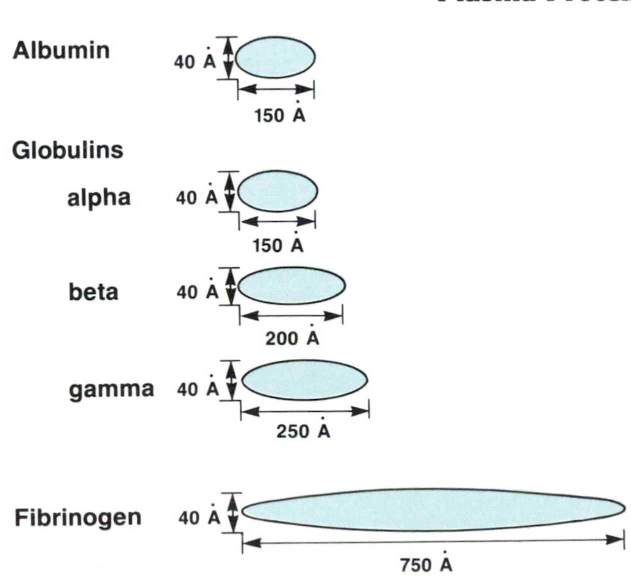

The plasma concentration of selected proteins can vary greatly in specific acute states such as tissue injury or infection, but, on the whole, plasma protein concentration is well controlled. The mechanisms of this control are not yet understood.

ERYTHROCYTES (RED BLOOD CELLS)

OVERVIEW

The **function** of erythrocytes is to **transport respiratory gases** between the lung and other tissues. This dictates their physical attributes:

- They are a 3-g/dL solution of **hemoglobin** (the major O_2 transporting protein) plus some **carbonic anhydrase** (a catalyst for CO_2 carriage).
- Their **exterior membrane** is **pliable.** They have no mitochondria, **no structural elements,** and no nuclei (in humans). These aspects give them remarkable **ductility** and allow them to recover again and again from deformations that occur as they are squeezed through capillaries.
- Their biconcave shape is conducive to surface exchange phenomena because it gives them **maximum surface area for a given volume.**

Red Cell Formation (Erythropoiesis) and Destruction

Prerequisites

In addition to the general need for vitamin B_{12} and folic acid to permit DNA synthesis for cell division, red cell formation **(erythropoiesis)** requires four important factors:

1. A stimulus
2. The presence of a normal, growth-promoting environment in the bone marrow
3. The presence of erythropoietin
4. The presence of iron

Stem cell conversion

Influences from local bone marrow cells guide the conversion of pluripotential stem cells to cells that are committed to the erythroid series (erythroid progenitor cells).

- **Erythroid progenitor cells** are **responsive** only **to erythropoietin.**
- Erythropoietin causes the formation of **erythroblasts,** the earliest morphologically recognizable members of the erythroid series.
- **Erythroblasts** are large cells whose prominent feature is a nucleus that occupies most of the cell interior. They **divide three to five times over a period of 6 to 8 days and mature into** the much smaller, nucleus-free **reticulocytes** that are released into the circulation.
- Within 24 hours the reticulocytes lose their trace amounts of RNA, and they then become true erythrocytes.

Hemoglobin (Hb)

Hb synthesis is the most significant intracellular event that takes place during erythrocyte formation. Hb binds O_2 and CO_2 reversibly, being responsible for most O_2 that is carried by blood and for about 25 percent of CO_2. (The remaining 75 percent is carried in the form of dissolved H_2CO_3 or HCO_3^-.)

Life cycle

Erythrocytes have a **life span of about 120 days,** all of them spent within the lumen of blood vessels. Aging cells undergo membrane changes that allow mononuclear phagocytes in the marrow, liver, and spleen to recognize and remove the deteriorating cells.

- Heme is dissociated from the globin portion and is oxidized. This separates Fe from the pigment portion.
- Transferrin, the iron-transporting protein, carries the iron back to the committed stem cells for incorporation into new erythrocytes.
- The pigment portion of heme is reduced to **bilirubin** and is excreted via bile into the gastrointestinal tract, giving stool its characteristic brown color.
- The globin chains are broken down into their amino acids and released to the body pool of amino acids.

ERYTHROCYTES (RED BLOOD CELLS)

Red Cell Formation (Erythropoiesis)

In the adult human about 200×10^9 red cells are formed each day in the marrow of the ribs, sternum, vertebrae, and pelvis. They, as well as the platelets and most leukocytes, originate from a single type of primitive cell, the **stem cell.**

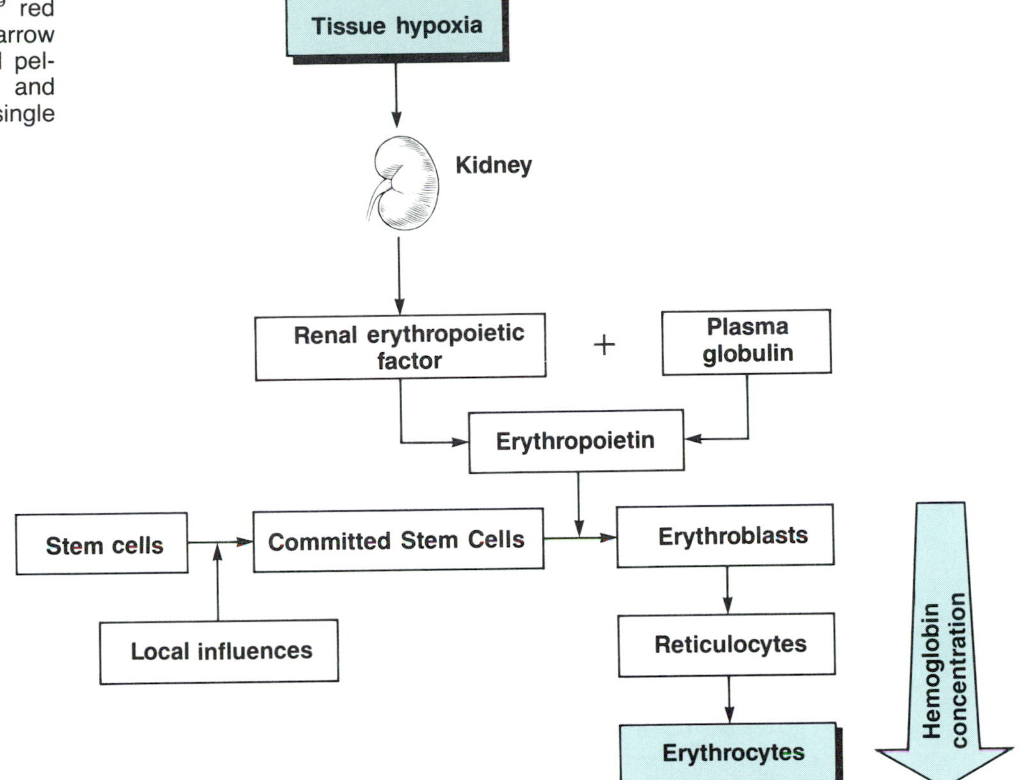

Hemoglobin (β-chain) and Oxygen Transport

Hemoglobin (Hb) consists of two pairs of chains and contains polypeptide portions (globin) as well as an iron-containing pigment (heme). Globin is an intertwined string of amino acids, and the heme ring is suspended between neighboring globin loops.

- **O_2 is carried between the iron of the heme moiety and a neighboring histidine residue.**
- **CO_2 is carried as a carbamino group at the NH_2 terminal.**

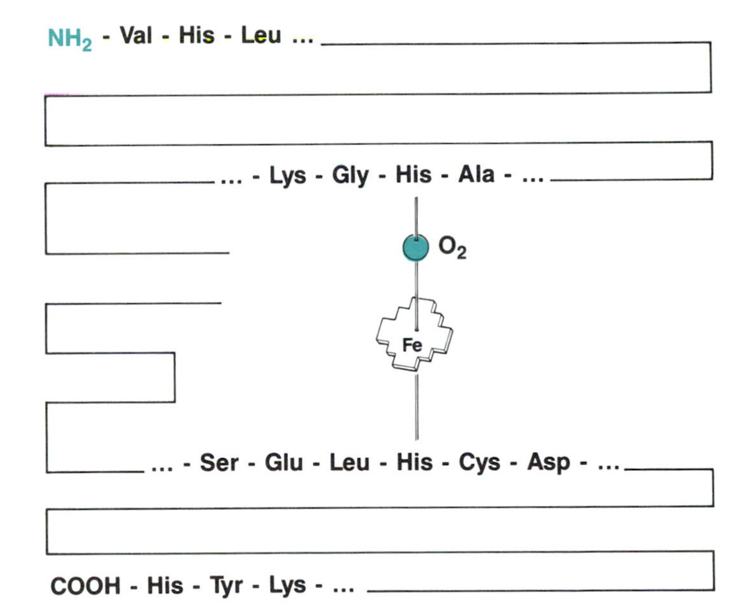

Hemoglobin and Carbon Dioxide Transport

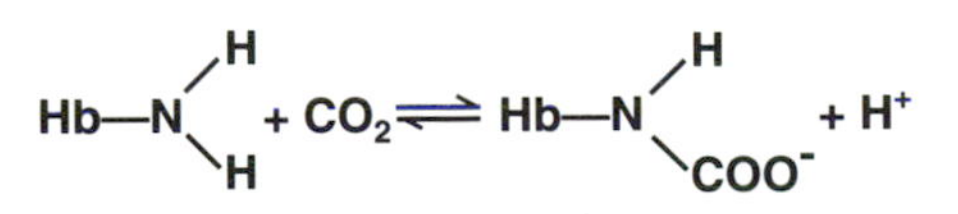

$$\text{Hb—N}\begin{smallmatrix}\text{H}\\ \text{H}\end{smallmatrix} + CO_2 \rightleftharpoons \text{Hb—N}\begin{smallmatrix}\text{H}\\ \text{COO}^-\end{smallmatrix} + H^+$$

GRANULOCYTES AND MONOCYTES

OVERVIEW

Groups of **leukocytes** (white blood cells) differ by their histological properties (microscopic appearance and affinity for certain stains), but they all have **one major function:** they are a **rapid and specific defense mechanism against infectious molecular agents and micro-organisms.**

Leukocyte Formation

Under the influence of unknown circulating "steering" factors, **stem cells** can be diverted from **self-renewal** toward a **committed line of progenitor cells.** These differ from stem cells in two ways:

1. Progenitor cells have lost the capacity for self-renewal.
2. Progenitor cells are not pluripotential, but are committed to producing (under the proper growth conditions) daughter cells of a particular type.

During maturation each cell line acquires distinctive properties such as

- packaging of enzymes into intracellular granules (in granulocytes);
- a contractile system that will permit locomotion (in granulocytes);
- incorporation of surface membrane receptors and surface membrane antigens.

Granulocytes and monocytes form the bulk of the leukocyte population.

Granulocytes

These mature in the bone marrow and arrive in the tissues fully differentiated for specific phagocytic tasks. They contain sophisticated mechanisms by which they can **develop rapidly from harmless circulating intravascular cells to specific phagocytic cells and killers of bacteria.** These mechanisms include

- an internal **microtubular skeleton** against which contractions can occur;
- a system of cytoplasmic **contractile proteins;**
- a system of **membrane receptors** for chemoattractive substances, for activated complement, or for immunoglobulins;
- systems of **primary granules** and **secondary (specific) granules.**

Primary granules

contain enzymes for the digestion of proteins and sugars as well as enzymes that catalyze scavenger actions of oxide radicals.

Secondary granules

contain agents that act in more specific ways rather than by general tissue destruction (e.g., agents that compete with bacteria for specific vital substances).

Monocytes

Monocytes form a pool of stationary or mobile **macrophages** in the tissues and continue to differentiate there. It is most likely that **local factors determine whether some macrophages specialize in antigen processing, others clear the blood of denatured proteins, and still others phagocytose micro-organisms.**

GRANULOCYTES AND MONOCYTES

Leukocyte Formation

1. Leukocytes are formed in the **bone marrow** from pluripotential **stem cells.**
2. Stem cells can either produce daughter stem cells **(self-renewal)** *or* they can differentiate during a few cell divisions into committed stem cells **(progenitor cells)** that give rise to specific blood cells.
3. Progenitor cells divide to form **blast cells,** the earliest morphologically recognizable precursors of the different end cells.

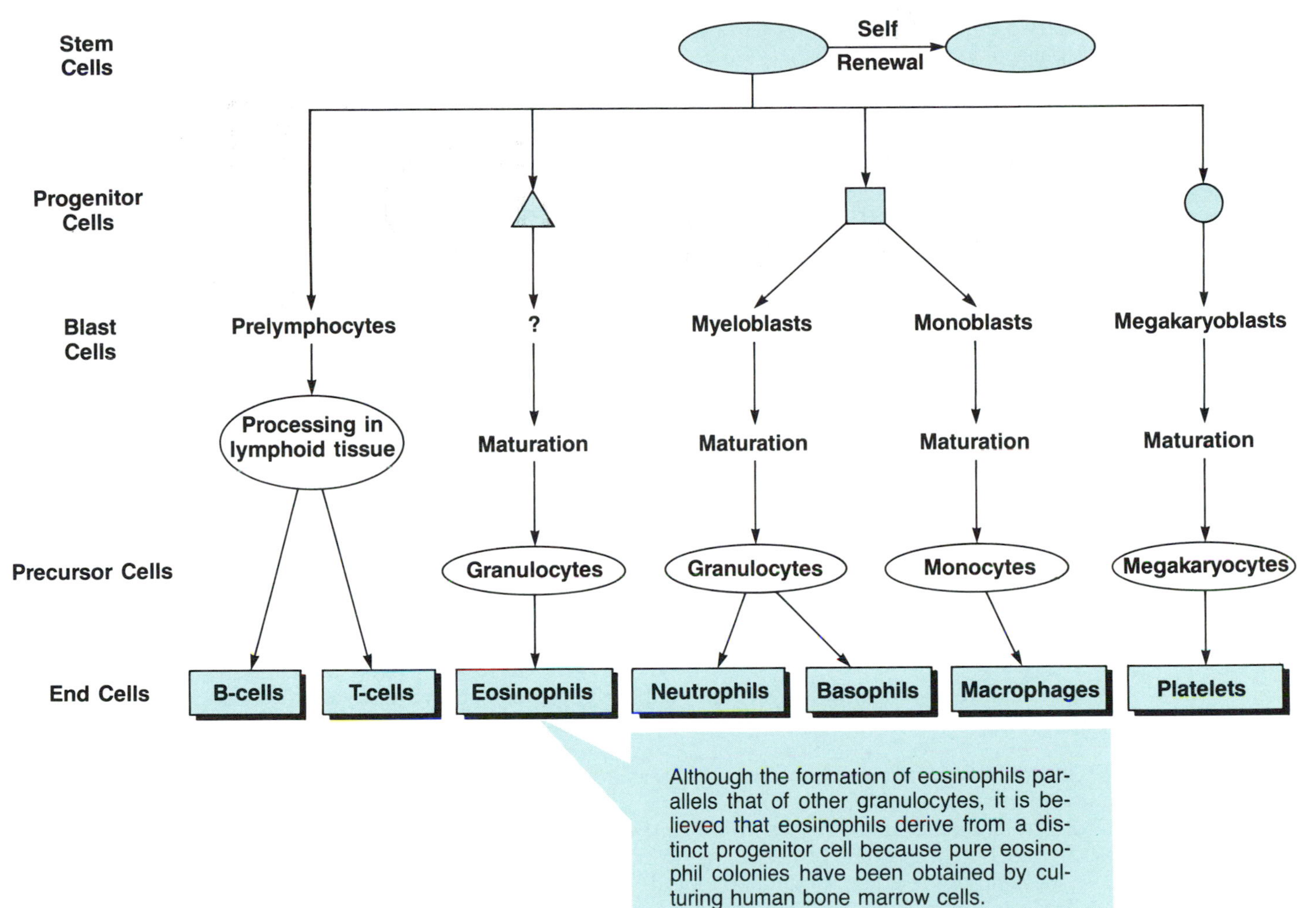

PHAGOCYTOSIS AND INFLAMMATION

Phagocytosis is **a process of immobilization, ingestion, and digestion of foreign agents by granulocytes and monocytes.** A vital first step in phagocytosis is the activation of the **complement** system.

Complement Activation

About 20 normally dormant, proteolytic plasma proteins make up the complement system. Two among them, named C1 and C3, can, via separate pathways, activate the system to full, cascading activity.

The classical pathway

The sequence of this path **begins with C1.** C1 activation is precipitated by physical attachment of C1 molecular filaments to antibody "bridges" engulfing a "foreign" membrane. The conformational change in C1 allows progression to proteolytic activity.

The alternative pathway

This sequence follows C3 activation. C3 molecules are sufficiently labile that a few are always spontaneously activated. If the activated few encounter a surface that has particular structural or immunological features, then the full cascade proceeds.

Both pathways result in the formation of a **membrane adherence complex.** This complex may either cause membrane lysis directly or act as a coupler between phagocytic cells and the membrane to be attacked. Such coupling is called **opsonization.**

Granulocyte Responses

In acute inflammation, granulocytes, accumulating locally in response to chemoattractants, form the initial defense. Macrophages follow later.

Chemoattraction

Chemoattractive agents interact with surface receptors on granulocytes to cause four effects:

1. Granulocyte accumulation in and egress from the microcirculation of the affected region
2. Emigration of granulocytes from blood vessels toward the noxious agent
3. Attachment of granulocytes to the opsonized surface of the agent (if it is large) or engulfing of the agent within a phagocytic vacuole
4. Fusion of granules with the opsonized membrane and release of granule contents

Lysis

Granulocytes release lysosomal proteases. They also cause the release of arachidonic acid (AA). AA oxidation yields leukotrienes, prostaglandins, and superoxide radicals.

Leukotrienes

- result from the oxidation of AA via the lypoxygenase pathway.

They are **potent inducers of leukocyte aggregation and chemotaxis.**

Prostaglandins and superoxide radicals (O_2^-)

- result from the oxidation of AA via the cyclooxygenase pathway.

Prostaglandins may act as local vasodilators, but their chief role may be to act as a local brake on the release of proteases. **O_2^-** quickly forms two metabolites, hydrogen peroxide H_2O_2) and hydroxyl radical (OH*). These have little inflammatory or bactericidal activity by themselves, but can react with other substances to form effective destroyers of bacteria.

PHAGOCYTOSIS AND INFLAMMATION

Complement Activation

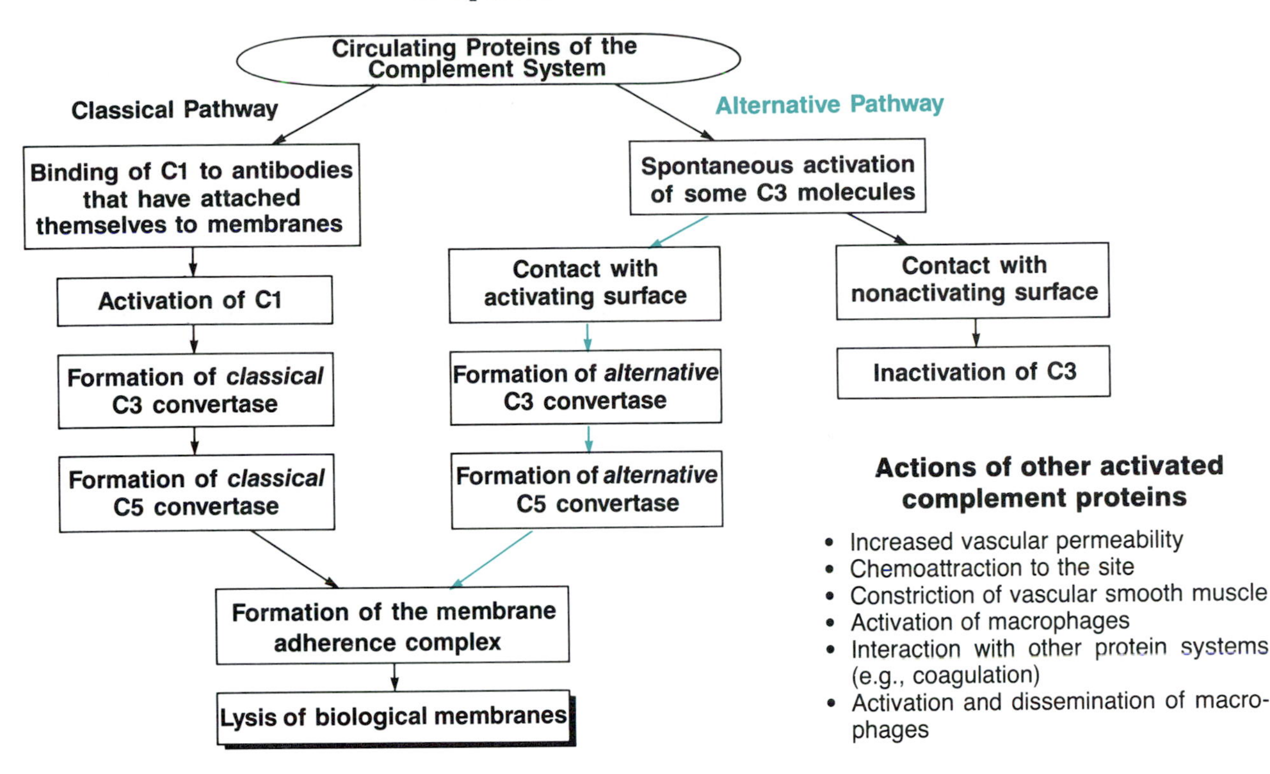

Actions of other activated complement proteins

- Increased vascular permeability
- Chemoattraction to the site
- Constriction of vascular smooth muscle
- Activation of macrophages
- Interaction with other protein systems (e.g., coagulation)
- Activation and dissemination of macrophages

Granulocyte Responses

1. ACCUMULATION and DIAPEDESIS

   Initiated by effects of chemoattractants on specific granules. Caused by release of a specific promoter.

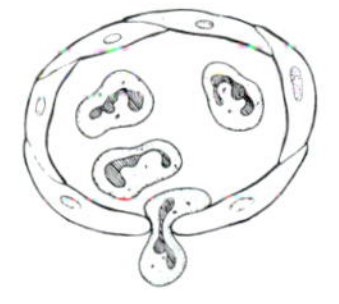

2. AMEBOID MOTION towards the noxious agent (CHEMOTAXIS)

   Initiated by low concentrations of chemoattractants.

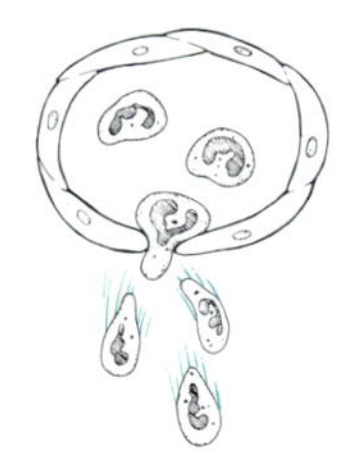

3. MEMBRANE ATTACHMENT

   Initiated by high concentrations of chemoattractants.

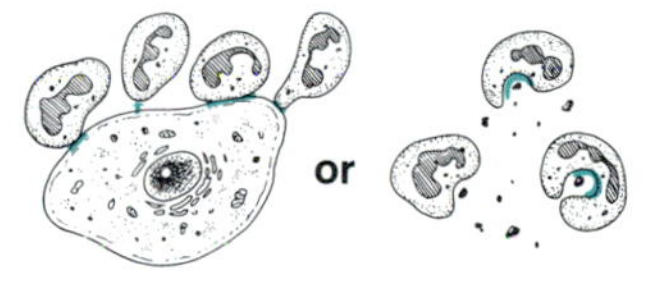

4. MEMBRANE FUSION and LYSIS

   Lysis occurs partly as a result of direct destruction by lysosomal proteases released from granules and partly as a result of surface receptor-initiated release of arachidonic acid (AA). The products of AA metabolism are leukotrienes, prostaglandins and superoxide radicals.

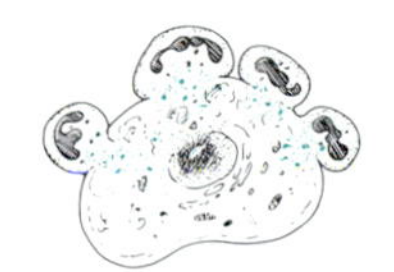

LYMPHOCYTES

Formation and Function

Lymphocytes arise from the bone marrow stem cells. Depending on the site of subsequent maturation, they develop into two classes that cannot be distinguished by light microscopy. Nevertheless, they have different immunological tasks:

- **B-cells** are responsible for **humoral immune responses.**
- **T-cells** are responsible for **cell-mediated immune responses.**

B-cells

Antigens can stimulate these cells to **form clones that synthesize a specific immunoglobulin antibody,** which is different from the antibodies synthesized by all other B-cell clones. The specificity of antibody synthesis is directed by sites of recognition and sites of binding located within the B-cell surface membrane.

T-cells

The surface sites of recognition and binding on T-cells respond to antigen as well as to the glycoproteins that couple antigen to the surface of the carrier cell. These glycoproteins are antigenic surface markers that are present on all cells. They belong to a single family of molecules whose synthesis is directed by a tight cluster of genes, generally named the **major histocompatibility complex (MHC)** or, specifically in humans, the **human leukocyte antigen (HLA).**

- **The nature of the HLA product on the surface of the carrier cell determines the nature of the T-cell response to the antigen.**

Two types of proliferative response occur. One yields a population of T-cells that modulates the processes by which B-cell clones are formed. The other yields a population of cytotoxic T-cells that can directly deactivate the antigen.

Immunoglobulins

Structure

The basic structural units are a pair of "heavy" amino acid chains and a pair of "light" amino acid chains.

- Papain cleaves immunoglobulin molecules at a point called the **"hinge."** The region to one side of the hinge is named the **Fc** region; that on the other side of the hinge is named the **Fab** region.
- Each chain consists of several **domains,** linked to neighboring domains by amino acids. Some domains occur in all immunoglobulins **(constant domains)** and some occur only in certain classes of immunoglobulins **(variable domains).**
- Within the variable domains, regions of **hypervariability** exist.
- The variable domains in the heavy chains designate the immunoglobulin as belonging to one of five types: IgG, IgA, IgM, IgD, or IgE.
- **The hypervariable regions within the variable domains on neighboring light and heavy chains form specific sites of antigen recognition and binding.**

Function

The unique function of immunoglobulins is to recognize and bind to other proteins. The purpose of forming such **antigen-antibody complexes** is either to precipitate antigen from solution or to attach antigen to phagocytic or cytotoxic cells for subsequent destruction.

- **Precipitation** requires that several molecules of antigen and antibody form a physical aggregate.
- **Attachment of complexes to other cells** requires that the receiving cell have surface receptors for the Fc portion of the antibody.

LYMPHOCYTES

Formation

Stem cells

Processing in Thymus → T-cells → Cell-mediated actions

Processing in Bone Marrow → B-cells → Antibodies

Function

B-Cells	T-Cells
B-cell surface recognition sites are complete immunoglobulin molecules.	T-cell surface recognition sites resemble the Fc region of a typical immunoglobulin.
B-cells can recognize freely circulating antigen.	T-cells can recognize only antigens that are presented to them on the surface of another cell.
B-cells respond to binding site occupation by synthetizing antibody.	T-cell responses depend on the nature of the coupler that holds the antigen on the surface of the presenting cell.

Immunoglobulins

Structure

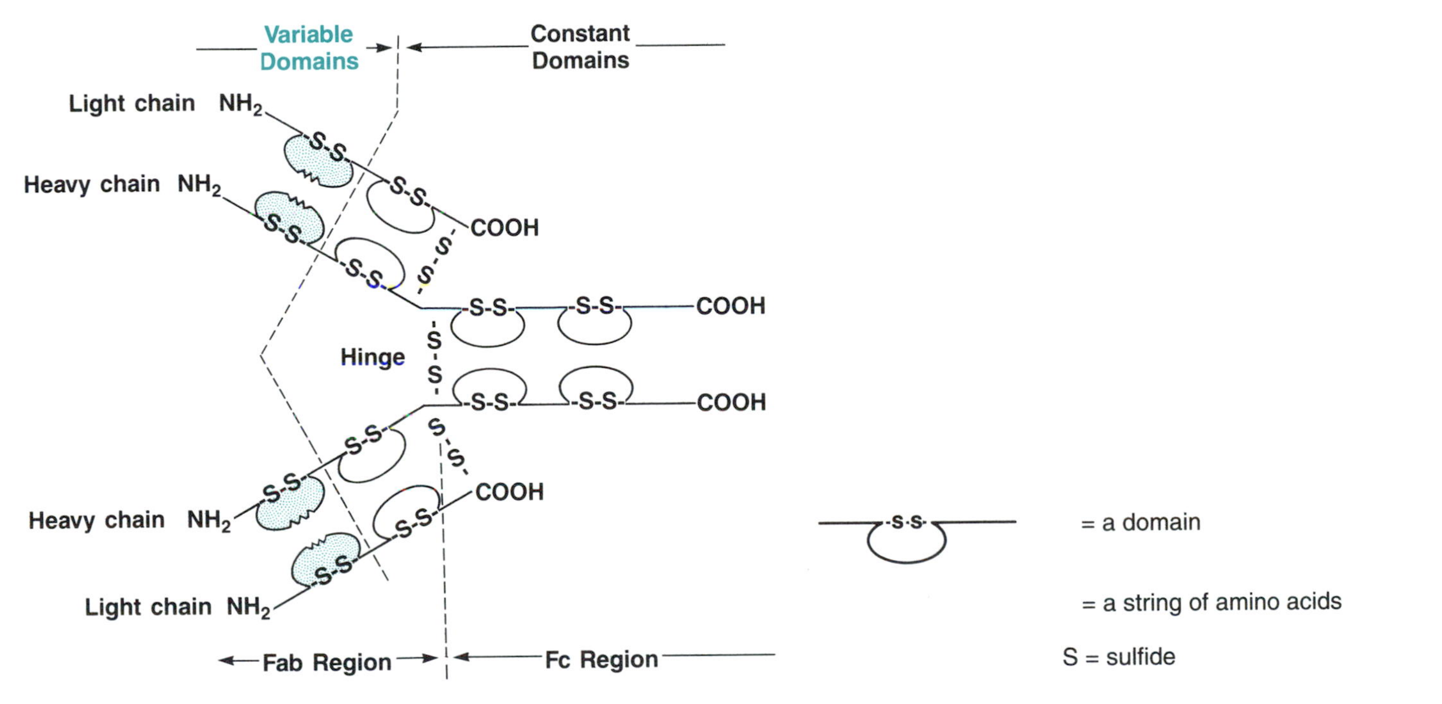

Function

Immunoglobulins have great mechanical flexibility in the *hinge* region. This allows

- Formation of physical aggregates (complexes) of antigen and antibody;
- Linking of a variety of antigen shapes (two are shown opposite) in a variety of linkage configurations (two are shown opposite).

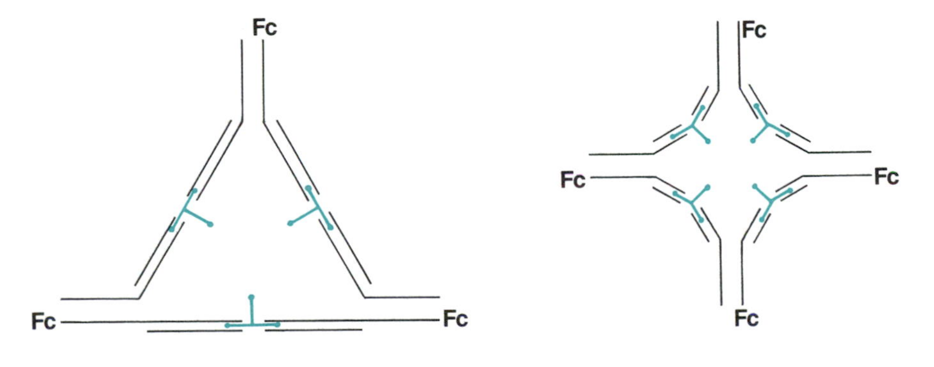

IMMUNITY

OVERVIEW

The immune system is a defense mechanism characterized by **recognition of nonself, specificity, and memory.** It has two basic components: **natural immunity and acquired immunity.** Natural immunity is bestowed by substances that are capable of acting directly and immediately on foreign matter (e.g., interferon, properdin, basic polypeptides). Acquired immunity is a normally dormant component of the immune system. It can be activated in response to specific stimuli. Passive activation (by the injection of previously activated components) is possible, but **the essence of the immune system is active acquired immunity, derived from circulating lymphocytes.**

Immune Reactions

Fully developed immune reactions involve antibody synthesis by B-cells (humoral response) as well as direct, cytotoxic T-cell responses.

B-cells

B-cell activation and cloning are under continuous control by T-cells:

- Helper T-cells promote cloning.
- Suppressor T-cells inhibit cloning by blocking helper T-cells.

T-cells

T-cell behavior is determined by the interaction between the T-cell and the HLA-determined glycoprotein binding sites on the antigen-presenting cell.

Humoral responses

These begin when a B-cell recognizes and binds antigen to its surface binding sites. However, **cloning occurs only after the B-cell and antigen have received a signal from an *activated* helper T-cell.**

Formation of activated helper T-cells is a two-step process:

1. Presentation of the antigen, bound to the surface of a presenting macrophage in conjunction with class II HLA products.

- **The presence of class II HLA products directs the formation of a helper T-cell.**

2. Activation and proliferation of the helper T-cell. This requires **interleukin 1,** a soluble factor derived from macrophages.

Activated helper T-cells are able to recognize antigen bound to the surface of B-cells in association with class II HLA products, and they are stimulated by this complex to secrete the factor(s) required for B-cell proliferation and antibody production.

Cytotoxic responses

These are initiated when precursors of killer T-cells recognize and bind **class I HLA surface markers** on a foreign cell.

Subsequent differentiation and proliferation occur only after a signal from activated helper T-cells. (Their formation is thought to be directed by class II HLA products.) Thus,

- **Recognition of coexisting class I incompatibilities (by precursors of killer T-cells) and class II incompatibilities (by activated helper T-cells) on the same cell causes the release of interleukin 2 from activated helper T-cells.**

This leads to the **formation and replication of activated killer T-cells** that are specific for the class I HLA surface incompatibility that initiated the response. Cytotoxic agents, whose precise nature is not yet known, will then destroy the foreign cell.

Immune Reactions

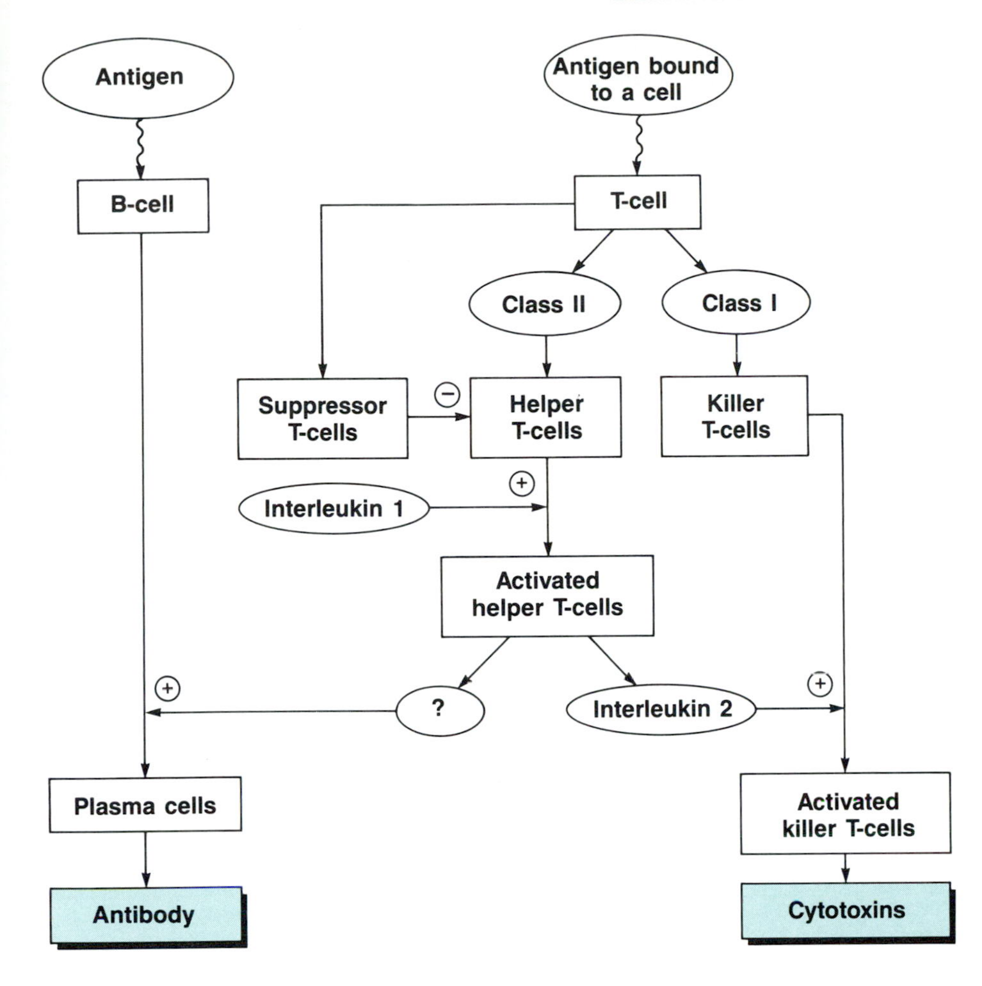

HLA surface glycoproteins
Genes within the HLA complex code for two different types of surface glycoproteins, named class I and class II products.

Class I products

- are present on all cells and lead to cytotoxic T-cell responses in an invaded organism.

Class II Products

- have more limited distribution;
- are present on B-cells *and*
- can be induced by soluble factors derived from macrophages and from epithelial cells;
- cause **proliferation** of a group of T-cells **(helper T-cells)** that regulate activation and cloning of B-cells.

PLATELETS

Platelets are disc-shaped cells without nuclei and with a diameter of 2 to 4 μm. They are **formed** in bone marrow **by budding from the cytoplasm of megakaryocytes,** and they spend their lifetime (estimated to be about 10 days) within the intravascular compartment, many of them being held in the red pulp of the spleen. Their **major function is in hemostasis.**

Structure

Surface membrane

The **exterior** of the encasing membrane **is covered with glycoproteins, plasma proteins, and receptors** for a variety of agents.

The **interior** of the covering membrane **adjoins a three-layered "unit membrane" that contains clot-promoting phospholipids, several different enzymes,** and filamentous structures that contain both **structural support elements and contractile proteins.**

The external membranes are invaginated to form a system of channels within the platelet interior.

Cytoplasm

The cytoplasm contains smooth endoplasmic reticulum, a Golgi apparatus, mitochondria, lysosomes, and a number of granules.

Function

Platelets normally circulate freely. Their **function is to aggregate in localized clumps at sites of vessel injury.** Whether or not platelets clump together depends entirely on the balance of prostaglandins on the exterior of the platelet surface membrane.

Membrane phospholipid metabolism

Phospholipases are released and activated when platelet surface receptors interact with their specific mate. They hydrolyze platelet membrane phospholipids (phosphatidylcholine and phosphatidylinositol) to yield prostaglandin endoperoxides (PGG_2 and PGH_2).

The chemical nature of subsequent metabolic products is determined by the lysosomal enzymes that are available:

- **Platelets** contain thromboxane synthetase. As a result, they **produce thromboxane A_2** (TXA_2), a powerful mediator of and specific stimulus for platelet aggregation and adhesion.
- **Endothelial cells** contain prostacyclin synthetase. Therefore, they **produce prostacyclin** (PGI_2), an inhibitor of platelet aggregation.

The equilibrium between TXA_2 and PGI_2 is shifted toward TXA_2 when the endothelial lining is damaged. Then platelets aggregate to form a **hemostatic plug.**

PLATELETS

Structure

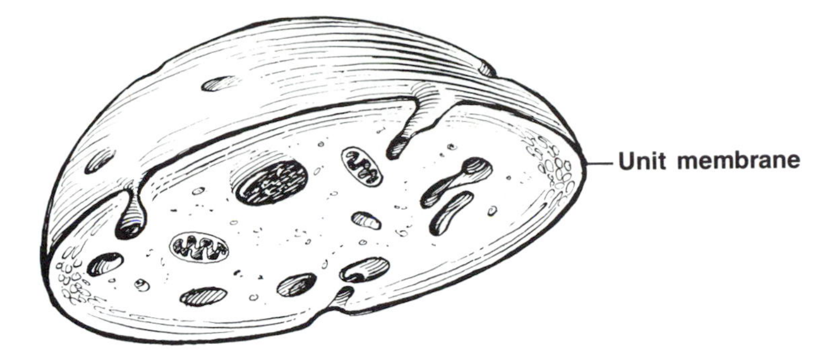

Function

Membrane phospholipid metabolism

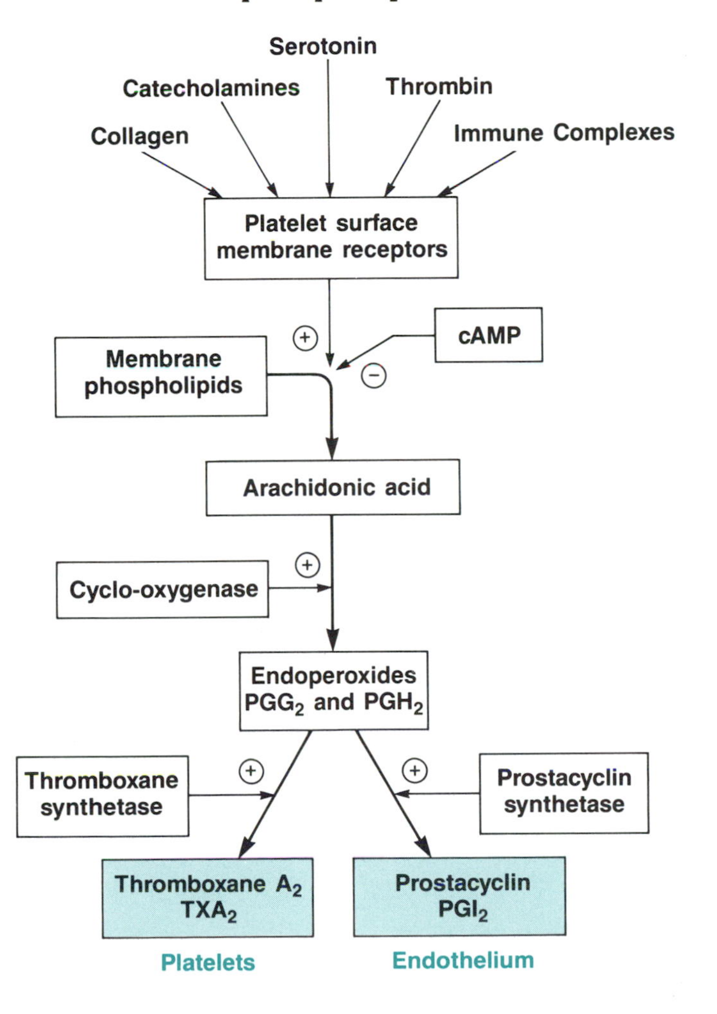

Plug formation

Damage of blood vessel endothelium is quickly followed by **platelet adhesion** to collagen in the exposed subendothelium.

The next step is **platelet activation,** leading to the **release reaction** (contraction of platelets and consequent release of granular contents into the system of surface channels). Among the sequelae of this reaction is the release of enzymes that are capable of converting fibrinogen to fibrin.

Platelet aggregation is due to fibrin threads that form on the platelet surface and hold neighboring platelets together.

The final step in plug formation is **consolidation of platelets** into a firm plug by an organized contraction of platelet actomyosin.

Plug formation

Platelet adhesion and aggregation

- Are promoted by TXA_2
- Are inhibited by PGI_2

TXA_2 and PGI_2 normally maintain an antagonistic equilibrium in which platelets are unattached and circulating. The equilibrium is altered by endothelial damage.

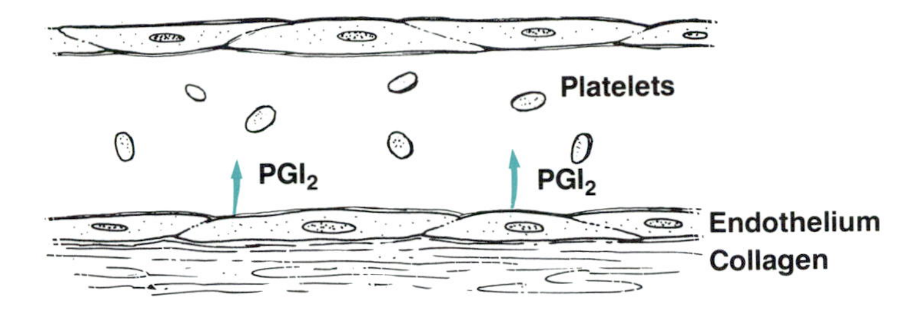

Adhesion

Mechanical attachment between platelet and collagen is provided by a plasma protein, the von Willebrand factor, that circulates as a complex with factor VIII coagulation protein.

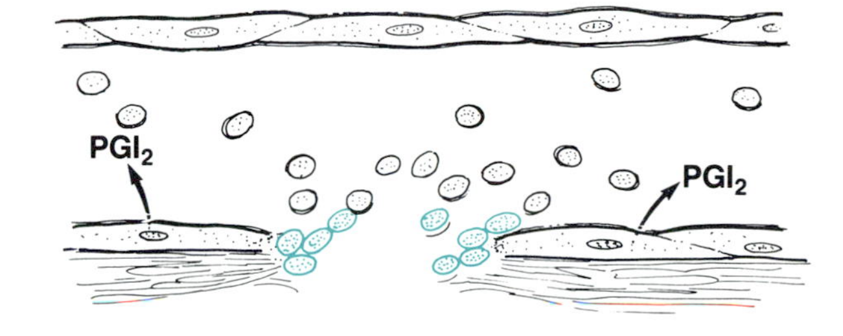

Activation

This series of events includes

- changes in platelet shape;
- formation of spicules and pseudopods that spread along the subendothelial fibers and intertwine with spicules from adjacent platelets;
- liberation and oxidation of arachidonic acid.

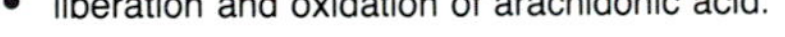

Aggregation

This occurs only if the surrounding medium supplies fibrinogen so that fibrin can be formed.

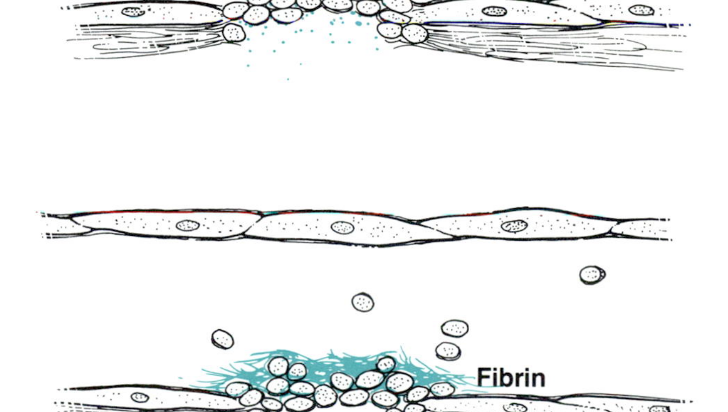

Consolidation

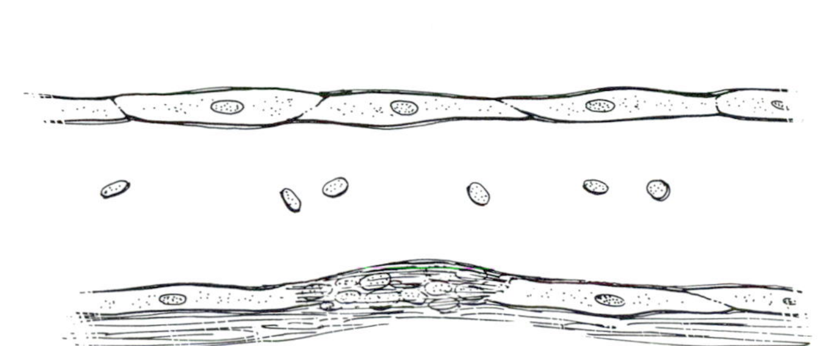

HEMOSTASIS

OVERVIEW

Prevention of blood loss after vessel injury involves three physiologic mechanisms:

1. **Vasoconstriction** in the affected area
2. Formation of a **platelet plug**
3. Formation of a firm **blood clot**

Vasoconstriction

Two aspects are involved:

- **Nervous factors** (initiated by pain or by the anticipation of pain)
- **Chemical factors** (tissue chemicals released by the injury as well as serotonin released from activated platelets)

Platelet Plug

Contact of circulating platelets with **collagen** fibers in the intima of the injured vessel leads to **platelet activation** and consequent formation of a hemostatic plug. It also activates the blood-clotting cascade.

Clot Formation

Blood vessel injury or contact between blood and an inappropriate surface will, within seconds, activate normally present plasma proteins and transform fluid, flowing blood into a gel in circumscribed areas.

- Activation by vessel trauma follows the **extrinsic pathway.**
- Activation by surface contact follows the **intrinsic pathway.**

Once initiated, the clotting process can be divided into three stages:

1. Development of thromboplastic activity
2. Activation of prothrombin
3. Conversion of fibrinogen to fibrin

Later the clot undergoes enzymatic digestion (fibrinolysis).

HEMOSTASIS

Clotting Cascade

- The objective of blood coagulation is to form stabilized fibrin threads from fibrin monomer.
- Fibrin monomer derives from plasma fibrinogen as a result of the enzymatic activity of thrombin.
- Thrombin must be generated from plasma prothrombin by the action of thromboplastic activity (prothrombinase).

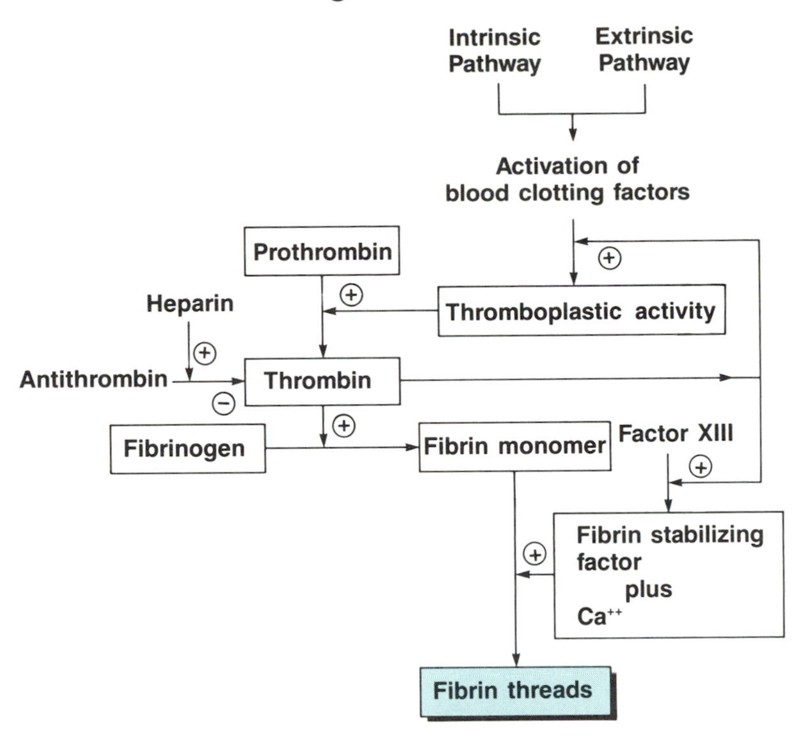

Fibrinolysis

- With time, fibrin is removed (and the clot is resorbed) by the proteolytic action of plasmin on fibrin.
- Plasmin results from the activation of plasminogen, a freely circulating inert proenzyme.
- The most potent physiological plasminogen activators are derived from cells (e.g., vascular plasminogen activator from vascular endothelium and urokinase from renal tubular epithelium).

Thromboplastic activity

This is a **"complex" of activated factor X, activated factor V, Ca^{++}, and platelet-derived phospholipids.**

The intrinsic pathway

- is initiated when blood contacts an **incompatible surface** (rough, nonwettable, or negatively charged);
- begins with the activation of factor XII (the Hageman factor).

The extrinsic pathway

- is initiated by blood vessel trauma;
- short-circuits many of the intrinsic activation steps and leads to thromboplastic activity after the activation of only one factor.

Ca^{++} is a vital requirement for normal progression of either path. Hence, agents that bind calcium (e.g., citrate, oxalate, EDTA) prevent blood clotting.

Prothrombin activation

Prothrombin is a circulating protein. Its synthesis (in the liver) requires **vitamin K** and can be inhibited by substances that compete with vitamin K (e.g., coumarin).

Prothrombin activation takes place on the surface of platelets when thromboplastic activity is present. It is prevented by antithrombin.

Conversion of fibrinogen to fibrin

This requires the proteolytic **action of thrombin on circulating fibrinogen.** Conversion is inhibited by heparin.

Once fibrin has been formed, it is further stabilized by strong **cross-linking.** That process requires Ca^{++} and is promoted by activated factor XIII (the fibrin stabilizing factor).

Thromboplastic Activity

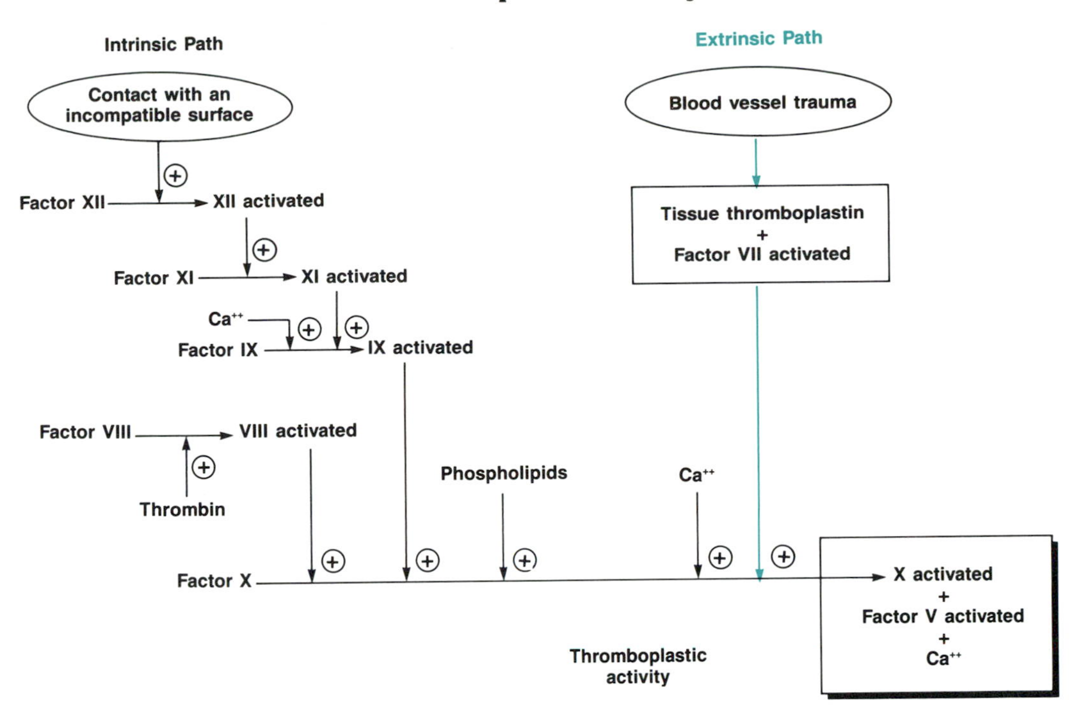

4

RESPIRATION

TRANSPORT OF OXYGEN AND CARBON DIOXIDE

OVERVIEW

The lungs provide an exchange interface with our body fluids. Gases are transported in blood in physical solution and in chemical combination with specific carrier agents.

Inspiration

Humidified air is brought to the alveoli by physiological mechanisms that decrease alveolar pressure below ambient atmospheric pressure.

Expiration

The pressure gradient is reversed, and water-saturated air leaves the lung.

Control mechanisms, responding to emotional and chemical stimuli (concentrations of O_2, CO_2, and H^+), adjust the **rate and depth of respiration.**

Interactions Between Gases and Blood

The amounts of O_2 and CO_2 that can be physically dissolved in blood represent only 5 percent of the total amounts carried.

TRANSPORT OF OXYGEN AND CARBON DIOXIDE

Overview

Within the lung a large number of terminal sacs, the **alveoli,** and the epithelium of their **enveloping capillary network** provide an **exchange interface** between air and blood.

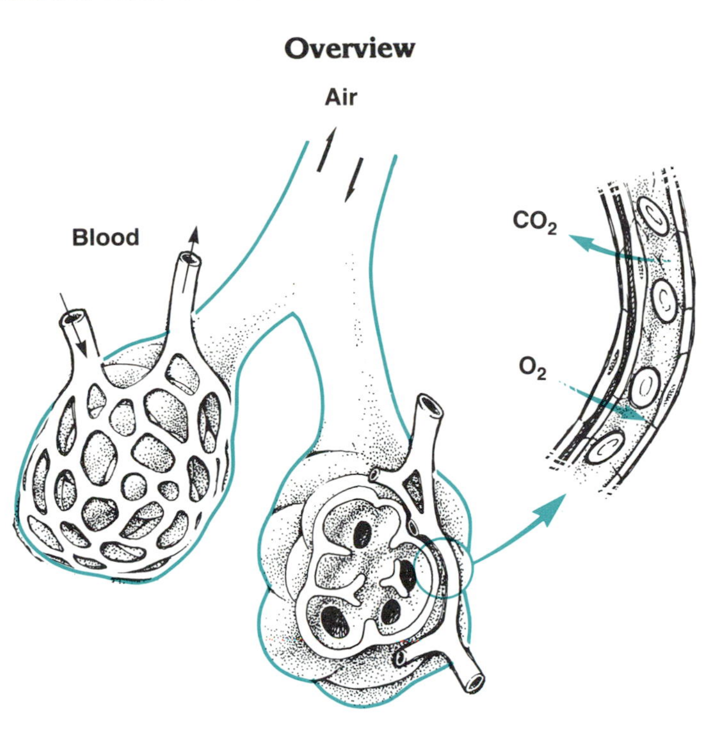

CO_2 and O_2 exchange between capillary and alveolus in response to concentration gradients. (Gas concentration is expressed as **partial pressure.**)

Interactions Between Gases and Blood

Partial pressures of respiratory gases

In a mixture of gases, the partial pressure of a gas is calculated as the product of total pressure and fractional contribution of the gas to the total mixture. For example, the fraction of O_2 in dry air is 21 percent. Hence, in dry air at atmospheric pressure (760 mm Hg), P_{O_2} is 0.21 × 760 = 160 mm Hg.

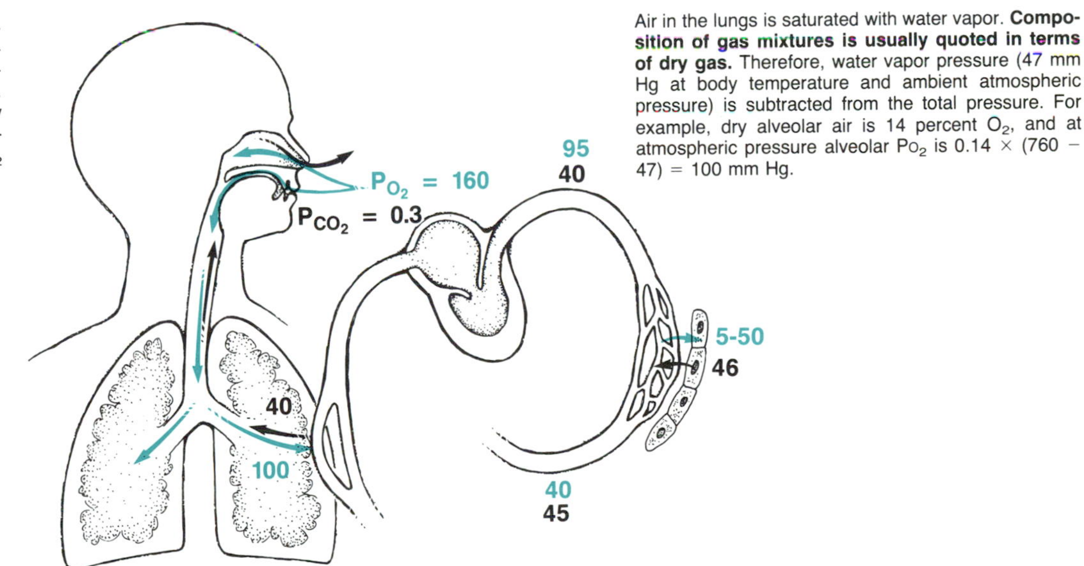

Air in the lungs is saturated with water vapor. **Composition of gas mixtures is usually quoted in terms of dry gas.** Therefore, water vapor pressure (47 mm Hg at body temperature and ambient atmospheric pressure) is subtracted from the total pressure. For example, dry alveolar air is 14 percent O_2, and at atmospheric pressure alveolar P_{O_2} is 0.14 × (760 − 47) = 100 mm Hg.

TRANSPORT OF OXYGEN AND CARBON DIOXIDE—CONT'D.

Gas Exchange

Carbon dioxide

The bulk of carbon dioxide is transported as HCO_3^- and H_2CO_3. Some transport (≈ 5 percent) occurs in chemical association with hemoglobin.

In the tissues

CO_2 is produced by cellular metabolism and enters tissue capillaries down its partial pressure gradient.

- **Carbonic anhydrase** (found **in red cells,** but not in plasma) catalyzes the combination of CO_2 with water to form H_2CO_3. A fraction of H_2CO_3 dissociates to HCO_3^- and H^+. (The H^+ ions are buffered by nonoxygenated hemoglobin.)
- CO_2 also reacts with the amine groups on hemoglobin. The combination of nonoxygenated hemoglobin, H^+, and CO_2 is called **carbaminohemoglobin.**

In the lungs

The processes described above are reversed because the partial pressure gradients cause diffusion of CO_2 out of the capillaries. This drives all binding reactions toward free CO_2.

Oxygen

The bulk of oxygen transport occurs by reversible chemical binding with hemoglobin.

In the lungs

- Oxygen moves down its partial pressure gradient, enters capillaries, and combines rapidly with hemoglobin (Hb) in red cells to form **oxyhemoglobin** (HbO_2).
- In comparison with carbaminohemoglobin, oxyhemoglobin is a **stronger acid and a weaker CO_2 binding agent.** As a result,
- Free H^+ is made available and CO_2 is released from carbaminohemoglobin.

In the tissues

- Oxygen moves down its partial pressure gradient and enters tissue cells. It leaves nonoxygenated hemoglobin behind.
- Nonoxygenated hemoglobin is then available to accept H^+ and CO_2 (forming carbaminohemoglobin).

TRANSPORT OF OXYGEN AND CARBON DIOXIDE—CONT'D.

Gas exchange

In the tissues

Capillary
HCO_3^-
Cl^-
Carbonic anhydrase
$HCO_3^- + H^+ \leftarrow H_2CO_3 \leftarrow H_2O + CO_2$
Red Cell
O_2 + Hb H^+ / NHCOO$^-$ $\leftarrow HbO_2$ $NH_2 + CO_2$
Carbamino Hb
CO_2
O_2
Tissue Cells

In the lungs

Capillary
Carbonic anhydrase
Cl^-
$HCO_3^- \rightarrow HCO_3^- + H^+ \rightarrow H_2CO_3 \rightarrow H_2O + CO_2$
Red Cell
O_2 + Hb H^+ / NHCOO$^-$ $\rightarrow HbO_2$ $NH_2 + CO_2$
Carbamino Hb
Alveolus
O_2
CO_2

The hemoglobin dissociation curve

Depending on Po_2, each Hb molecule can bind up to four molecules of O_2 in an easily reversible manner. In the lungs, at a Po_2 of 100 mm Hg, the binding reaction is 97 percent complete (i.e., the Hb saturation is 97 percent). In the tissues, at a Po_2 near 20 mm Hg, the reaction is only about 20 percent complete.

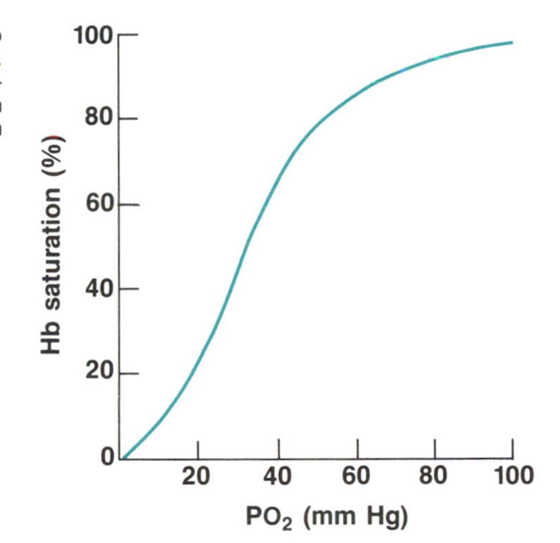

The plateau at the upper end signifies a stable blood-oxygen content despite wide variations in alveolar Po_2. The steep portion between 20 and 60 mm Hg signifies the extent of O_2 unloading in the tissues without large changes in Po_2.

Increasing temperature, Pco_2, or $[H^+]$ will shift the Hb dissociation curve to the right. That is, each of these conditions causes Hb to begin to give up O_2 at higher values of Po_2. This facilitates release of O_2 to the tissues.

MECHANICS OF BREATHING

OVERVIEW

The lungs are filled and emptied as a result of changes in the volume of the thoracic cavity.

Inspiration

Two factors cause expansion of the chest cavity:

- Its vertical size is increased by active **contraction of the diaphragm.**
- Its cross section is increased in part by contraction of the external intercostal muscles, sometimes assisted by accessory muscles.

Expiration

Expiration is **passive** in quiet breathing (driven by elastic recoil of the lungs and chest wall), but may be assisted by active contraction of intercostal and abdominal muscles.

- Inspiratory muscles continue to contract during part of expiration. Their gentle opposition to elastic recoil makes the expiration time almost double the inspiration time.

Changes in Pressure and Volume During the Respiratory Cycle

Compliance and **resistance** in the components of the respiratory system influence the changes in pressures and lung volume during respiration.

Compliance

This is **a measure of the elastic force** that aids or opposes a change in the volume of a distensible organ. It is defined as a ratio:

$$\text{Compliance} = \frac{\text{Change in volume}}{\text{Change in transmural pressure}}$$

- **It is the slope of a pressure/volume curve.**

Both **lung compliance** and **chest wall compliance** influence the pressure-volume curve of the system.

Lung compliance

- Is influenced by **tissue elements,** degree of **tissue hydration, lung blood volume, air space geometry,** and **surface forces** at the air-fluid interface of the alveoli.
  - Major contributing tissue elements are fibers of elastin or collagen.
  - Distended interstitial fluid spaces and engorged blood vessels reduce lung compliance.
  - Surface forces influence lung compliance because the pressure required to maintain an air bubble at a given size within a fluid medium is directly related to the surface tension of the air-fluid interface (Laplace's Law).

Chest wall compliance

- Is influenced by **chest geometry, composition of the chest wall,** and **mass of abdominal contents.**
  - **Anything that inhibits motion of the chest wall reduces its compliance.**

MECHANICS OF BREATHING

Changes in Pressure and Volume During the Respiratory Cycle

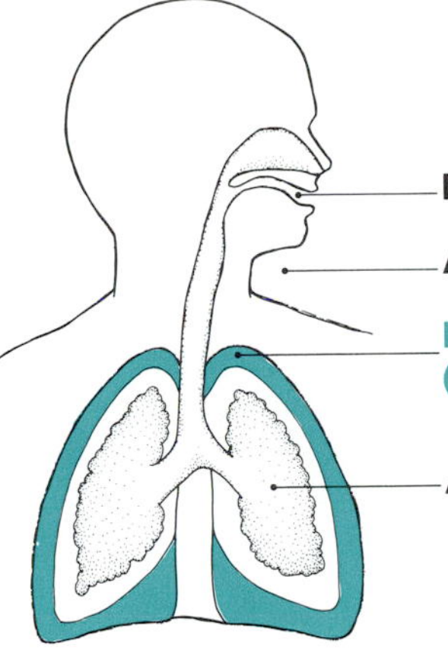

- The intrapleural space connects the mechanical forces of the lung to those of the chest wall.
- At the end of a quiet expiration,

$$P_{pl} = -5 \text{ cm } H_2O \text{ and } P_{alv} = 0$$

i.e., the lungs are being stretched by a pressure gradient of 5 cm H_2O and the chest cavity is being contracted by a pressure gradient of 5 cm H_2O.

Static pressure-volume characteristics

Static pressure-volume curves can be determined in experimental settings where the expanding pressure is maintained until all air flow stops. Under conditions of zero air flow,

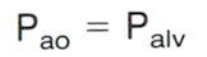

$$P_{ao} = P_{alv}$$

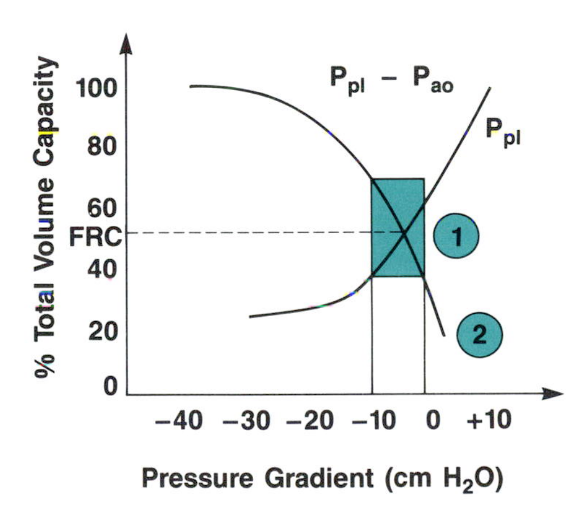

- [P_{pl} − P_{ao}] represents the force acting on the elastic components of the lungs.
- [P_{pl} − atmospheric P] represents the force acting on the elastic components of the chest wall.

(1) Expiration is never to the point where all air is removed from the lung. It normally ends at a point where the net elastic force arising from the collapsing lung is exactly balanced by the net elastic force with which the chest cavity resists further volume reduction. This point is represented by the intersection of the two lines. The volume at that point is called the **functional residual capacity.** The intrapleural pressure at that point is about −5 cm H_2O.

(2) Lung volume does not collapse to zero at zero pressure gradient because the small airways collapse before the alveoli do. This leaves a **residual volume.**

☐ indicates the intrapleural pressure range traversed during normal, quiet breathing. In this range lung compliance and chest wall compliance are each near 200 mL/cm H_2O.

- Both curves normally show hysteresis, less pressure being required to maintain a given expansion during deflation than during inflation. Hysteresis is not demonstrated here.

Dynamic pressure-volume characteristics

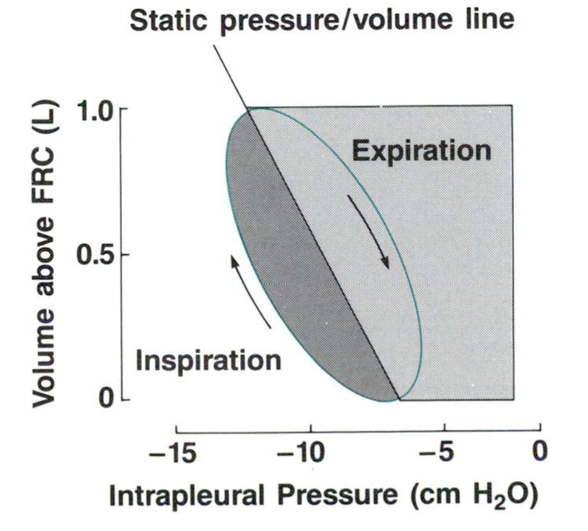

When pressure-volume curves are determined while there is air flow, then [P_{ao} − P_{pl}] represents not only the forces arising from elastic components, but also the forces required to overcome **airway resistance to flow** and **tissue resistance to movement.**

- During inspiration, pressure in excess of static pressures is required at each volume to overcome resistance.
- During expiration, less pressure than predicted by static conditions is required because some energy is supplied by elastic recoil.
- Area on a pressure-volume curve represents work done. Therefore, several components of the **work of breathing** can be identified:

▪ represents work done by respiratory muscles to overcome tissue resistance and airway flow resistance. This work is lost as heat.

☐ represents work done to overcome elastic forces in the lungs and chest wall. This work is stored as elastic recoil energy.

Resistance

This is **a measure of the pressure gradient required to move air along the airways at a certain rate of flow.** It is expressed as

$$\text{Resistance} = \frac{\text{Pressure gradient}}{\text{Flow}}$$

- Three sources of resistance must be overcome during breathing: (1) **tissue resistance in the chest wall,** (2) **tissue resistance in the lungs,** and (3) **flow resistance in the airways.**
- Chest wall resistance contributes up to 20 percent of total resistance.
- Lung tissue resistance contributes up to 15 percent of total resistance.
- **Airway resistance is the most important resistive component,** and it is most subject to increase with disease. It increases greatly if air flow becomes **turbulent,** and it increases dramatically with small decreases in **airway diameter.**

Surface Tension, Pressure, and the Shape of Alveoli

Alveoli are open at one end and are interconnected with other alveoli. This creates the potential for two problems:

1. Because the surface tension is fixed by the fluid medium, the pressure inside small alveoli should be higher than that inside large alveoli. Therefore, small alveoli should empty into large alveoli.
2. Because the air can escape from alveoli when surface tension shrinks them, no opposing internal pressure can build up and they should collapse.

These potentially fatal problems are avoided by **lung surfactant** and **anatomic interdependence.**

Surfactant

This is a secretion of some cells lining the alveoli.

- Its surface tension is only one-fifth that of extracellular fluid. As a result, **the pressure gradients required to expand the lung are lower than they would be without surfactant.**
- It has the unusual property of reducing surface tension more in small alveoli than in large alveoli. As a result, **small alveoli can be maintained by the same hydrostatic pressure as can large alveoli** and there is no tendency for small alveoli to empty into large alveoli.

Anatomic interdependence

Alveoli are anatomically interdependent in two respects:

1. Alveolar outflow tracts are interconnected.
2. Pulmonary interstitial pressure influences the pressure gradient across the alveolar walls.

◆ Interconnected outflow tracts are a liability because they create potential paths of low flow resistance to adjacent alveoli. These paths are not normally functional, however, because there are no gradients in pressure among neighboring alveoli.

◆ Alveoli do not collapse at the end of expiration, when P_{alv} is zero, because pulmonary interstitial fluid pressure is -5 cm H_2O. That is, the pressure inside alveoli is greater than that outside alveoli.

- **The ultimate factor preventing alveolar collapse is that the lung as a whole cannot collapse because of close mechanical coupling between its outer lining (the visceral pleural membrane) and the inner lining of the thorax (the parietal pleural membrane).**

Surface Tension, Pressure, and the Shape of Alveoli

Surface forces have great influence on lung behavior because breathing amounts to periodic inflation and partial deflation of a very large number of small bubbles.

Normal Bubbles

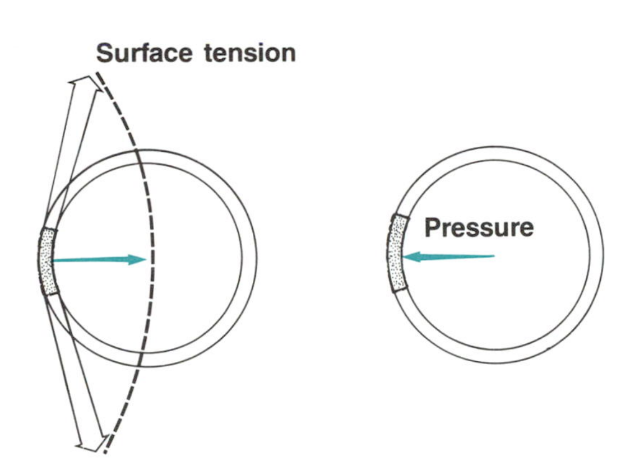

In an air bubble lined with water, **surface tension** creates **an inwardly directed force** that tends to collapse the bubble. The size of the bubble depends on the **opposing hydrostatic pressure** of the air within the bubble.

At equilibrium,

$$\text{Pressure} \propto \frac{\text{Surface tension}}{\text{Radius}}$$

The magnitude of surface tension per unit of surface area is fixed for a given liquid. As a result, more hydrostatic pressure is required to maintain an air bubble of small radius than to maintain one of large radius.

Alveoli

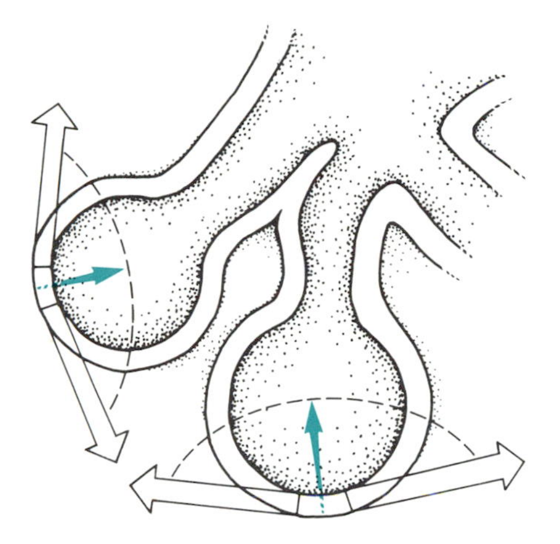

Alveoli are open at one end and are interconnected with other alveoli. They would collapse except for the facts that

- the pressure in small alveoli is the same as the pressure in large alveoli;
- a force other than internal pressure (alveolar pressure) prevents them from collapsing.

Lung Volumes

The volume of air that is moved into or out of the lungs during a breath (tidal volume) can be increased or decreased by calling on inspiratory or expiratory reserve volumes. However, not all of the tidal volume is available to ventilate the alveoli because some of it is used to fill the **deadspace.**

Deadspace

This volume (about 0.15 L) normally consists of the anatomic volume of those respiratory structures that are not involved in respiratory exchange processes. The physiologic deadspace (measured by a gas equilibration technique) may be slightly different from the anatomic deadspace.

Lung Volumes

At steady state, the volume of the lungs is such that collective alveolar surface tension is exactly balanced by intrapleural suction. This lung volume is called the functional residual capacity (FRC).

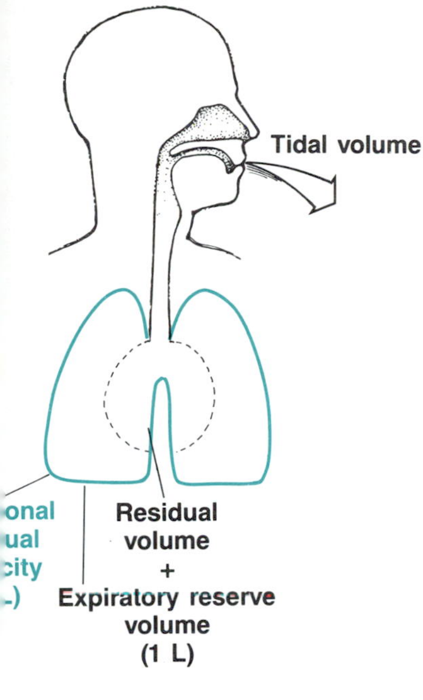

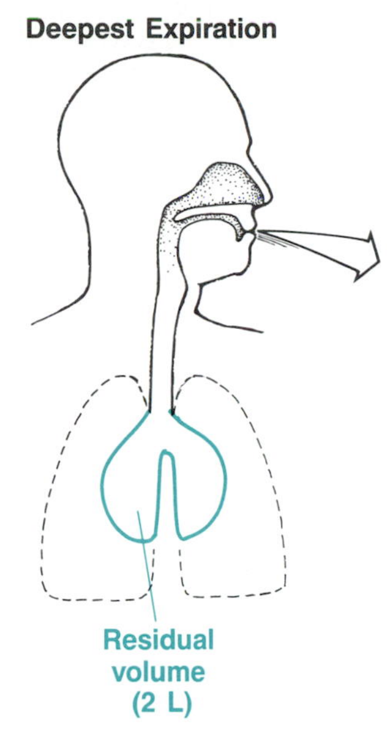

Functional residual capacity (FRC)

The steady-state volume of the lungs at the end of a normal, quiet expiration. It is the **sum of residual volume and expiratory reserve volume.**

Residual volume

At the end of a **forced expiration,** involving contraction of the internal intercostal muscles, lung volume can be decreased below FRC, down to the residual volume. **Residual volume is not zero because small airways collapse before all alveoli do.** This traps air distal to the point of collapse.

Expiratory reserve volume

The difference between FRC and residual volume.

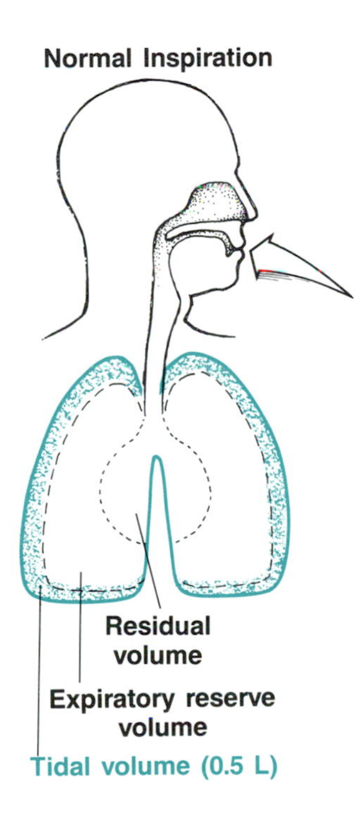

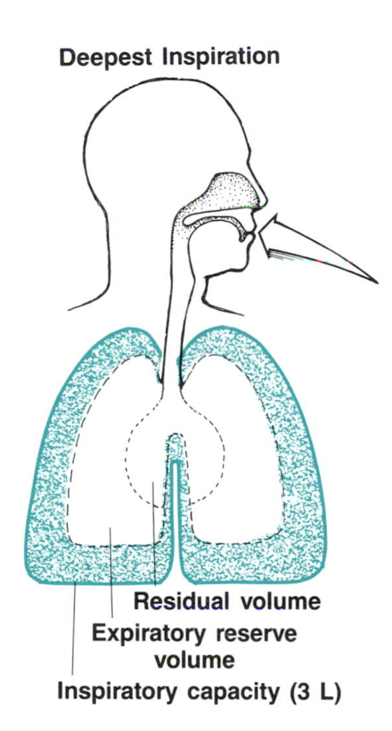

Tidal volume

The volume of air moved into or out of the lungs during one breath. It is normally 0.5 L, but it can be increased to 4 L by invading both the expiratory reserve volume and the inspiratory reserve volume. **The maximal tidal volume is called the vital capacity.**

Inspiratory capacity

During a **forced inspiration** (requiring more forceful contractions of the usual inspiratory muscles as well as the use of accessory muscles such as the scalene muscles and the sterno-mastoids), lung volume can be increased by the inspiratory capacity. At the end of such a forced inspiration, the total lung volume is at its maximum, which is FRC + inspiratory capacity.

Inspiratory reserve volume

The difference between maximal lung volume and lung volume at the end of a normal inspiration.

PULMONARY BLOOD FLOW

Effects of Gravity

The pulmonary vascular bed, like the systemic vascular bed, has all of the cardiac output flowing through it. However, pulmonary vascular resistance is less than one-fifth of systemic resistance. Consequently, **mean pulmonary arterial hydrostatic pressure is only about 15 mm Hg (20 cm H_2O) above atmospheric pressure.** This value is so low that the effects of gravity on local hydrostatic pressure are significant:

- Pulmonary vessels that are located in the apex, more than 10 cm above heart level, have an intravascular hydrostatic pressure near 0 mm Hg.
- Vessels near the base of the lungs have higher pressure.

The consequences of this vertical intravascular hydrostatic pressure gradient are as follows.

- **Blood flow and blood volume in certain regions of the lungs can be affected by fluctuations in respiratory pressure and by postural changes.**
- In regions where alveolar pressure exceeds pulmonary venous pressure, blood flow is determined by the difference between pulmonary arterial pressure and alveolar pressure, not by the difference between arterial pressure and venous pressure.
- Near the base of the lungs, the elevated hydrostatic pressure tends to expand blood vessels and, thereby, tends to increase the local blood volume.

Chemical Effects

In most vascular beds, hypoxia causes pronounced vasodilatation. However, **pulmonary arterioles constrict when *alveolar* Po_2 is low.**

- This has protective significance in that perfusion is directed toward well-oxygenated regions of the lungs.
- Low Po_2 in the blood that perfuses the arterioles does not cause vasodilatation.

Ventilation and Perfusion in the Lung

The amount of O_2 extracted from, and the amount of CO_2 delivered to, each alveolus vary directly with its blood flow. Therefore, **adequate gas exchange in the lungs requires that ventilation and blood flow (perfusion) be matched** to each other. Both decrease from base toward apex in the upright lung, but **the change in ventilation with height is not the same as that in perfusion.** As a result, near the base of the lungs (compared to the apex),

- the ventilation/perfusion ratio is lower;
- the extent of capillary-alveolar gas diffusion is lower;
- pulmonary venous Po_2 is lower and pulmonary venous Pco_2 is higher.

These **unequal vertical gradients in ventilation and perfusion are responsible for the observation that Po_2 in pulmonary venous blood is generally lower than alveolar Po_2.**

PULMONARY BLOOD FLOW

Effects of Gravity

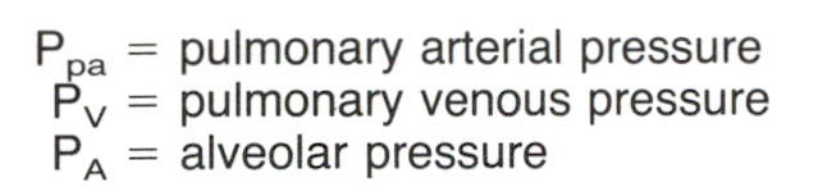

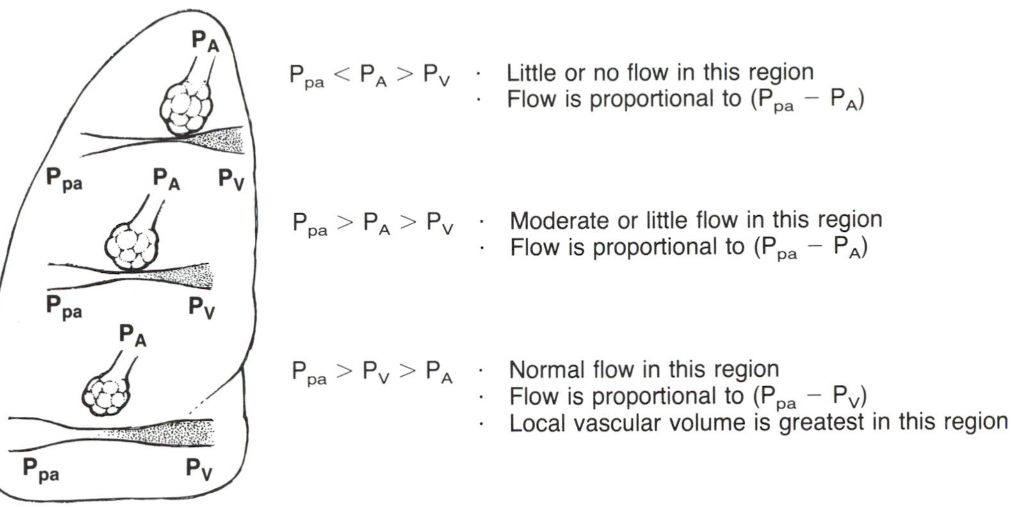

Ventilation and Perfusion in the Lung

As a result of gravitational effects, pulmonary blood flow (lung perfusion) in the normal upright lung is smallest at the lung apex and increases from apex to base.

Ventilation (measured per unit of lung volume) also increases from apex to base. The reason is that alveoli near the base have smaller resting volumes, but expand relatively normally on inspiration.

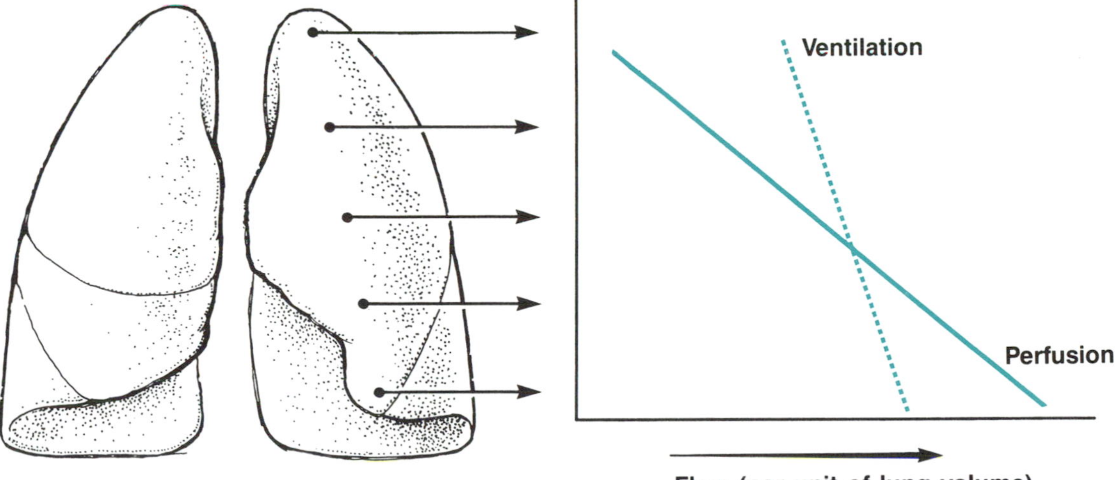

Flow (per unit of lung volume)

CONTROL OF RESPIRATION

The basic aim of respiratory regulation is to maintain arterial Po_2 and pH within their normal limits. Under resting conditions this can be accomplished by the basal respiratory rate and depth, which are set by central nervous mechanisms.

◆ Emotional factors, as well as mechanical and chemical information from the periphery, modify the basal rate in such a way that ventilation (rate and depth of respiration) and arterial pH are homeostatically linked.

Neuronal Connections

The rate and depth of the basic respiratory rhythm are set by medullary neurons that spontaneously generate a program of motor instructions to the muscles of respiration. The basic program is modulated by other central nervous areas for a variety of purposes:

- The hypothalamus accomplishes whole-body autonomic integration and adjusts breathing for moment-to-moment needs.
- The forebrain initiates breathing patterns that have emotional connotations (e.g., gasps of surprise or languorous sighs).
- The motor cortex issues modifications to the breathing program for the purposes of generating speech and volitional control over breathing.
- The cerebellum participates in breathing modulations that are associated with postural changes.

Respiratory adaptations to peripheral needs are accomplished through three classes of receptor afferents.

Chemoreceptors

These are located centrally (in the medulla) as well as peripherally (in the carotid and aortic bodies). They respond to changes in Po_2, Pco_2, or pH of tissue fluid or blood and elicit homeostatically appropriate reflex changes in ventilation.

Irritant receptors

These are located throughout the airways and lungs. They respond to mechanical or chemical irritation and elicit complex reflexes such as sneezes and coughs.

Stretch receptors

These are found in the lungs and in the respiratory muscles. They monitor lung volume and, in general, cause both inhibition of inspiration and promotion of expiration upon lung inflation (the Hering-Breuer reflex).

◆ Their role is easily demonstrated in animals, where severing of afferent stretch receptor paths leads to a characteristic change in breathing pattern (slow, deep breathing that is thought to be regulated by chemical factors). Such a change does not occur in humans, suggesting that mechanical factors are a less important regulatory influence in us.

Chemical Regulators

The link between ventilation and body needs is normally provided by Pco_2, but Po_2 and $[H^+]$ are also important chemical regulators. These three agents behave in an interactive, not an algebraically additive, manner.

◆ **At normal Po_2, ventilation is directly related to Pco_2, and the relationship is governed mostly by medullary chemoreceptors.**

◆ At normal Pco_2, ventilation is affected little by changes in Po_2 until Po_2 falls to near 50 mm Hg. At that point ventilation increases steeply with further small decreases in Po_2. Thus, **peripheral chemoreceptors provide a last-resort defense mechanism during hypoxemia.**

CONTROL OF RESPIRATION

The Neuronal Connections

Neurons in the **medulla** *(nucleus tractus solitarius* and *nucleus retroambiguus)* generate an **automatic breathing rhythm** and convey the appropriate action potential patterns to spinal motor neurons that drive the primary and accessory muscles of respiration.

Three classes of receptor afferents provide central nervous system input that is capable of influencing ventilation.

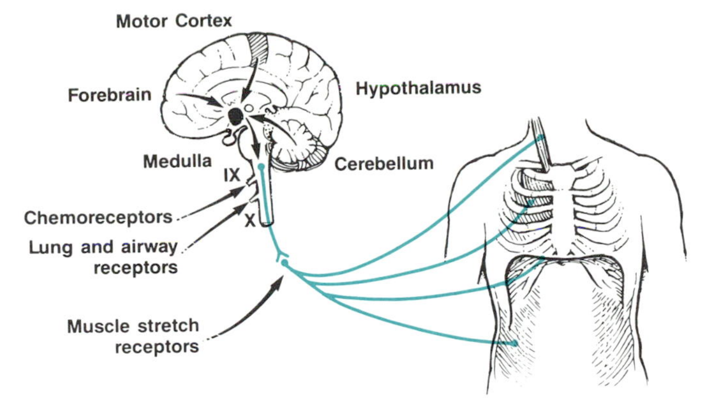

A variety of physiological activities influence respiration via information from other central nervous areas. Their influence can be direct, via paths to the spinal motor neurons, or indirect, via the hypothalamus-medulla.

Effects of CO_2, H^+, and O_2 on Chemoreceptors

- Medullary chemoreceptors do not respond significantly to changes in Po_2 within physiological ranges.
- Peripheral chemoreceptors (located in the carotid and aortic bodies) respond to changes in both Pco_2 and Po_2.
- Chemoreceptors do not respond to CO_2 per se, but react to the H^+ concentration of their extracellular environment. Their rate of response is related either inversely to ambient Po_2 or directly to the rate at which H^+ are formed by the stimulus.
- The coupling mechanisms between either $[H^+]$ or Po_2 and action potential frequency are not known.

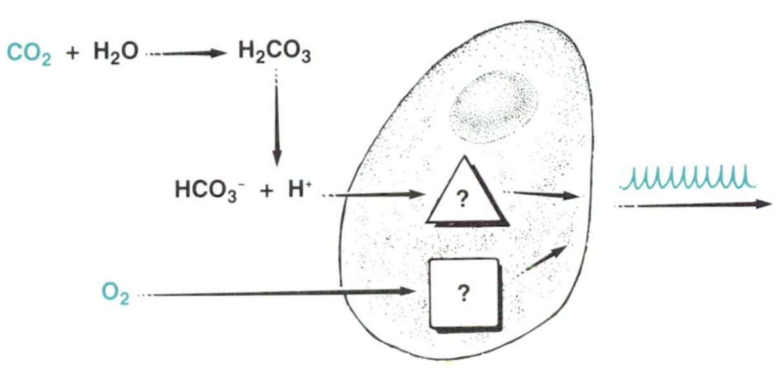

Effects of Arterial Blood Gases on Ventilation

Carbon dioxide (at different Po_2)

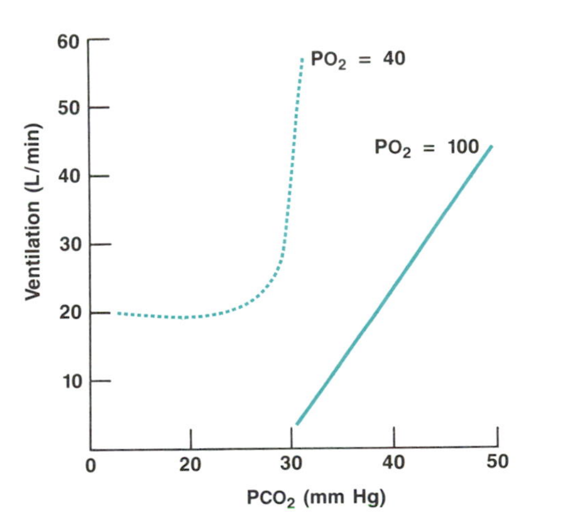

At normal Po_2 there is an almost linear relationship between ventilation and arterial Pco_2.

In conscious humans, about 75 percent of the increased drive under circumstances of elevated Pco_2 originates in the central chemoreceptors; the remainder derives from peripheral chemoreceptors.

Oxygen (at different Pco_2)

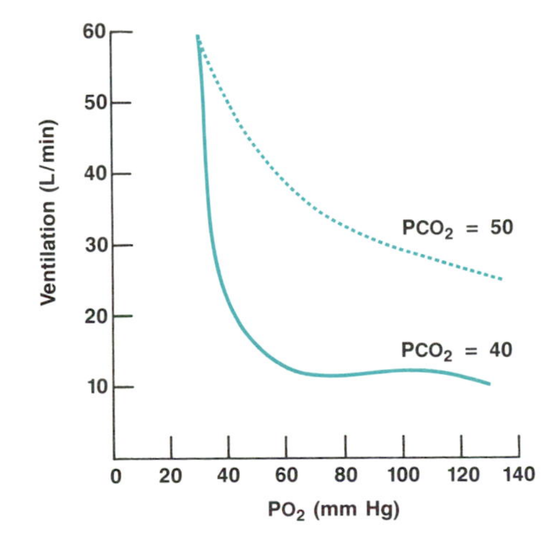

Ventilatory responses to Po_2 are due to peripheral chemoreceptors:

- At normal Pco_2 they show little effect on ventilation in the face of decreasing oxygen tension until arterial Po_2 falls to near 50 mm Hg.
- If Pco_2 increases simultaneously with the fall in Po_2, then the stimulatory effect of falling oxygen tension is markedly enhanced.

5

CARDIOVASCULAR PHYSIOLOGY

CARDIOVASCULAR PHYSIOLOGY

OVERVIEW

The cardiovascular system consists of the **heart,** the **blood vessels,** and the **regulatory mechanisms** that govern their functions.

The system circulates blood in a continuous loop. Its purposes are

- **to provide for all tissues an adequate supply of nutrients and**
- **to remove from all tissues the products of their metabolic activity.**

The **left ventricle** of the heart is generally taken as the starting point of the continuous loop. Its function is to **pump blood into the root of the aorta.** From the aorta, its progression of flow is as follows:

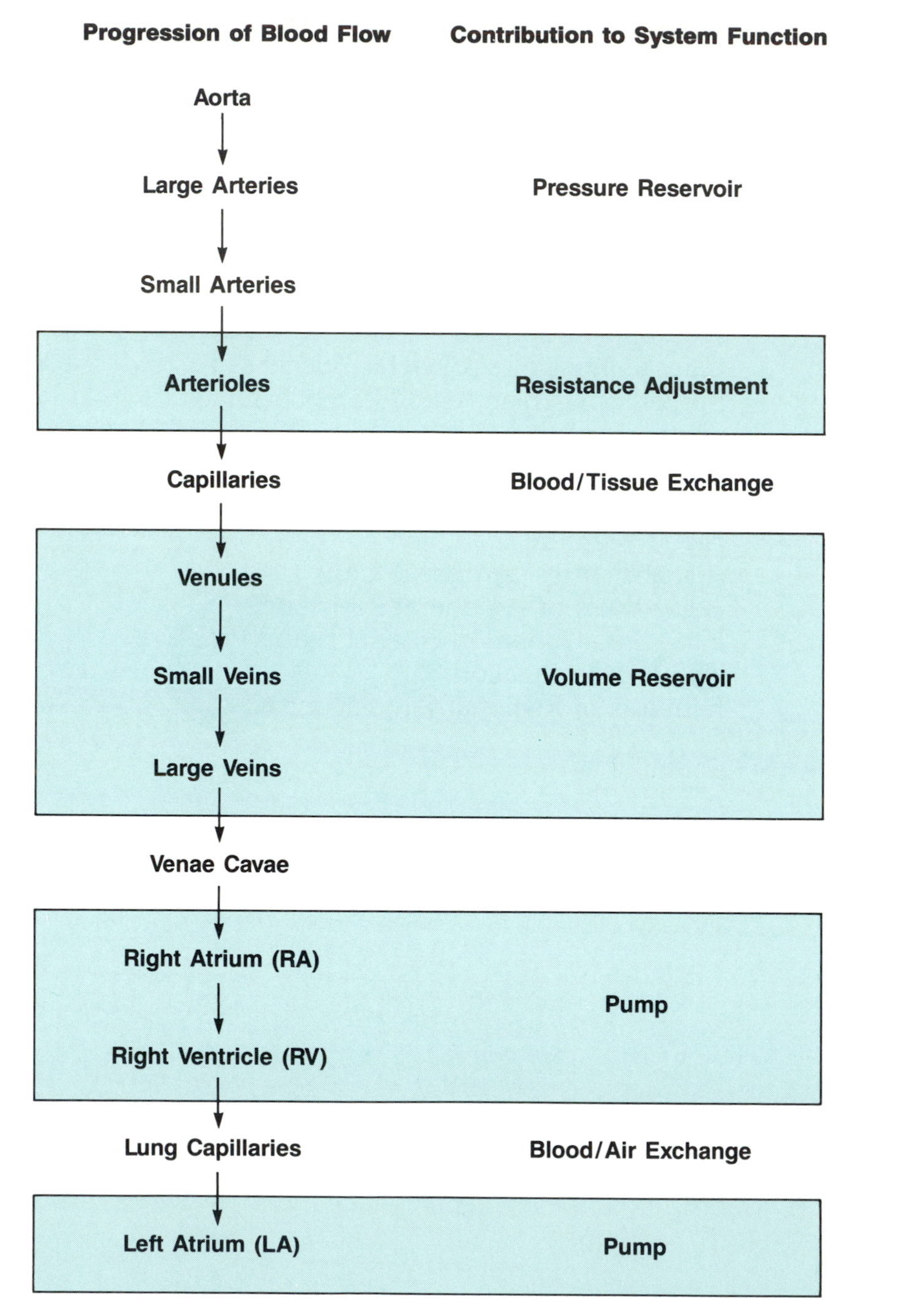

OVERVIEW

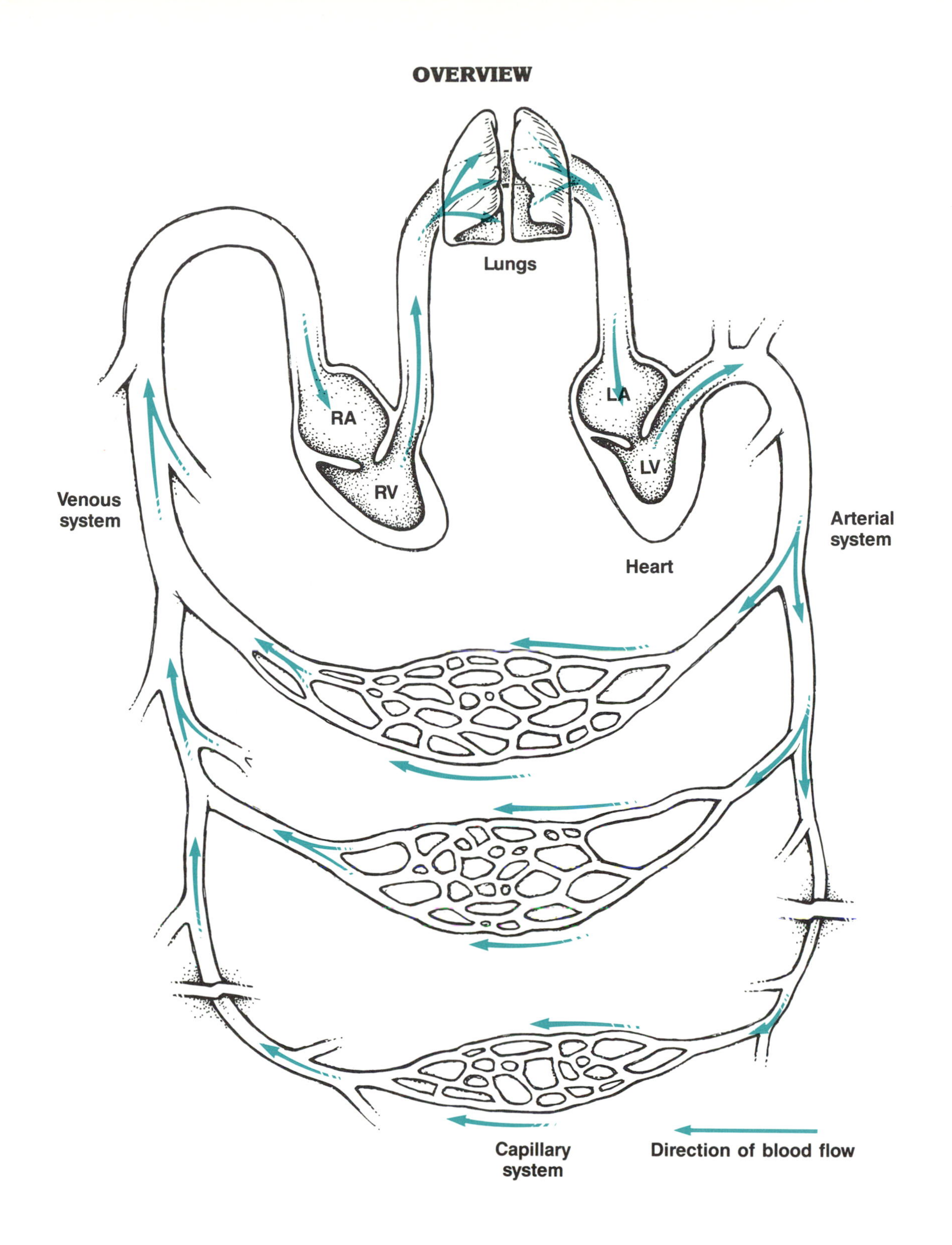

Lungs
RA
RV
LA
LV
Venous system
Arterial system
Heart
Capillary system
Direction of blood flow

PRESSURES AND FLOWS

While the Heart Is Stopped

The total volume of blood within the cardiovascular system is slightly greater than the system can hold without stretch. As a result,

- **there exists blood pressure even when the heart has been stopped;**
- it is called **mean circulatory filling pressure.**

The magnitude of mean circulatory filling pressure is

- about **7 mm Hg** in humans;
- **the same at every point** in the cardiovascular system;
- **influenced mainly by blood volume and blood vessel stiffness** (compliance).

While the Heart Is Beating

The heart takes blood from the venous side and adds it to the arterial side. **This must lower the pressure below 7 mm Hg on the venous side and raise it above 7 mm Hg on the arterial side.**

At steady state the **magnitude of the arterial-venous pressure gradient is just great enough to maintain a steady average flow** through the system, **against the flow resistance** that is offered by the blood vessels.

- The flow out of the heart is called **cardiac output.**
- The flow returning to the heart is called **venous return.**

Cardiac output

This is the average flow in the system. It is related to the pressure gradient:

$$\text{Cardiac output} = \frac{\text{Mean aortic pressure} - \text{Mean right atrial pressure}}{\text{Total peripheral resistance}}$$

Venous return

Venous return can differ from cardiac output for only a few minutes because the heart and the lungs have restricted abilities to sequester and supply blood.

- It is **profoundly altered by external pressures (respiratory pump, muscle pump) and by gravity (orthostasis, postural hypotension).**

Respiratory pump

- Decrease in thoracic pressure during inspiration is transmitted to thoracic venous pressure, but veins away from the chest do not experience the same pressure decrease on inspiration.
- The change in pressure gradient from extrathoracic to thoracic veins causes increased venous return during inspiration.

Muscle pump

- Muscle activity compresses veins and massages blood toward the heart in veins that are equipped with valves.

Orthostasis

- Changes in posture cause changes in hydrostatic pressure within blood vessels, but not outside blood vessels (because the outside is not a continuous column of fluid).
- Arteries are strong vessels, but veins are distended (or compressed) by changes in transmural pressure.
- During venous distension, venous return is momentarily altered as blood is pooled in the venous system.

PRESSURES AND FLOWS

Pressures in the cardiovascular system are always given with reference to atmospheric pressure. Thus, a vascular pressure of 25 mm Hg means 25 mm Hg *above atmospheric pressure.*

While the Heart Is Stopped

The hydrostatic pressure that exists in the cardiovascular system during the first few minutes after the heart has been stopped is called the **mean circulatory filling pressure.** It is near 7 mm Hg.

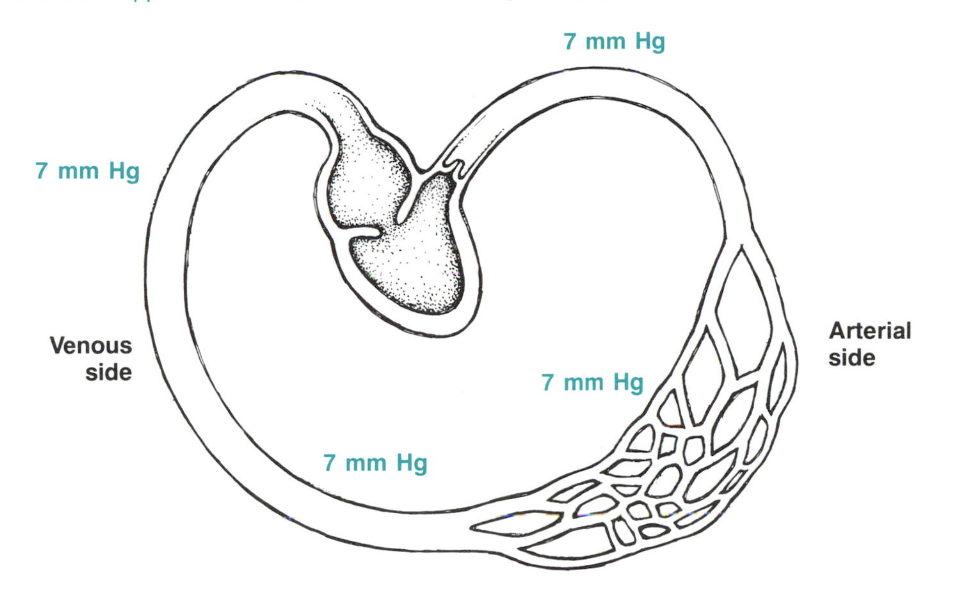

While the Heart Is Beating

The beating of the heart establishes a **pressure gradient** in the cardiovascular system, pressure on the arterial side exceeding that on the venous side.

The pressure gradient maintains flow through the blood vessels.

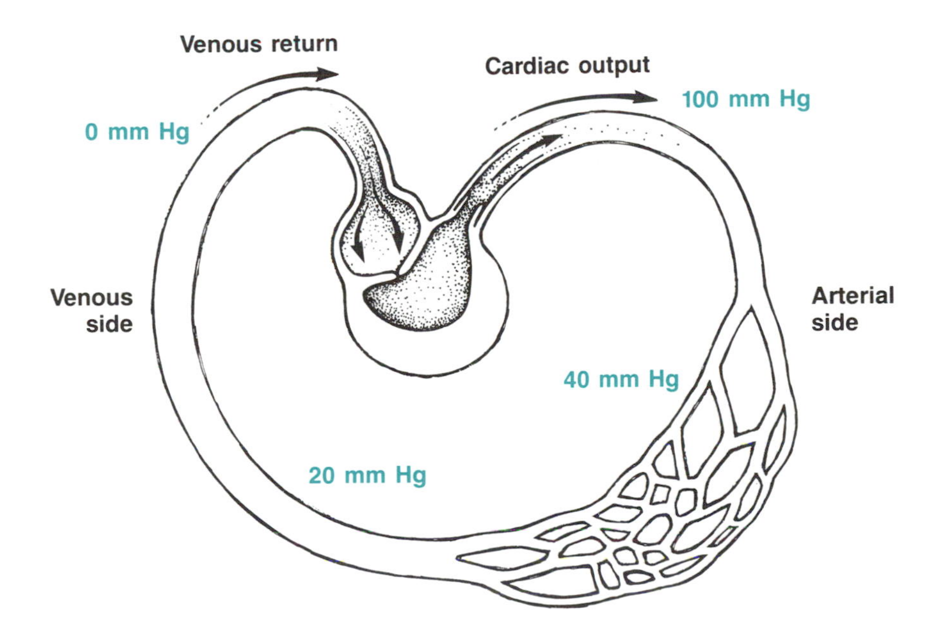

CARDIAC MUSCLE VS. SKELETAL MUSCLE

Structural Differences

Cardiac Muscle	Skeletal Muscle
• Intercalated discs	
• Cross connections between neighboring fibers	• Nerves to individual fibers
• Large T-tubules	• Small T-tubules
• One T-tubule per sarcomere	• Two T-tubules per sarcomere
• Small longitudinal sarcoplasmic reticulum (LSR)	• Well-developed longitudinal sarcoplasmic reticulum
• Numerous mitochondria	• Few mitochondria

Functional Differences

Cardiac Muscle	Skeletal Muscle
• Pacemaker cells are present ↓ Automaticity	• Stable resting membrane potential ↓
	• Stimulus required to generate AP
• Passive Ca^{++} transport helps to shape the action potential	• Ca^{++} transport has negligible role in shaping the action potential
• Long-lasting action potential	• Action potential has short duration
• Duration of action potential relative to duration of muscle twitch makes tetanization impossible	• Tetanization is possible for maximum force
• Aerobic metabolism only	• Aerobic plus anaerobic metabolism

CARDIAC MUSCLE VS. SKELETAL MUSCLE

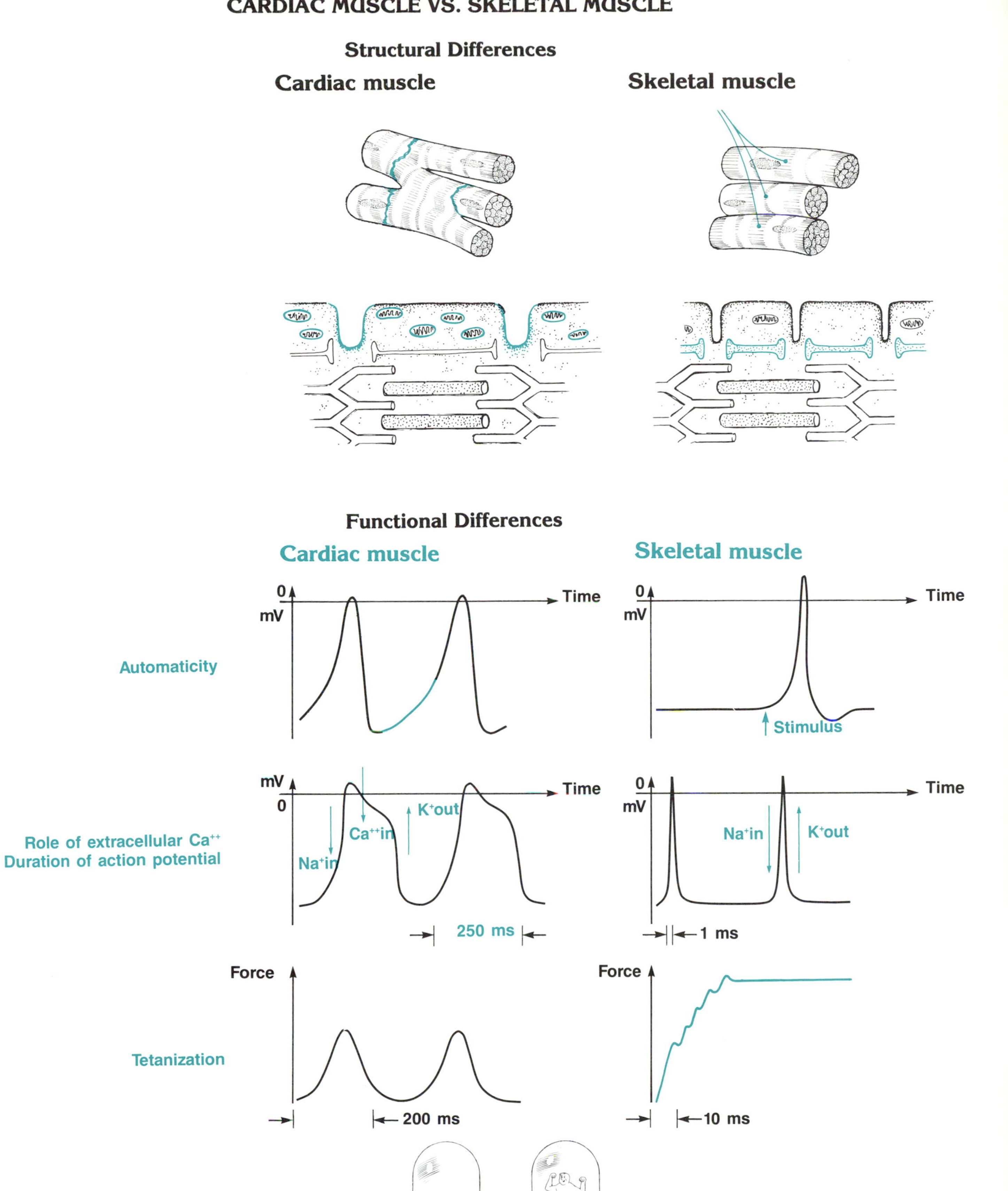

SPREAD OF CARDIAC ACTION POTENTIALS AND GENESIS OF ECG

Two aspects of cardiac anatomy help to create, just before each heartbeat, a large, organized electrical field in and around the heart:

- The specialized conduction systems of the heart (cross connections between muscle fibers and the His-Purkinje system) ensure that large numbers of cells depolarize in synchrony.
- The spread of depolarization along relatively fixed, predetermined paths ensures that the orientation of the electric field with respect to the body surface changes little from beat to beat.

The temporal changes in magnitude and direction of the electrical fields that accompany cardiac depolarization and repolarization can be detected as potential differences between any two points on the body surface. A number of surface points have become conventional for the attachment of recording electrodes. Recording potential differences between them results in **bipolar lead ECGs (leads I, II, and III).**

Cardiac Electrophysiologic Event	Corresponding ECG Deflection
0–70 ms: Action potentials that were spontaneously generated by the dominant pacemaker (S-A node) spread through atrial tissue	**P-wave:** • Upward deflection in lead I • Smaller upward deflection in lead II • Upward, biphasic, or downward deflection in lead III
70–150 ms: A-V node delay	**P-R segment:** • Normally isoelectric • Measured from the end of the P-wave to the onset of the QRS complex (Note that it is **not the same as the P-R *interval*,** which is defined as the interval from the onset of P to the beginning of the QRS complex.)
150 ms: Depolarization of upper interventricular septum (Note that the tip of the septum does not depolarize until later.)	**Q-wave** (defined as the first negative deflection preceding R): • Downward deflection in lead I • Small or absent in lead II • May be absent in lead III because the Q vector would produce an upward deflection in lead III
152–165 ms: Depolarization of both ventricles	**R-wave:** • Large upward deflection in leads I, II and III
180–200 ms: Depolarization of last portions of ventricle and septal tip	**S-wave** (defined as the first negative deflection following R): • Downward deflection in leads I, II, and III
250–600 ms: Repolarization of ventricle	**T-wave:** • Upward deflection in leads I, II, and III

It is also conventional to record ECGs at a number of surface points with respect to a reference that is obtained by voltage division among other surface points. This results in **unipolar ECGs (aV_R, aV_L, aV_F, and precordial leads V_1 through V_6 are typical unipolar leads).**

SPREAD OF CARDIAC ACTION POTENTIALS AND GENESIS OF ECG

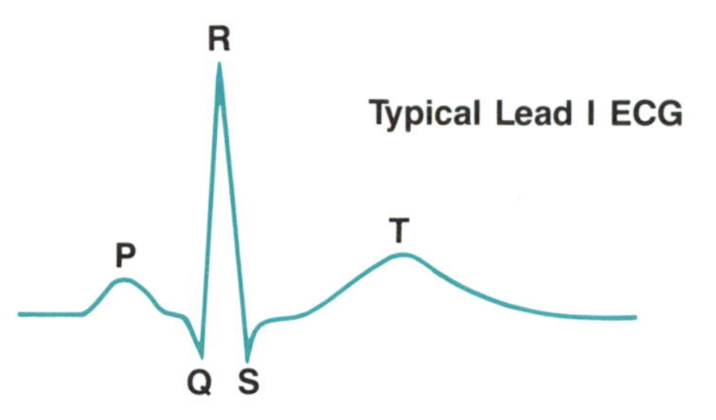

Spread of Depolarization Wave

Through Atria

0-70 milliseconds (ms)

S-A node

LV

RV

Through Ventricles

150 ms

152 ms

155-165 ms

180 ms

Vector Representation and ECG Lead Projection

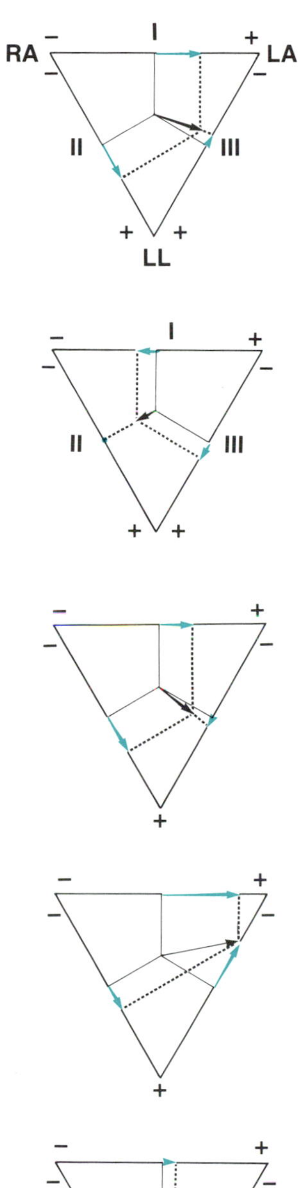

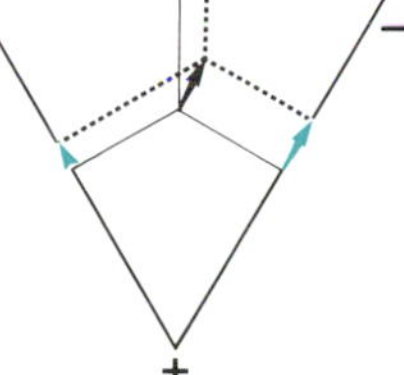

HEMODYNAMICS OF THE CARDIAC CYCLE

The cardiac cycle can be considered in three sequential phases: **atrial contraction, ventricular contraction,** and **significant ventricular filling.**

Left Atrial Contraction

- **The aortic valve is closed** (because of the pressure gradient).
- **The mitral valve is open.**
- Atrial pressure rises **("a" wave)** as atrial muscle contracts.
- Blood is added to the left ventricle. As a result, left ventricular pressure rises with "a," and left ventricular volume increases to the level of left **ventricular end-diastolic volume** (LVEDV).

Left Ventricular Contraction

- Ventricular pressure rises above atrial pressure, and the **mitral valve closes.**
- **The aortic valve is still closed** because of the pressure gradient.
- Ventricular pressure continues to increase during **isovolumetric** (constant ventricular volume) **contraction,** pushes against the mitral valve, and causes the **"c" wave** in the atrial pressure trace.
- There is not yet a flow between ventricle and aorta.
- Continuing ventricular contraction pulls the fibrous valve skeleton toward the apex of the heart, causing the **"x" descent** in the atrial pressure.
- **When ventricular pressure just exceeds aortic pressure, the aortic valve opens.**
- **Ventricular volume decreases** rapidly as the **stroke volume** is ejected.
- Ventricular pressure, as well as aortic pressure and aortic flow, continues to rise until the ventricle begins to relax.
- **The atrium fills throughout ventricular contraction** and initial ventricular relaxation, causing atrial pressure to rise (after the "x" descent).

Initial ventricular relaxation

As the ventricle begins to relax, ventricular pressure falls and the rate of ventricular ejection decreases markedly.

- Ventricular pressure falls below aortic pressure because the ventricles relax rapidly while blood inertance and vessel compliance retard pressure decline in the aorta.

◆ **During this phase there continues to be flow from each ventricle in the direction of the energy gradient (not the pressure gradient!).**
- After the aortic valve closes, there is a clear separation between aortic pressure and ventricular pressure. This separation begins in the period of **isovolumetric relaxation.**

Significant Ventricular Filling

- When the falling ventricular pressure intersects the rising atrial pressure ("v" wave), the **mitral valve opens.**
- **Ventricular volume increases rapidly** as the ventricle fills in response to a gradient created between the full atrium and the rapidly relaxing ventricle.

◆ **At normal heart rate, most ventricular filling occurs in this phase.**
- Rapid ventricular relaxation is responsible for the descent in ventricular pressure ("y" descent in atrial pressure) even while ventricular volume increases rapidly.

HEMODYNAMICS OF THE CARDIAC CYCLE

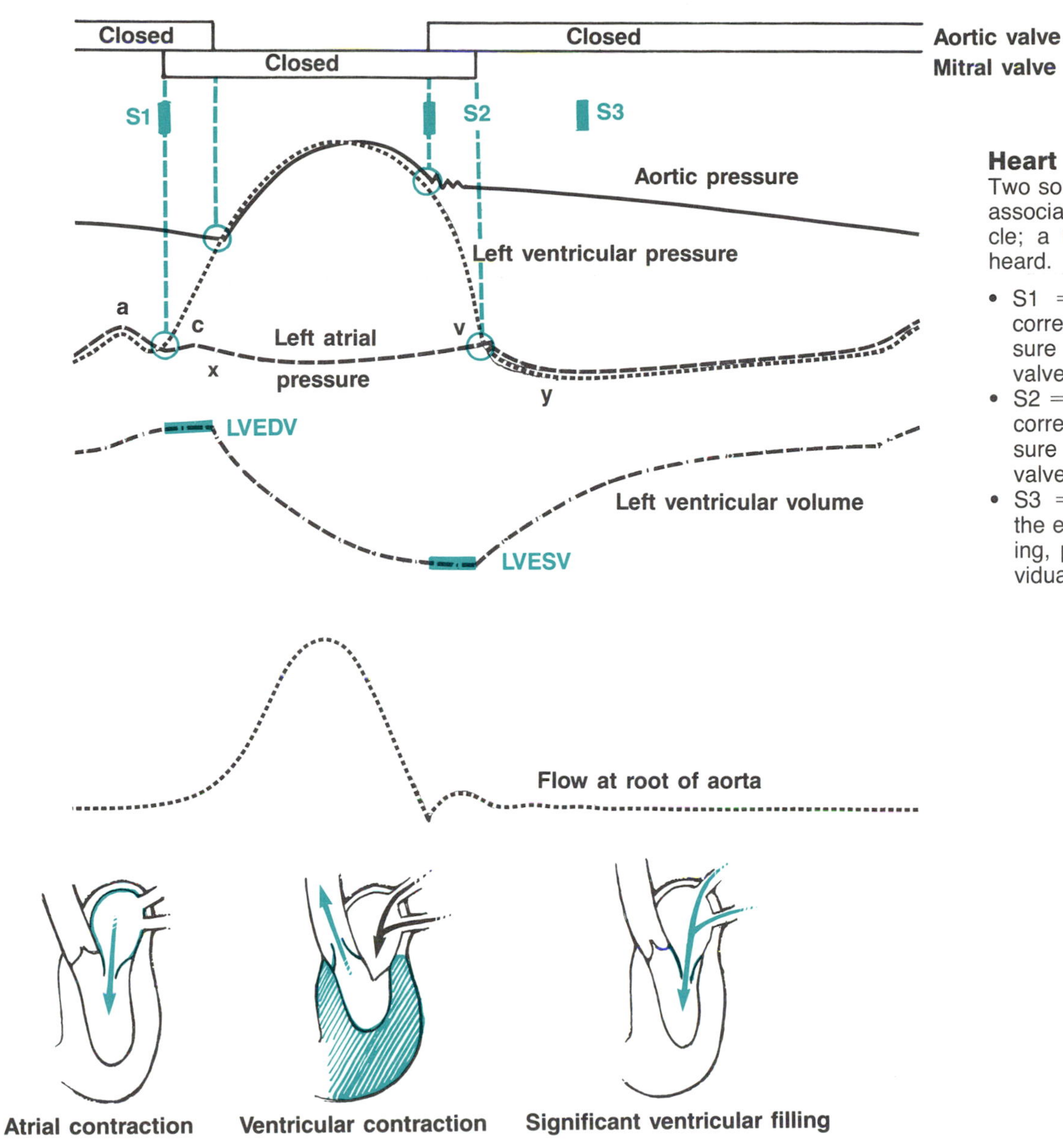

Heart sounds

Two sounds are most commonly associated with the cardiac cycle; a third can sometimes be heard.

- S1 = The first heart sound corresponds in timing to closure of mitral and tricuspid valves.
- S2 = The second heart sound corresponds in timing to closure of aortic and pulmonic valves.
- S3 = can often be heard at the end of rapid ventricular filling, particularly in young individuals.

PERIPHERAL VASCULAR MECHANICS

Mechanisms That Ensure Steady Flow Despite Intermittent Pump Action

Arterial elasticity

Elasticity is a mechanical property of all materials.

- It is measured by a modulus of elasticity.
- It **describes the tendency of a material to return to its resting state after it has been deformed by a force** and after the force has been removed.

Blood vessel elasticity

- derives mostly from **collagen;**
- provides elastic recoil to **propel flow in diastole.**

Viscoelasticity

- derives mostly from **relaxed vascular smooth muscle;**
- provides smoothing-out of fluctuations in flow and pressure.

Mechanisms That Permit Preferential Distribution of Flow

$$\text{Blood flow} = \frac{\text{Driving pressure}}{\text{Resistance}}$$

Resistance in blood vessels

Resistance to flow in a single vessel is influenced most significantly by **vessel diameter.**

Resistance in vascular beds

Two patterns of blood vessel connections are possible to form vascular beds:

Blood vessels in series

- The flow occurs sequentially from one vessel to the next.
- $R_{Total} = R_1 + R_2 + R_3$.

◆ **Vascular beds that show a "series" arrangement of blood vessels can change their resistance only by precise adjustments in the diameter of individual vessels.**

Blood vessels in parallel

- Flow is divided and occurs simultaneously in several vessels.
- $1/R_{Total} = 1/R_1 + 1/R_2 + 1/R_3$.

◆ **Vascular beds that show a "parallel" arrangement of blood vessels can change their resistance by vasomotion (total closing or total opening of selected blood vessels).**

- Opening of previously closed blood vessels is called **recruitment.**

PERIPHERAL VASCULAR MECHANICS

Mechanisms That Ensure Steady Flow Despite Intermittent Pump Action

Arterial elasticity

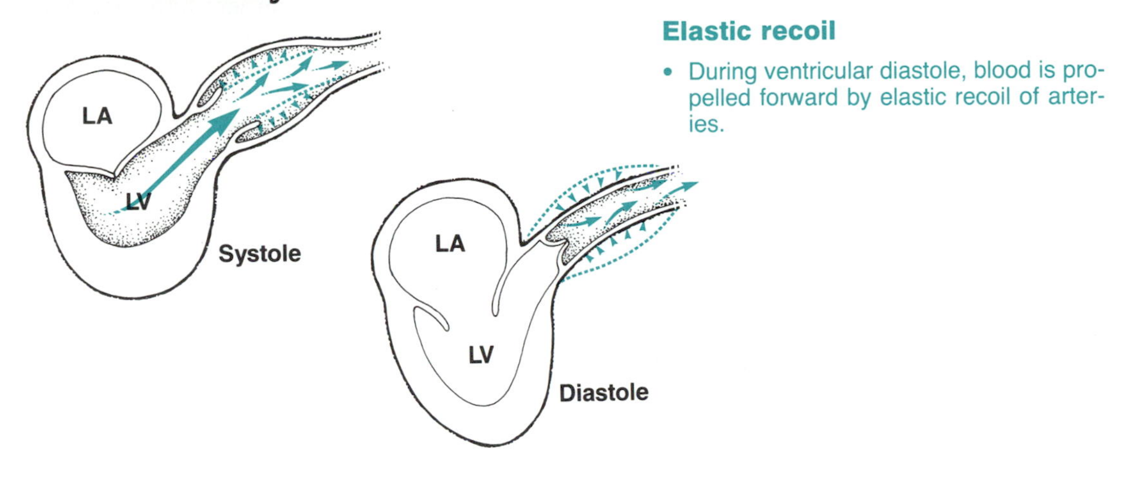

Elastic recoil

- During ventricular diastole, blood is propelled forward by elastic recoil of arteries.

Smooth muscle viscoelasticity

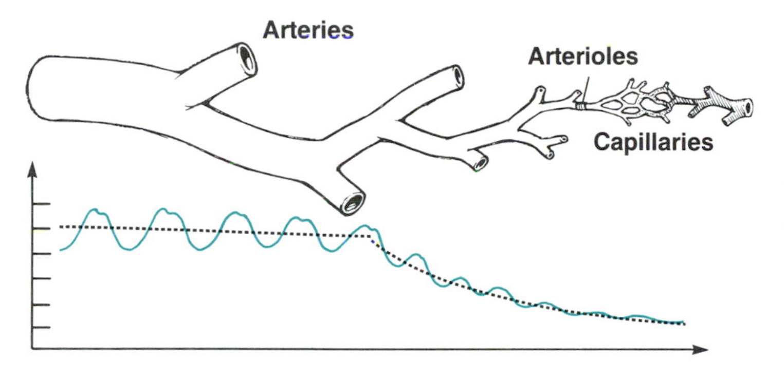

Arterioles

- offer high flow resistance (this decreases the mean pressure in more distal vessels);
- offer high viscoelasticity (this removes pressure fluctuations).

Mechanisms That Permit Preferential Distribution of Flow

Resistance in blood vessels

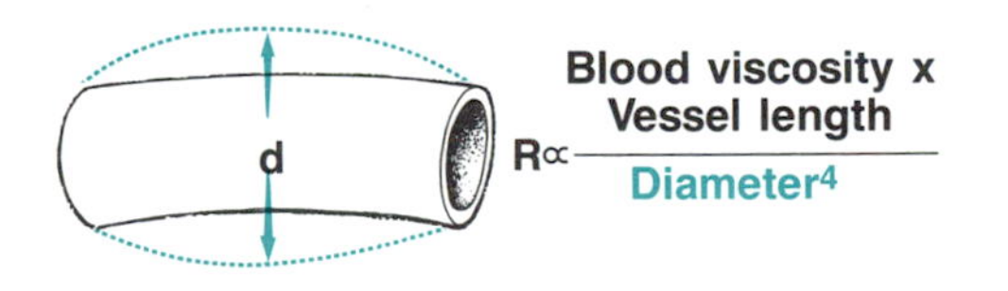

$$R \propto \frac{\text{Blood viscosity} \times \text{Vessel length}}{\text{Diameter}^4}$$

Resistance in vascular beds

Blood vessels in series

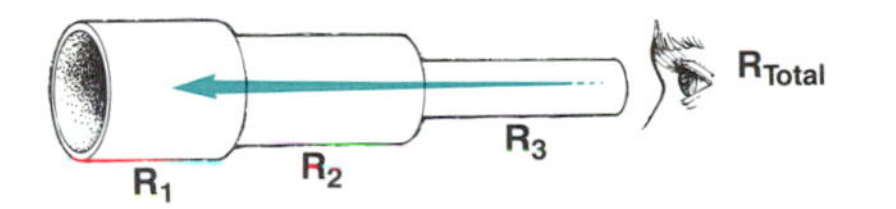

Increasing the resistance in one vessel decreases flow through all vessels.

Blood vessels in parallel

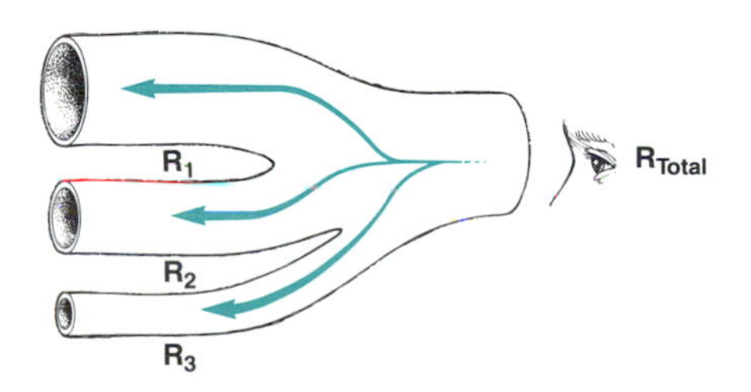

Increasing the resistance in one vessel increases flow through all other vessels.

BLOOD PRESSURE

Effects of Gravity on Blood Pressure

Within a continuous column of fluid and with respect to a measurement reference point within the column,

- **pressure changes with distance from the reference point;**
  - it increases for points closer to the absolute center of gravitational attraction (center of the earth);
  - it decreases for points farther away from the absolute center of gravitational attraction.

Effects of Gravity on the Total Energy Per Unit Volume of Blood

For a volume of blood (of density ρ), flowing at velocity v,

$$\text{Total energy} = \text{Hydrostatic energy} + \text{Gravitational energy} + \text{Kinetic energy}$$
$$= (P_{HEART} + \rho gh) + \rho gh + \tfrac{1}{2}\rho v^2$$

- **The net effect of gravity on total energy is zero.**
- Since blood flows from regions of higher energy to regions of lower energy (*not necessarily* from regions of higher pressure to regions of lower pressure), the position of the unit volume with respect to the measurement reference point has no steady-state effects on blood flow. (There may, however, be *transient* effects after changes in position—e.g., orthostasis.)

BLOOD PRESSURE

Effects of Gravity on Blood Pressure

- All pressures are expressed relative to atmospheric pressure.
- The tricuspid valve is taken to be the relative gravitational reference point for all cardiovascular pressures. This is the point to which all measuring devices are nulled no matter what the body position may be.

In Large Veins (mm Hg) | In Large Arteries (mm Hg)

In Large Veins (mm Hg)	In Large Arteries (mm Hg)
−35	50
−10	75
0	100
20	118
	130
45	140
115	200

Ref Ref

($P_{Heart} - \varrho gh$)

P_{Heart}

($P_{Heart} + \varrho gh$)

The pressure measured at any point in the circulation is the sum of P_{HEART} and gravitational forces (ρhg). (P_{HEART} at any point is what remains after flow resistance losses have been subtracted from the pressure created by heart action at the root of the aorta.)

For example, at the elevated wrist,

ABP = (100 − 5) + blood density × gravitational constant × distance from reference
= 50 mm Hg

VBP = 10 + blood density × gravitational constant × distance from reference
= −35 mm Hg

Note that with respect to the reference point, the distance is −h for all points that lie away from the absolute center of gravitational attraction and +h for all points that lie closer to the absolute center of gravitational attraction.

Effects of Gravity on the Total Energy of Blood

The total energy of a unit volume of blood at any point consists of the sum of hydrostatic energy ($P_{HEART} + \rho gh$), gravitational energy (ρgh), and kinetic energy ($\frac{1}{2}\rho v^2$).

For example, at the elevated wrist ($P_{HEART} + \rho hg$)
= 95 + (−45) on the arterial side
= 10 + (−45) on the venous side

That is, the hydrostatic energy of the volume of blood has decreased because, with respect to the cardiovascular reference point (tricuspid valve), the volume is located within a continuous column of fluid, but farther from the absolute center of gravitational attraction.

Gravitational energy = +45 on both the arterial and the venous sides.

That is, the volume of blood has acquired additional gravitational potential energy by virtue of being located farther from the absolute gravitational reference point (center of the earth).

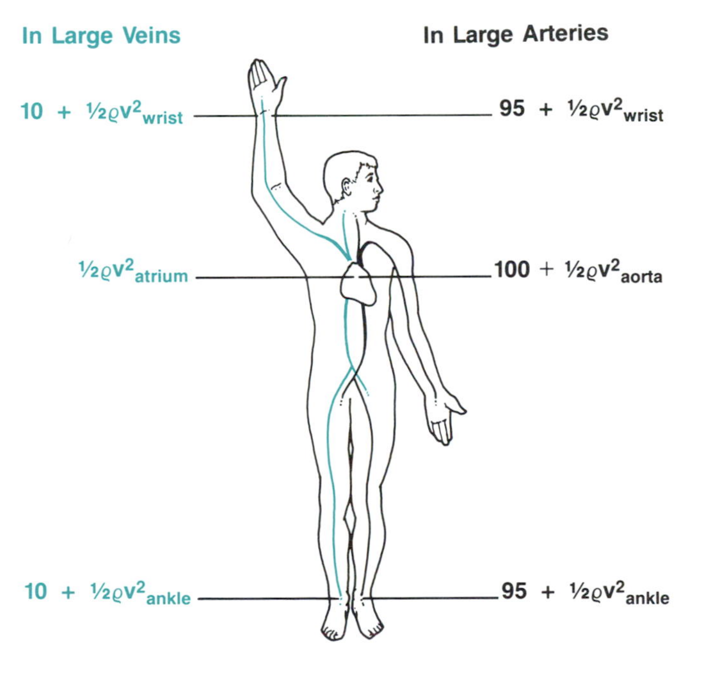

At the elevated wrist

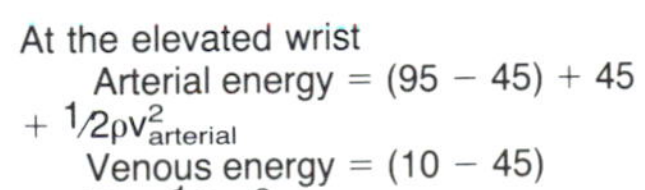

Arterial energy = (95 − 45) + 45 + $\frac{1}{2}\rho v^2_{arterial}$

Venous energy = (10 − 45) + 45 + $\frac{1}{2}\rho v^2_{venous}$

The Relationship Between Pressure and Wall Tension

Laplace's Law can be viewed with slightly different perspectives depending on whether the area of interest is a blood vessel that is being distended by internal pressure or a cardiac chamber in which pressure is being generated as a result of wall tension.

The Arterial Pressure Pulse

Systolic blood pressure

This is the highest pressure reached during a cardiac cycle. Given a diastolic arterial blood pressure upon which the effect of cardiac action is superimposed, **systolic arterial blood pressure is determined by**

- **cardiac performance,**
  - determining the rate of inflow into the arterial system during systole;
- **aortic compliance;**
- total peripheral resistance,
  - determining the rate of outflow from the arterial system and, thereby, the net volume added to the arterial system during systole (its contribution is less than 20 percent of that due to cardiac performance).

Diastolic blood pressure

This is the lowest pressure reached during a cardiac cycle. Given a systolic level from which the pressure begins to decline, **diastolic arterial blood pressure is determined by**

- **total peripheral resistance,**
  - determining the rate of peripheral outflow from the arterial system during diastole;
- **heart rate,**
  - determining the point at which the steady decline of pressure in the diastolic interval is halted.

Pulse pressure

This is the difference between systolic and diastolic pressures.

The Relationship Between Blood Pressure and Tension in Vessel Walls

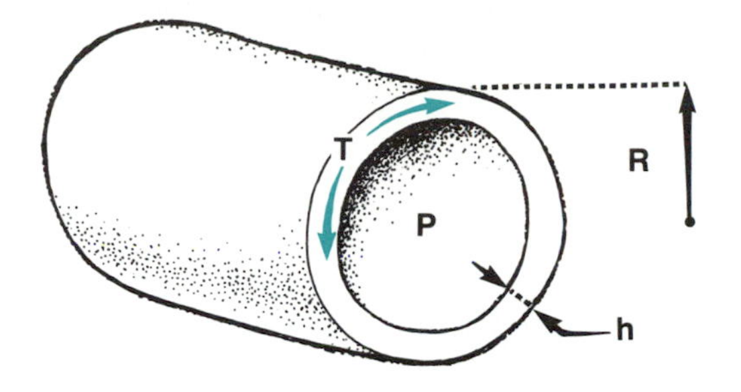

Two Points of View

$T = (P \times R)/2h$

Effects of pressure on wall tension

The tension created in the wall of a blood vessel by local pressure is given by

$$T = (P \times R)/2h$$

Effects of wall tension on pressure

The pressure generated in the ventricle as a result of ventricular wall tensions is given by

$$P = (2h \times T)/R$$

The Arterial Pressure Pulse

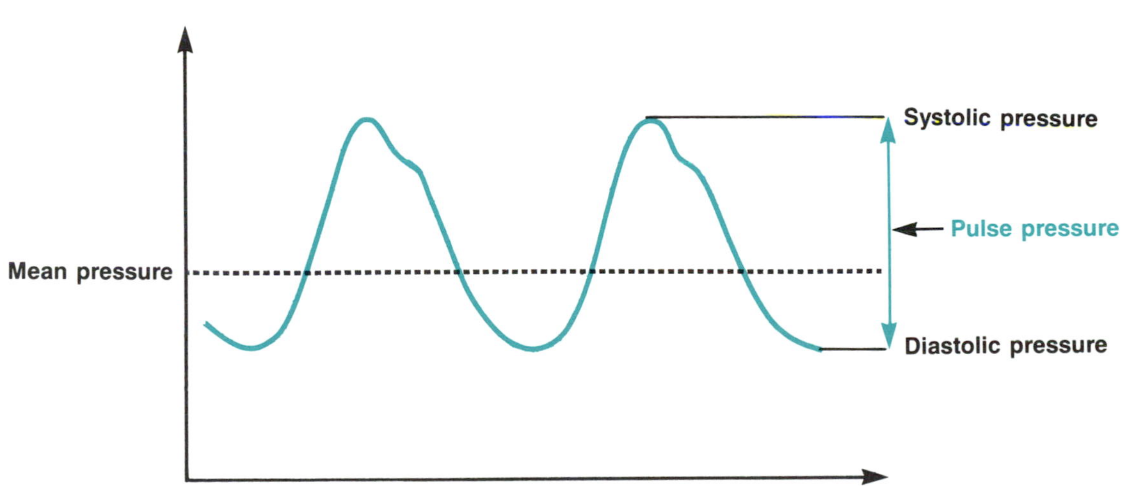

Pulse pressure

is determined mostly by cardiac performance and aortic compliance (total peripheral resistance contributes only 10 to 20 percent).

Mean blood pressure = The average pressure during a cardiac cycle
= Diastolic + ⅓ pulse pressure (for aortic pulse shape)

ADJUSTMENT OF CARDIAC OUTPUT

OVERVIEW

With each beat, each ventricle ejects a **stroke volume** of about 70 mL (in adults). The total volume of blood ejected per minute from each ventricle (i.e., the cardiac output) is the product of stroke volume (SV) and **heart rate** (HR):

$$HR \times SV = CO$$

- **Cardiac output is adjusted by changes in heart rate or stroke volume.**

Mechanisms for Changing Heart Rate

Each heartbeat is initiated by electrical activity in the dominant pacemaker cells, which are normally located in the region of the SA node. The dominant pacemaker generates a spontaneous action potential whenever its membrane potential reaches threshold. Therefore,

- **heart rate is determined by the interval between successive points of intersection of the pacemaker potential with the threshold potential.**

The duration of that interval depends on only three factors, and a change in any of them will change heart rate:

- **Slope** of the pacemaker potential
- Level of threshold potential
- Level of minimum repolarization potential

Short-term mechanisms

Minute-to-minute changes in HR are achieved mainly by changing the slope of the pacemaker potential (by nerves or hormones).

Long-term mechanisms

These are related to changes in extracellular electrolyte concentrations:

- Changes in $[K^+]_{extracellular}$ have marked effects on the minimum level to which pacemaker potentials will fall during repolarization.
- Changes in $[Ca^{++}]_{extracellular}$ influence the threshold level (the threshold is made more negative by decreasing $[Ca^{++}]_{extracellular}$).

Mechanisms for Changing Stroke Volume

Stroke volume is determined by the completeness of systolic ventricular emptying. Systolic emptying, in turn, varies directly with **myocardial performance.**

Four factors influence myocardial performance:

- Ventricular **preload**
- Ventricular **afterload**
- Heart **rate**
- Ventricular **contractility**

ADJUSTMENT OF CARDIAC OUTPUT

Mechanisms for Changing Heart Rate

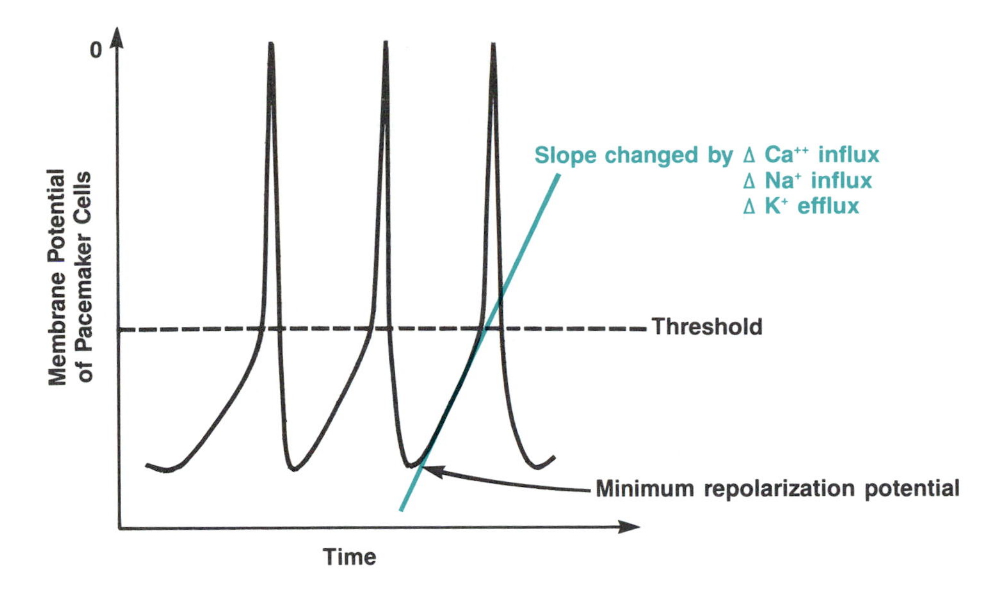

Mechanisms for Changing Stroke Volume

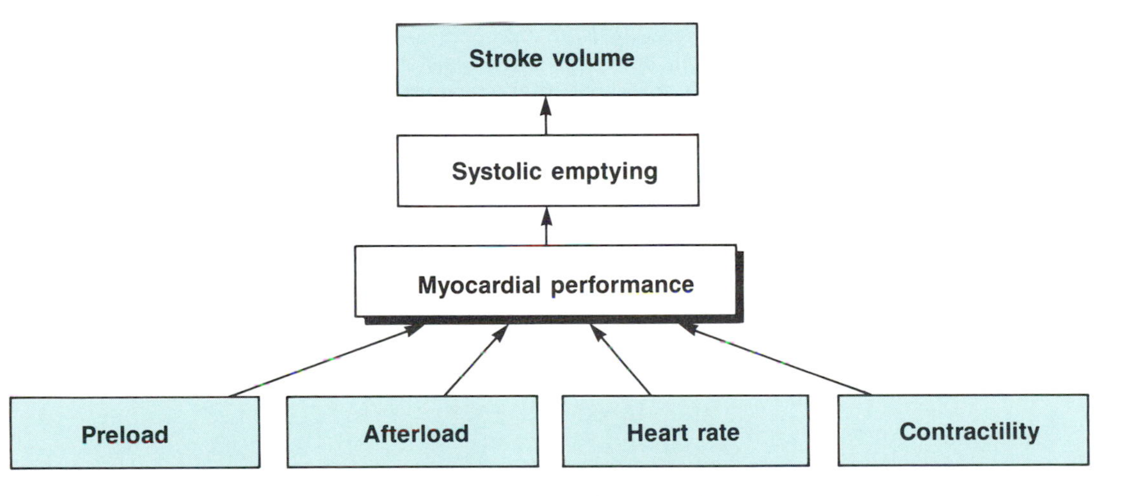

Preload

- determined by the degree of stretch of **ventricular muscle fibers** before they contract.
- Changes in stretch are thought to alter the extent of **alignment between potential myosin cross bridges and potential actin binding sites.**
- The relationship between preload and myocardial performance is expressed by the **Frank-Starling Law** of the heart.
- **Preload depends on**
  —filling time (inversely proportional to **heart rate**),
  —filling pressure **(atrial pressure),**
  —**ventricular compliance.**

Afterload

- Afterload is determined by **resistance to ventricular emptying.**
- Increased resistance increases systolic ventricular pressure and, thereby, the degree of compression of subendocardial blood vessels.

It is thought that **metabolites,** accumulated during this period of compression, alter **coronary blood flow** and, hence, alter myocardial performance.

- The effect of afterload on myocardial performance is often called the **Anrep effect.**

Heart rate

- **Heart rate affects contractility** by changing the duration of the diastolic interval and, thereby, the period of time during which **intracellular Ca^{++}** can be sequestered.
- If the interval is short, then little sequestration occurs and free intracellular $[Ca^{++}]$ is high when the next beat occurs.
- The number of active actomyosin cross bridges is directly related to intracellular Ca^{++}.
- The effect of heart rate on performance is called the **Bowditch effect** or the **Treppe effect.**

Contractility

- This parameter, which cannot be measured directly, is influenced most profoundly by changes in the **biochemical milieu** within the sarcomere.
- In physiological settings it is influenced mostly by **sympathetic discharge, circulating adrenalin, and coronary blood flow.**

ADJUSTMENT OF PERIPHERAL RESISTANCE

The average perfusion pressure of all vascular beds is determined by flow (cardiac output [CO]) and resistance to flow (total peripheral resistance [TPR]):

$$CO \times TPR = ABP$$

Vascular Resistance in Any Tissue

Vascular resistance

- is changed by changing the degree of constriction of **vascular smooth muscle;**
- **represents a balance between**
  - **locally driven needs for matching tissue blood flow to tissue metabolism and**
  - **centrally directed need for survival of the organism.**

Central Nervous and Local Interactions

- Regional resistance is adjusted in such a way that **a given tissue receives the blood flow that is required for the maintenance of its metabolic activity.**
- **Total peripheral resistance** is adjusted in such a way that **mean arterial blood pressure is regulated at a desired level.**

Central nervous regulation of resistance

This is governed by receptors that respond to arterial blood pressure. As a result,

- tissues that are metabolically active have low resistance to blood flow because in them vasodilator actions of metabolites counteract centrally directed sympathetic vasoconstrictor influences;
- blood vessels in tissues that are not active are constricted by sympathetic nerves.

ADJUSTMENT OF PERIPHERAL RESISTANCE

Resistance to Flow

In single blood vessels

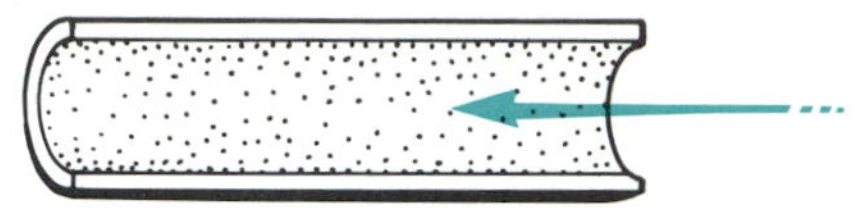

Resistance is determined by the diameter of the blood vessel.

- This is determined by smooth muscle tone.

In vascular beds

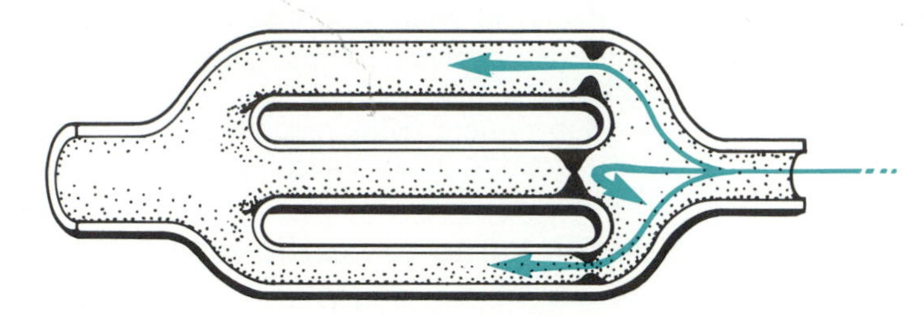

Resistance is determined both

- by the diameter of individual blood vessels (smooth muscle tone) and
- by the number of blood vessels that are open to flow. (This is determined by the setting of precapillary, smooth muscle sphincters: a vessel is open only when its precapillary sphincter is open.)

Influences on Vascular Smooth Muscle

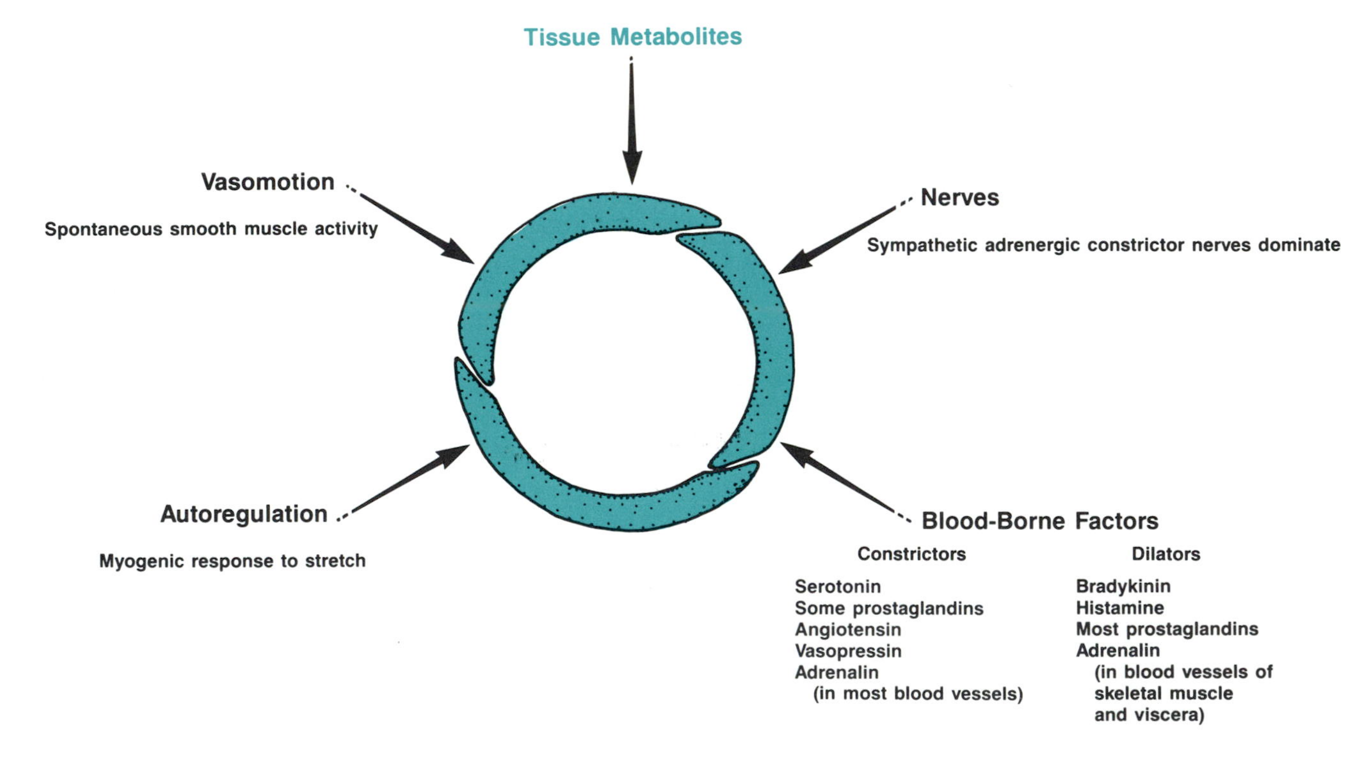

INTEGRATED CONTROL OF ARTERIAL BLOOD PRESSURE

Overall Scheme

In any tissue, blood flow is governed by metabolism because local metabolic factors ensure adequate perfusion by adjusting local vascular resistance.

If a local change in flow or volume is large enough to have systemic effects, then the effects will express themselves as a change in the activity of **cardiopulmonary stretch receptors.**

Subsequent **reflex responses** in heart rate, stroke volume, and total peripheral resistance are designed to accomplish five specific aims.

Specific aims of cardiovascular regulation

1. To adjust cardiac output to levels sufficient to meet whole-body metabolic needs
2. To adjust total peripheral resistance in such a way that it and the existing cardiac output result in the appropriate arterial blood pressure
3. To adjust regional blood flows to meet local needs while protecting cerebral and coronary blood flow
4. To adjust total and regional venous capacitance for transient augmentation of venous return
5. To adjust the rate of net fluid transfer across the walls of the capillary bed by altering the ratio of precapillary resistance to postcapillary resistance.

Cardiovascular regulation is summarized in the following mnemonic structure:

$$\begin{array}{r}\text{HR} \times \text{SV} = \text{CO} \\ \times \\ \text{TPR} \\ \| \\ \text{ABP}\end{array}$$

Reflex Regulation

Some cardiovascular responses are of an intrinsic nature, requiring no nerves or hormones. Examples are responses to changes in preload and afterload. However,

- **the bulk of cardiovascular regulation is governed by receptors that respond to stretch;**
- chemoreceptors normally play a small role except via their effects on ventilation (respiratory pump).
- The cardiovascular reflex centers are located in the brainstem (pons-medulla), and integration of cardiovascular function with other physiological systems is achieved via tracts from the hypothalamus to the brainstem. Emotional correlates derive from limbic and cortical tracts (see Autonomic Nervous System, Chapter 8).
- The effector mechanisms for cardiovascular regulation are
  - pacemaker cells in the SA nodal region,
  - cardiac muscle, and
  - vascular smooth muscle.

INTEGRATED CONTROL OF ARTERIAL BLOOD PRESSURE

Reflex Regulation

Afferent fibers from the stretch receptors travel to the pons and medulla

- in the vagus nerves (X),
- in the aortic depressor nerves (ADN, a branch of the vagus nerves), and
- in the carotid sinus nerves (CSN, a branch of the glossopharyngeal nerves [IX]).

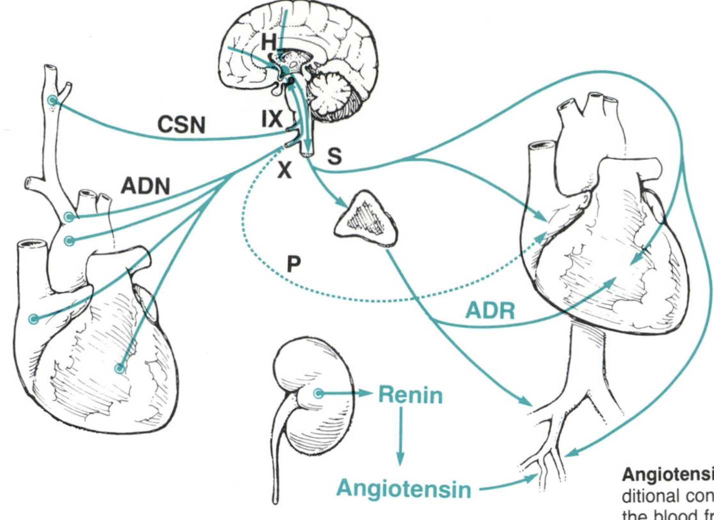

Efferent fibers from the pons-medulla reflex center travel

- via parasympathetic cholinergic fibers (P) to the heart only;
- via sympathetic adrenergic fibers (S) to the heart as well as to all blood vessels.
- Parasympathetic cholinergic fibers do go to some blood vessels, but their number is too small to affect whole-body cardiovascular function. They cause interesting things to happen, however that (see Human Sexual Response).

Angiotensin II, a potent vasoconstrictor, provides additional control of peripheral resistance. It is formed in the blood from **renin** that is released from renal afferent arteriolar (juxtaglomerular) cells in response to decreased local stretch.

Example: Summary of Responses to Hypotension

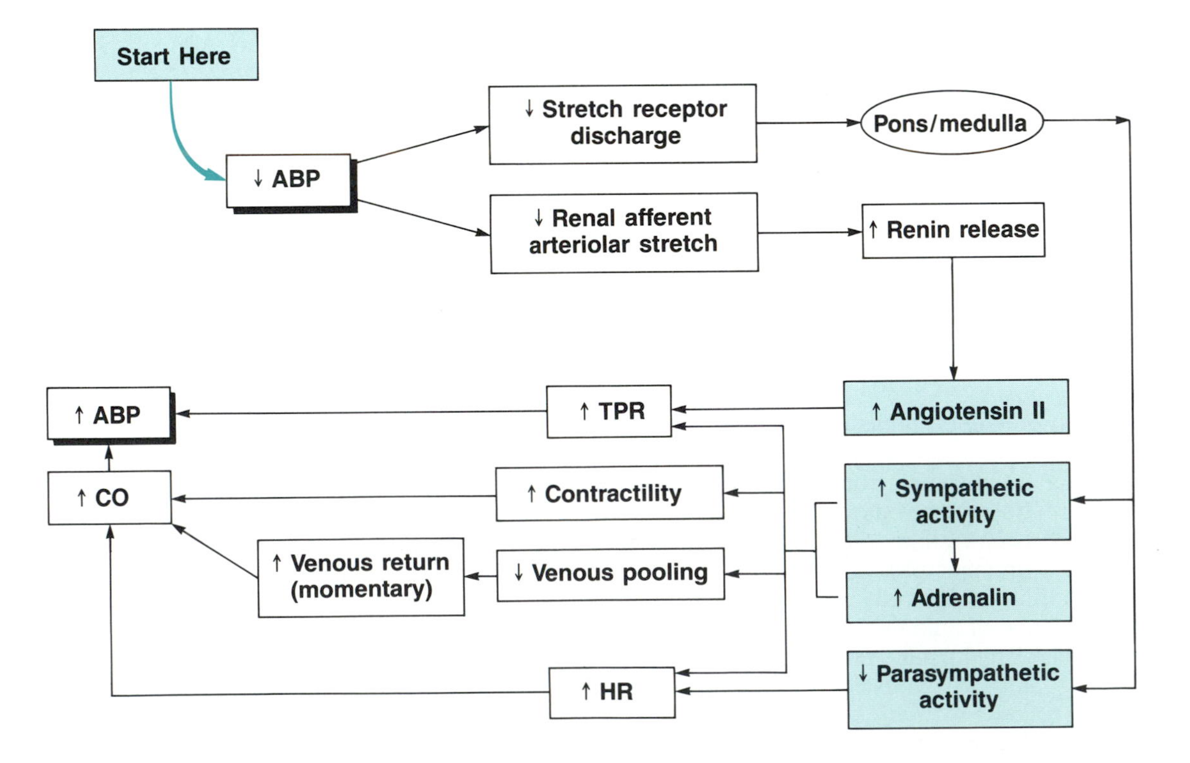

THE MICROCIRCULATION

Structure

Arterioles and metarterioles

These vessels contain smooth muscle units and, therefore, serve as controlling elements. Capillaries often do not directly join arterioles, but branch from metarterioles that form preferential flow channels.

Capillaries

Smooth muscle fibers form the **precapillary sphincter** at the junction between capillary and metarteriole. Beyond that, capillaries contain only endothelial cells with varying sizes and numbers of junctions between them.

- **Nonfenestrated (continuous) capillaries** (found in brain and muscle) have few endothelial junctions, and their clefts are small.
- **Fenestrated capillaries** (found in renal glomeruli and in the splanchnic bed) have more endothelial junctions.
- **Discontinuous capillaries** (found in liver, spleen, and bone marrow) have many and large endothelial junctions.

Venules

Smooth muscle units reappear at this end of the microvascular unit and permit these vessels to act as **elements that control flow.** In addition, their permeant structure allows them to participate in the **exchange functions** of the microcirculation.

Terminal lymphatics

Lymphatic microvessels are anchored to surrounding tissue by microfilaments. They exhibit **spontaneous contractile activity.** Larger lymphatic vessels have valves; this permits **unidirectional pumping** of lymph.

Arteriovenous anastomoses

These muscular bypass channels are found in some tissues—most prominently in those occasionally requiring a **blood flow far in excess of metabolic needs** (e.g., the skin).

Function

Endothelial micromilieu

Capillary endothelial cells synthesize many factors. Their functions range from modulation of flow (e.g., endothelin) to modulation of hemostasis (e.g., prostacyclin, PGI_2). This rich area of cardiovascular physiology lies beyond the scope of this book.

Exchange of substances across the capillary endothelium

Substances move across capillary endothelium at rates that are determined by **concentration gradients, pressure gradients, and endothelial permeability.**

- Lipid-soluble substances move through endothelial cytoplasm; other substances move through cell junctions or via pinocytotic vesicles.
- Loss of fluid to the interstitium is an inevitable consequence of the existence of a hydrostatic pressure within the leaky capillaries. Homeostasis requires that the lost fluid be recaptured.

◆ Balance of fluid exchange across capillary endothelium is maintained by two mechanisms:
1. A hydrostatic pressure gradient tends to push fluid out of the capillary, and an opposing protein-osmotic (i.e., oncotic) pressure gradient tends to draw fluid into the capillary.
2. Excess interstitial fluid is removed by lymphatic uptake.

THE MICROCIRCULATION

Structure

A microcirculatory unit consists of an **arteriole,** perhaps a small number of **metarterioles,** several **capillaries,** a **venule,** perhaps an **arteriovenous anastomosis** (a-v shunt), and several terminal **lymphatics.**

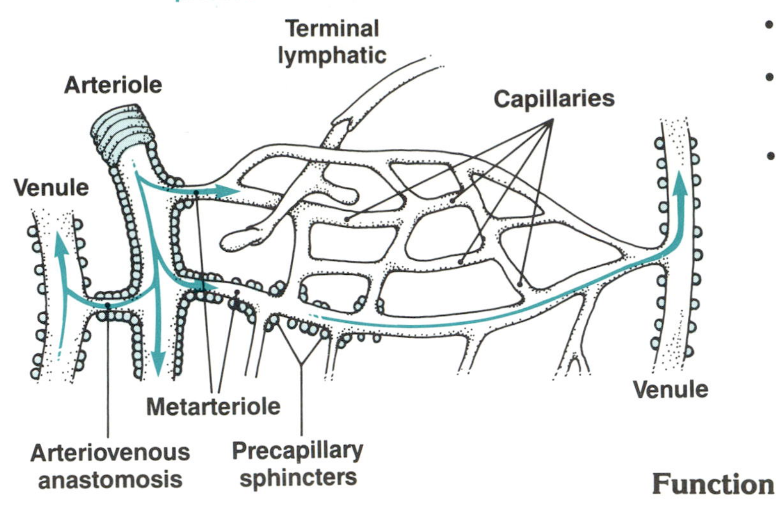

- The tone of precapillary sphincters determines the fraction of regional flow that passes through true capillaries.
- The tone of smooth muscle in arteriole, metarteriole, and venule determines the total flow through the microvascular unit.
- The tone of precapillary (i.e., arteriolar and metarteriolar) smooth muscle relative to postcapillary (i.e., venular) smooth muscle determines the hydrostatic pressure within capillaries.

Function

Exchange of fluid across the capillary endothelium

The rate of fluid loss is governed by transmural gradients in hydrostatic and oncotic pressures.

Transmural Pressures When the Precapillary Sphincter Is Open

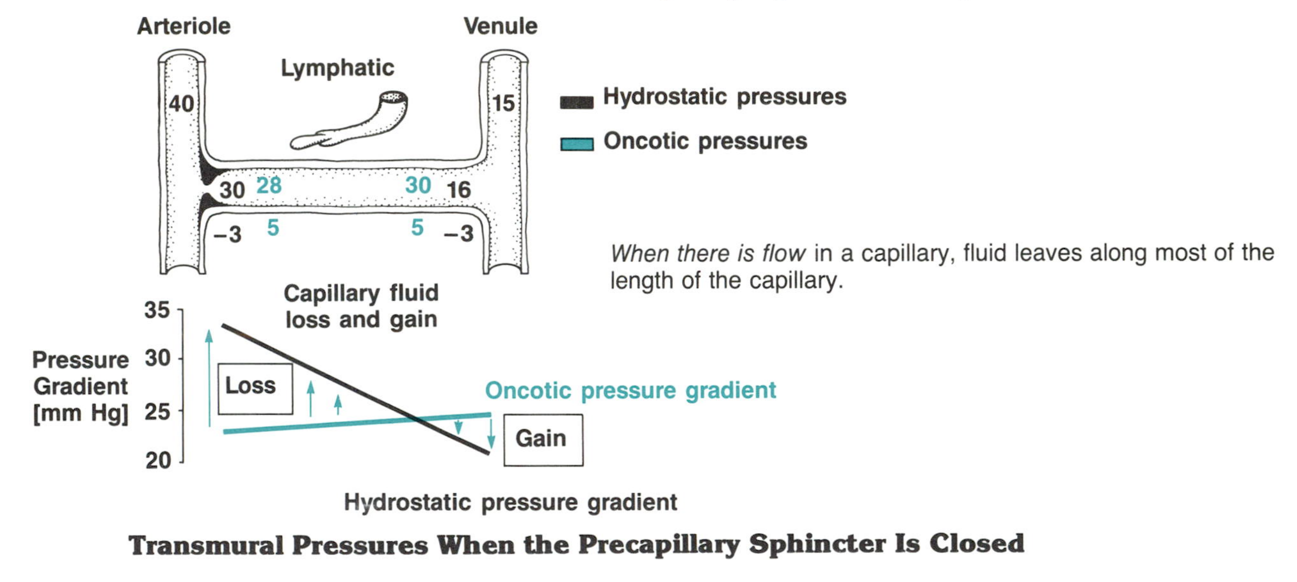

When there is flow in a capillary, fluid leaves along most of the length of the capillary.

Transmural Pressures When the Precapillary Sphincter Is Closed

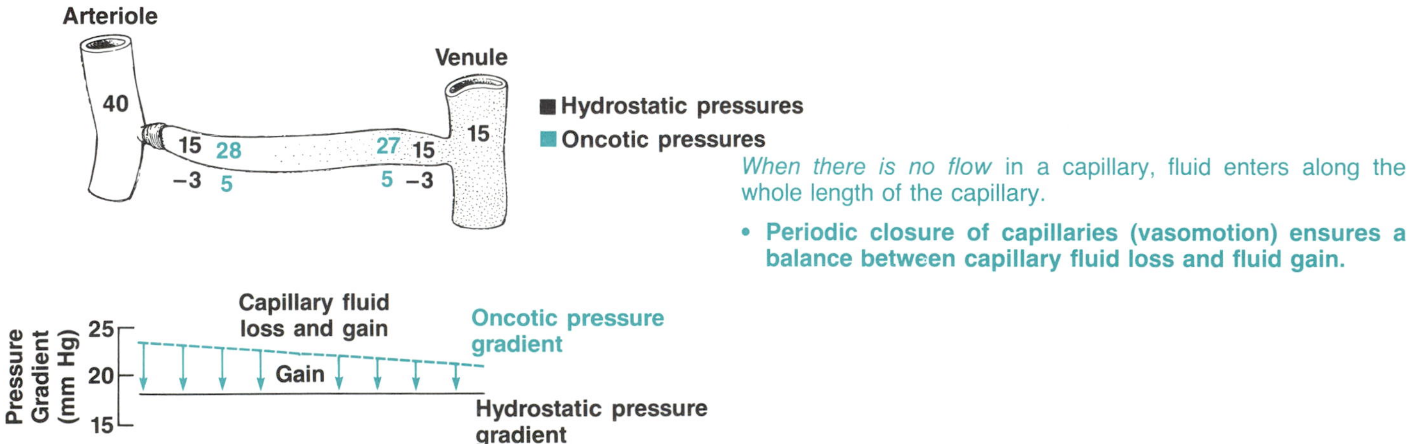

When there is no flow in a capillary, fluid enters along the whole length of the capillary.

- **Periodic closure of capillaries (vasomotion) ensures a balance between capillary fluid loss and fluid gain.**

FEATURES OF REGIONAL CIRCULATIONS

Cerebral Circulation

- Overall blood flow is well autoregulated in the 60- to 160-mm Hg arterial blood pressure range, but **regional cerebral flow varies with regional metabolic activity.**
- Although blood vessels are innervated, control by metabolic factors predominates:

◆ **Cerebral blood vessels are very sensitive to Pco_2** (which causes vasodilatation via H^+ formed when CO_2 combines with H_2O), but less sensitive to plasma [H^+] because H^+ cannot get through the blood-brain barrier.

- Most cerebral capillaries are of the nonfenestrated type and form an effective barrier against many substances (the **blood-brain barrier**). It is breached in only a few areas of the brain.

Coronary Circulation

- Tissue pressure provides strong mechanical impediment to flow during systole, particularly in subendocardial vessels. Therefore,

◆ **blood flow is maximal during ventricular diastole.**

- Control of flow is driven by local **metabolic factors** including tissue O_2 demand.
- **Tissue O_2 consumption is determined mostly by the cardiac muscle tension that must be developed for ejection of stroke volume.**
- O_2 extraction is normally high. As a result, **increased O_2 demand is met mostly by increased coronary flow.**

Cutaneous Circulation

- The cutaneous circulation is characterized by **a capillary plexus subserving tissue nourishment and a venous plexus subserving body temperature regulation.**
- Sympathetically controlled arteriovenous anastomoses permit a large blood flow through the venous plexus if it is required for thermoregulation.
- "Active" vasodilator influence is derived from chemicals that are liberated secondarily to cholinergic activation of sweat glands (bradykinin might be such a chemical).

Splanchnic Circulation (GI Tract, Liver, Spleen, and Pancreas)

- The major circulatory function is that of a **blood reservoir.**
- Dilator influence derives from adenosine (purinergic nerves), gastrin, cholecystokinin, and others.
- Constrictor influence derives from strong sympathetic influence; however, sympathetic constriction cannot be maintained (owing to autoregulatory escape).
- Most **liver blood flow** is provided by the **portal vein;** some is provided by the hepatic artery. Higher-order branches of these vessels meet in the liver sinusoids to form the capillary network of the liver. Sinusoids are lined with highly permeable endothelium. **The liver contains about 15 percent of total body blood volume. About half of this can be rapidly expelled by sympathetic constrictor effects.**
- The **spleen** contains blood of high hematocrit and acts as an **erythrocyte reservoir.** Its effectiveness in this regard varies among species.

Skeletal Muscle Circulation

- Blood vessels in ***resting* skeletal muscle** are a major locus of **peripheral resistance adjustments** in response to neural stimuli.
- Vessels in ***exercising* skeletal muscle** are controlled by local **metabolic factors.**

Pulmonary Circulation

Pulmonary vascular pressures at any level from artery to veins are lower than comparable systemic pressures. Nevertheless, the entire cardiac output passes through this vascular bed, requiring a **low resistance to flow.**

Hydrostatic implications of low vascular pressures

Orientation relative to gravity (upright vs. supine) and alveolar pressure (inspiration vs. expiration) has marked effects on flow distribution in the lung. In an upright individual,

- intravascular pressure ($P_{HEART} - \rho gh$) is low at the apex. As a result, the vasculature is compressed by alveolar pressure (which, in this region, is greater than intravascular pressures during all phases of respiration), and **the apex is poorly perfused.**
- Alveolar pressure is greater than (or nearly equal to) venous pressure in many regions of the lung. Hence, in those regions
  - **alveolar pressure is an important determinant of blood flow;**
  - respiratory fluctuations in alveolar pressure have significant influence on blood flow.

Fluid exchange across pulmonary capillaries

The dynamics of fluid exchange are governed by the same factors as in peripheral capillaries, but there are quantitative differences:

- Pulmonary capillary hydrostatic pressure is only about 7 mm Hg, and pulmonary interstitial hydrostatic pressure is about −8 mm Hg.
- Pulmonary capillaries are more permeable to protein than are systemic capillaries. Therefore, the oncotic pressure of pulmonary interstitial fluid is near 14 mm Hg. (Pulmonary plasma oncotic pressure is 28 mm Hg, identical to systemic plasma.)
- **The net pressure forcing fluid out of patent capillaries is only about 1 mm Hg.**

Structural factors contributing to low pulmonary vascular resistance

Two structural features contribute:

- The capillary bed is very large. (This makes the effective diameter also very large in comparison with the effective diameter of the systemic circulation.)
- Arterioles and venules have little smooth muscle. This has implications for the control of pulmonary vascular resistance.

Control of pulmonary vascular resistance

- Sympathetic and parasympathetic effects are present but weak.
- **Pulmonary vascular resistance is controlled mostly by passive influences** such as the effect of gravity on intravascular pressure and the compressive effect of alveoli on blood vessels.
- Hypoxia and hypercapnia cause vasoconstriction in the lung. This shifts blood flow away from poorly ventilated alveoli.

6

BODY FLUIDS AND ELECTROLYTES

Body Fluid Compartments
Filtration, Reabsorption, and Secretion
Renal Clearance
Renal Handling of Ions and Water
Urinary Concentration and Dilution
Renal Regulation of Extracellular Volume and Osmolality
The Role of the Kidney in Acid-Base Balance

BODY FLUID COMPARTMENTS

OVERVIEW

Body water exists within anatomically defined compartments. At the largest level of division it is classed as **intracellular fluid** and **extracellular fluid,** which are separated from each other by the **cell membrane.** Extracellular fluid is further subdivided into **interstitial fluid** and **intravascular fluid** (plasma), which are separated from each other by the **capillary endothelium.**

Electrolyte Composition of Body Fluid Compartments

Intracellular fluid

- **K^+** and **$protein^{n-}$** are the predominant ions (i.e., many proteins exist as negatively charged ions).

Extracellular fluid

- **Na^+, Cl^-,** and **HCO_3^-** are the predominant ions.

Intravascular fluid (plasma)

Plasma has a **higher $protein^{n-}$ concentration** than does interstitial fluid. This has two consequences:

- Because the protein molecules are charged, charge equality demands that all diffusible ions (including Na^+, Cl^-, and HCO_3^-) be distributed at slightly unequal concentrations between interstitial fluid and plasma (this is known as the **Gibbs-Donnan phenomenon**).
- Plasma osmolality exceeds interstitial osmolality by about 1 mOsm/kg. Although this difference is too small to be measured, its water-attracting effect **(oncotic pressure)** is equivalent to a hydrostatic pressure of 25 mm Hg.

BODY FLUID COMPARTMENTS

Fluid Intake and Fluid Output

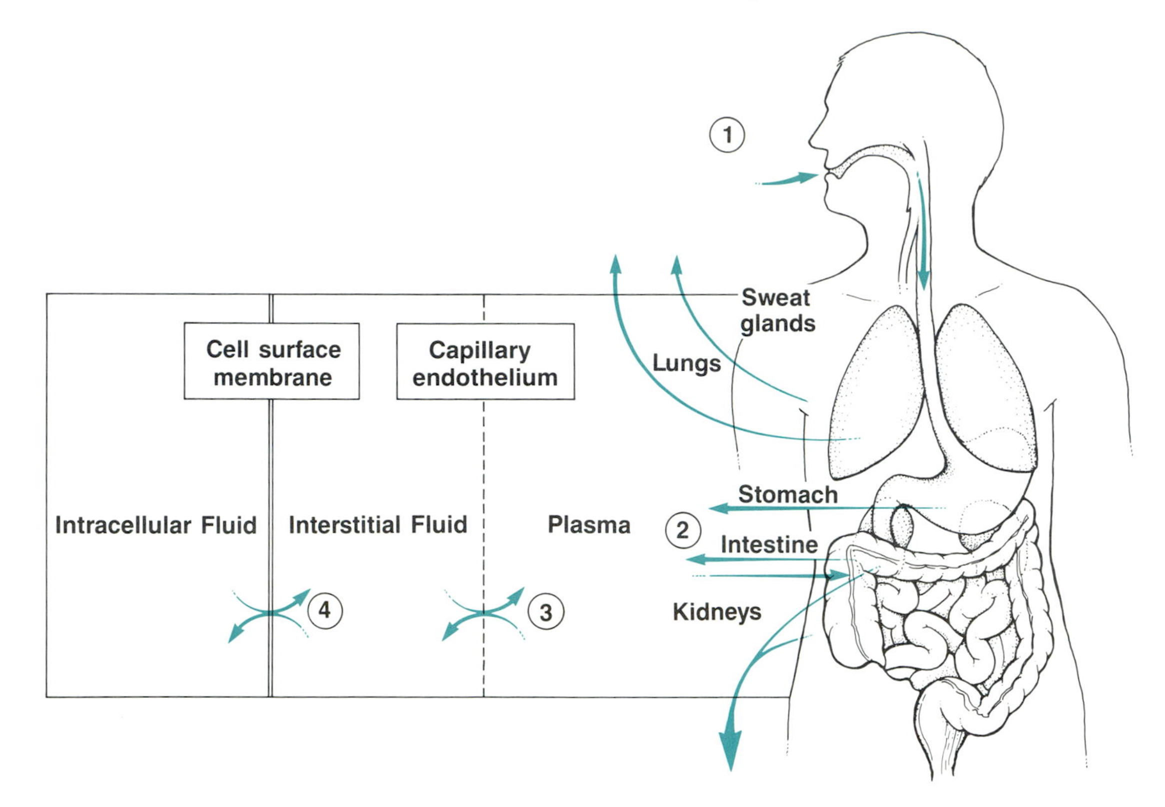

① Water and electrolytes normally enter by mouth.
② Most of the intake enters plasma via the intestinal wall and its adjacent blood vessels. Some of the intake travels through the GI system and leaves with stool.
③ From the plasma, some water leaves via the lungs, some water and electrolytes leave via sweat glands and via the kidney, and the remaining water exchanges with interstitial fluid across the capillary membrane.
④ Interstitial fluid exchanges with intracellular fluid across the cell membrane.

Forces Determining Exchanges Between Body Fluid Compartments

Fluids

Across the cell membrane

- Net fluid movement occurs instantaneously in response to an **osmotic gradient.**

Across the capillary endothelium

- Net fluid movement occurs in response to a gradient in hydrostatic pressure (ΔP) or a gradient in oncotic pressure ($\Delta\pi$).

Electrolytes

Across the cell membrane

Net electrolyte movement across the cell membrane occurs in response to changes in one or more of the following:

- Gradient in electrical potential
- Concentration gradient
- Rate of active transport processes

Across the capillary endothelium

There is free diffusion of all ionic species except $protein^{n-}$.

Whole-Body Balance and Distribution of Fluids and Electrolytes

Two principles determine content and distribution:

- **The kidneys determine how much of the water intake and electrolyte intake are to be left in the body.**
- **The physicochemical forces just outlined determine how to distribute throughout the body the materials left behind by the kidneys.**

Forces Determining Exchanges Between Body Fluid Compartments

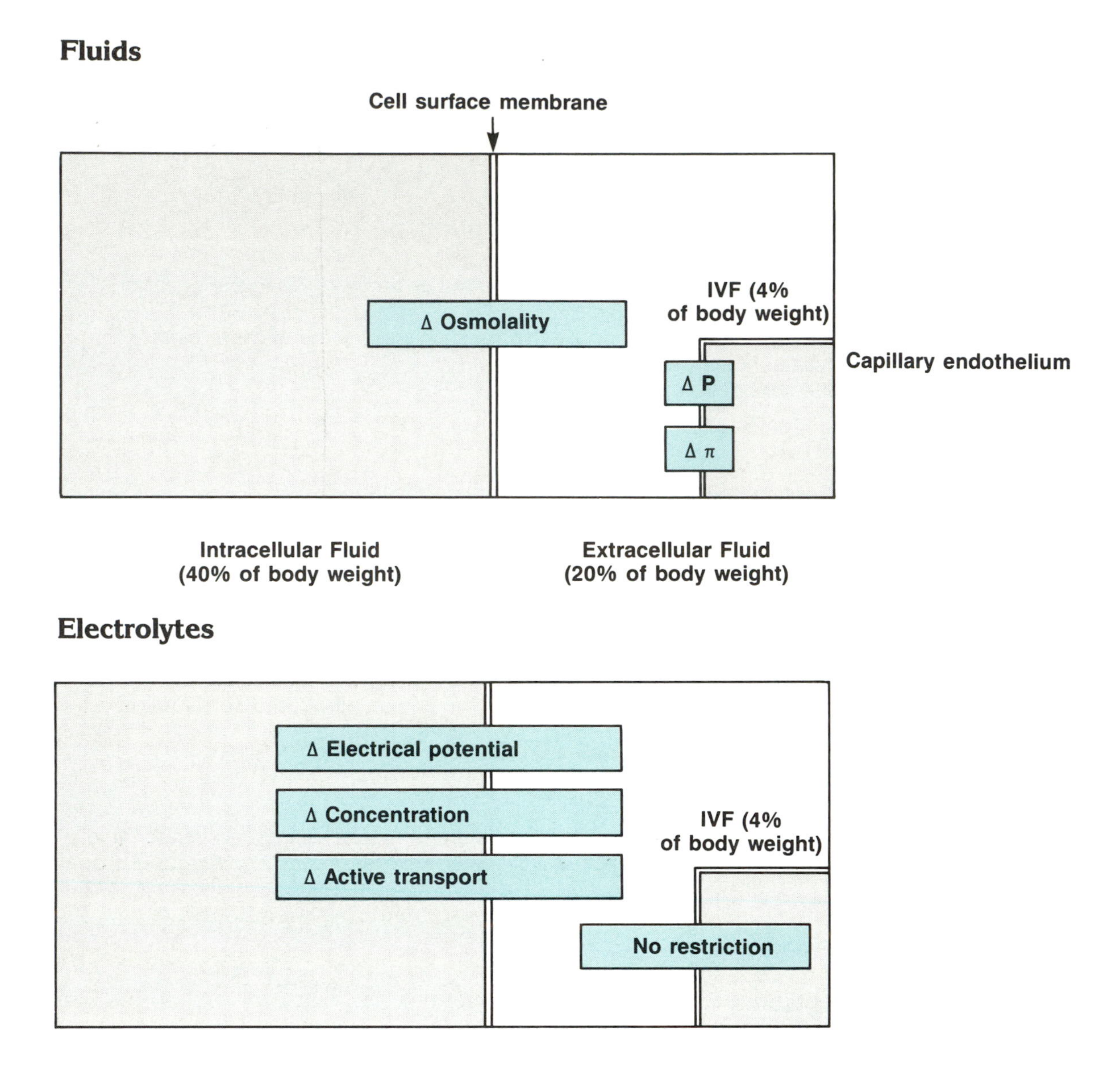

FILTRATION, REABSORPTION, AND SECRETION

OVERVIEW

Nephrons produce urine from blood by three mechanisms: filtration, tubular reabsorption, and tubular secretion.

Filtration

Glomerular filtration is driven by an **imbalance between hydrostatic pressure and protein-osmotic pressure gradients** across the glomerular capillary wall. Glomerular filtration pressure is of the same magnitude as that in capillaries of skeletal muscle.

Tubular Reabsorption

Ninety-nine percent of the filtered fluid is reabsorbed as it moves along the nephron.

- **Solute reabsorption** occurs by **active or passive** mechanisms. Some of these establish osmotic gradients across the tubular wall.
- **Fluid reabsorption** is driven by **osmotic gradients.**

Active reabsorption

This requires **direct expenditure of metabolic energy.** There are two kinds of active reabsorptive mechanisms: those exhibiting a transport maximum (T_m-limited) and those exhibiting a gradient-time maximum.

T_m-limited transport (e.g., glucose)

- This occurs via **a limited number of carriers.**
- The reabsorption maximum (T_m) is reached when all carriers are occupied.
- **If substance in excess of T_m is presented in the filtrate, all of the excess will be excreted.**

Gradient-time-limited transport (e.g., Na^+)

- During the time of contact between tubular fluid and tubular cell, only a certain gradient in concentration or in electrical potential can be established across the tubular epithelium.
- The magnitude of the transepithelial gradient varies inversely with the rate of flow of tubular fluid.

Passive reabsorption

This occurs in the direction of gradients in concentration or electrical potential. Energy is not expended directly on the transported substance, although energy must have been used to establish the electrochemical gradient.

Tubular Secretion

Secretion is analogous to tubular reabsorption, but transport is into tubular fluid rather than out of it. The most important examples of secreted substances in humans are creatinine, H^+ and K^+.

FILTRATION, REABSORPTION, AND SECRETION

The basic functional unit of the kidney is the nephron and its associated peritubular capillary network.

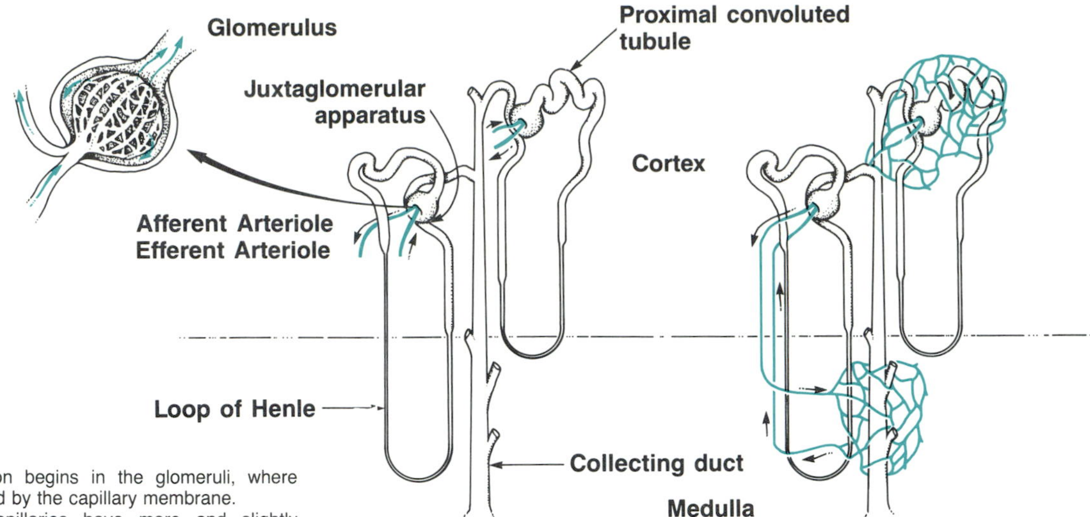

Filtration

- Urine formation begins in the glomeruli, where blood is filtered by the capillary membrane.
- Glomerular capillaries have more and slightly larger pores than skeletal muscle capillaries. This accounts for the large volume of filtrate being produced in the kidneys each day (about 180 L in a healthy adult human).

Cortical nephrons are shorter than juxtamedullary nephrons. Hence, they offer less total surface area for solute-solvent exchange than do juxtamedullary nephrons.

The early part of each distal tubule makes close contact with the afferent arteriole of the same nephron at the juxtaglomerular apparatus.

Efferent arterioles of **cortical nephrons** divide and form capillaries that surround the convoluted tubules.

Efferent arterioles of **juxtamedullary nephrons** form long, straight capillary vessels (the descending vasa recta) that run parallel to the loops of Henle for variable distances into the medulla. At their target depth they break up into a network of vessels that empty into the ascending vasa recta.

- Glomerular filtration yields capsular fluid, whose composition resembles that of interstitial fluid.
- As this fluid moves along the nephron, solutes are reabsorbed from it to the peritubular blood vessels or secreted into it from tubular cells.

Tubular Reabsorption

The handling of Na^+, Cl^-, water, and urea in the nephron demonstrates the interdependence of active and passive mechanisms:

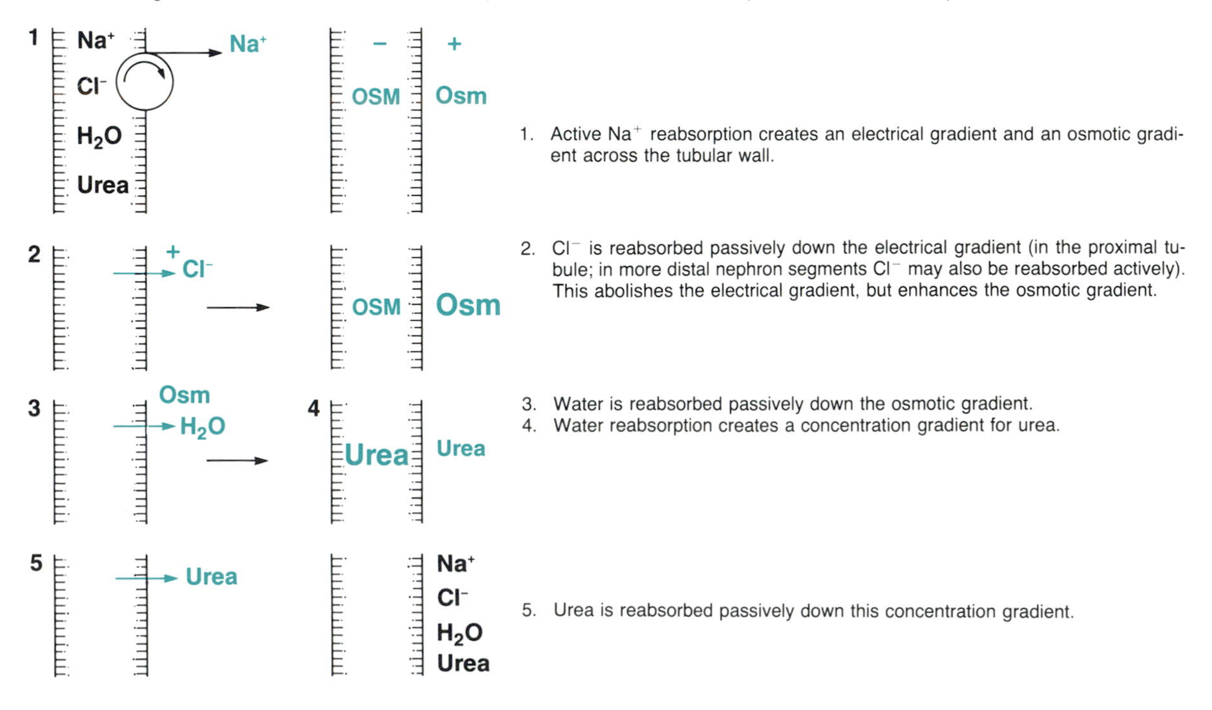

1. Active Na^+ reabsorption creates an electrical gradient and an osmotic gradient across the tubular wall.
2. Cl^- is reabsorbed passively down the electrical gradient (in the proximal tubule; in more distal nephron segments Cl^- may also be reabsorbed actively). This abolishes the electrical gradient, but enhances the osmotic gradient.
3. Water is reabsorbed passively down the osmotic gradient.
4. Water reabsorption creates a concentration gradient for urea.
5. Urea is reabsorbed passively down this concentration gradient.

RENAL CLEARANCE

Clearance of Inulin

The concept of renal clearance is important mainly because there exists a substance, namely inulin, that

- is not secreted into the nephron, but
- is freely filtered through the glomerular capillaries and
- is not reabsorbed from the nephron.

As a consequence of these features of the renal handling of inulin,

- at steady state the amount of inulin appearing per minute in the urine is equal to the amount of inulin filtered per minute through all glomeruli.
- **The clearance of inulin (C_{In}) is identical to the rate of glomerular filtration (GFR), and C_{In} is used as a measure of GFR.**
- The clearance of any substance can be used as an indication of the manner in which the substance is handled by the nephron; that is,
  - if $C_X > C_{In}$, then X is both filtered *and* secreted;
  - if $C_X < C_{In}$, then X is first filtered and then reabsorbed.

RENAL CLEARANCE

Definition

For any substance that is excreted in urine, the total amount excreted during 1 minute can be considered to have come from a hypothetical volume of plasma that is now completely cleared of the substance. This hypothetical minute-volume is defined as the **renal clearance** of the substance.

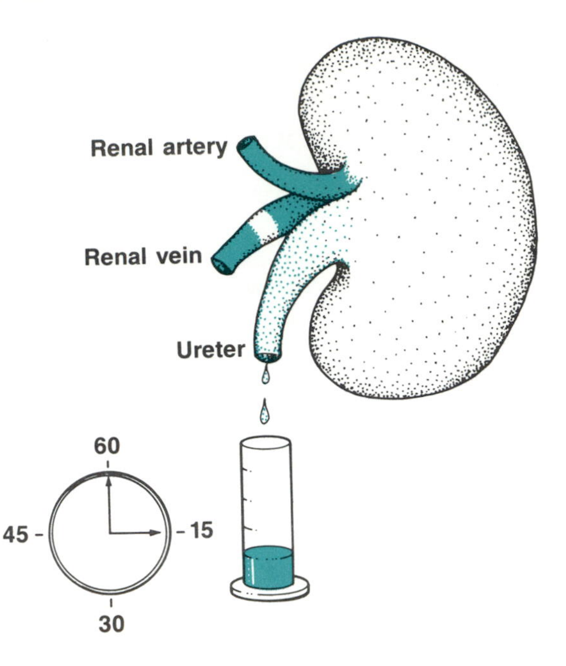

Calculation

$$\text{Clearance} = \frac{\text{Urine concentration}}{\text{Plasma concentration}} \times \text{Urine flow rate}$$

$$\text{i.e., } C_X = \frac{U_X}{P_X} \times V$$

Renal Handling of Inulin

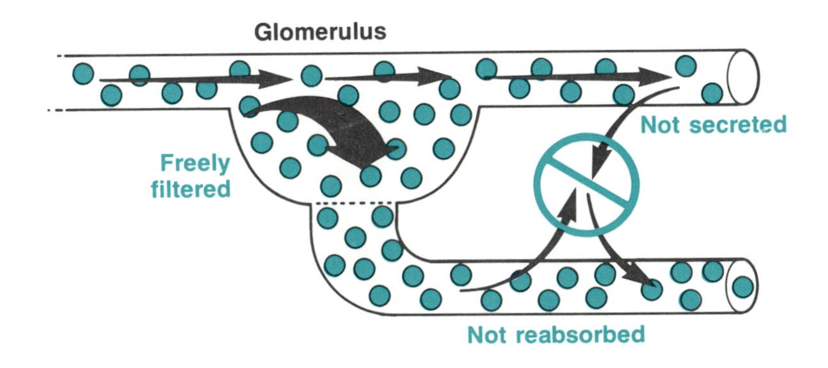

RENAL HANDLING OF IONS AND WATER

Reabsorption occurs in three steps: from tubular lumen into tubular cells, from there to the peritubular interstitial space, and then to the peritubular capillaries.

Proximal Convoluted Tubule

Solute reabsorption and secretion

Solute transport occurs by **active transport,** by **Na^+ cotransport,** and by **passive transport.**

Active transport mechanisms

- Driven directly by ATP
- Responsible for the transport of Na^+ and K^+. (Much of this occurs in the lateral intercellular spaces and creates in them a region of high osmolality.)

Na^+ gradient-coupled cotransport

- Driven by the Na^+ concentration gradient across the cell surface membrane on the brush border side
- Responsible for reabsorption of glucose and amino acids
- Responsible for secretion of H^+

Passive transport

- Driven by electrochemical gradients across the cell surface membrane. (Many of these are created or enhanced by mechanisms of active transport or cotransport.)
- Responsible for reabsorption of Cl^- and urea

Water

Water always follows passively down the osmotic gradients created by solute transport. As a consequence of water transport, interstitial hydrostatic pressure increases and interstitial protein osmotic pressure decreases. The resulting change in capillary transepithelial forces favors capillary uptake of interstitial fluid.

RENAL HANDLING OF IONS AND WATER

The energy turnover of the kidney is concerned mostly with the reabsorption of Na^+. As a consequence, 180 L of glomerular filtrate are reduced each day to 1 to 2 L of urine. The amount and composition of the urine depend on the requirements for homeostasis of body fluids, body electrolytes, and acid-base composition.

Proximal Convoluted Tubule

Epithelial structure

Proximal tubular cells have a large surface area for reabsorption and show evidence of high metabolic activity:

- An extensive brush border on the lumenal side
- Many infoldings on the interstitial side
- Large lateral intercellular spaces
- Many mitochondria
- High concentration of protein carrier molecules in the brush border

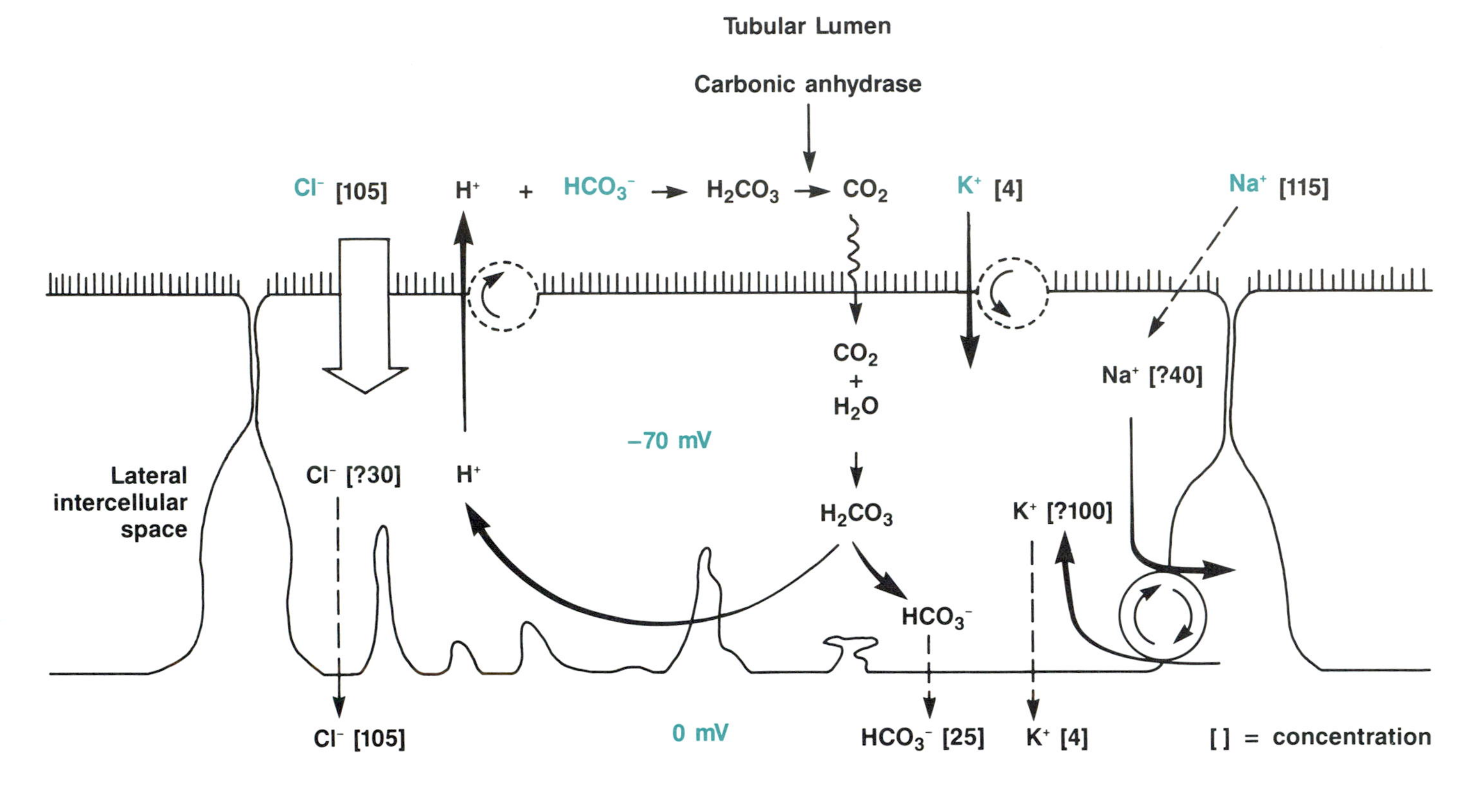

Chloride

In the proximal tubule, Cl^- is reabsorbed by passive mechanisms, but details are not yet known.

Bicarbonate

H^+ is secreted into the lumenal fluid and combines with filtered HCO_3^- to form H_2CO_3, which dissociates because carbonic anhydrase is present. The resulting CO_2 diffuses into the tubular cell, where it forms H^+ and HCO_3^-. H^+ is secreted into the lumen, and HCO_3^- diffuses passively into the peritubular interstitium.

Sodium

Na^+ enters the cell passively on the lumenal side, down concentration and electrical gradients. It is pumped actively into the lateral intercellular spaces.

Potassium

K^+ enters the cell mostly by active transport on the lumenal side because the electrical gradient is not sufficiently large to cause net transport against the concentration gradient. There is net outward transport on the interstitial side because the concentration gradient is larger than the opposing electrical gradient and because the active, inwardly directed pump is relatively slow.

RENAL HANDLING OF IONS AND WATER—CONT'D.

Loop of Henle

Thin segment

Cells in this portion of the nephron are thin. They have no brush border and few mitochondria.

Descending portion

- Highly permeable to water and moderately permeable to most solutes

Ascending portion

- Less water-permeable than the descending portion

Thick segment

Cells in this portion resemble those of the proximal convoluted tubule, but lack the brush border.

Distal Convoluted Tubule, Collecting Tubule, and Medullary Collecting Duct

Reabsorptive mechanisms in these portions are similar to those acting in earlier segments of the nephron.

Although only 10 to 20 percent of the filtered load reaches the distal convoluted tubule (and less than that reaches more distal portions), **these segments of the nephron are responsible for the fine adjustment of solute handling and water handling to physiological needs.** Such adjustment is accomplished in part by the hormones aldosterone and vasopressin.

RENAL HANDLING OF IONS AND WATER—CONT'D.

Distal Nephron

Distal convoluted tubule

- Reabsorption of solutes and water occurs by mechanisms that are similar to proximal tubular mechanisms.
- There is net secretion of K^+ and H^+ into the tubular lumen at this site, the ratio depending on acid-base status.
- Both Na^+ and K^+ transport rates are regulated by aldosterone.
- Water follows down the osmotic gradients.

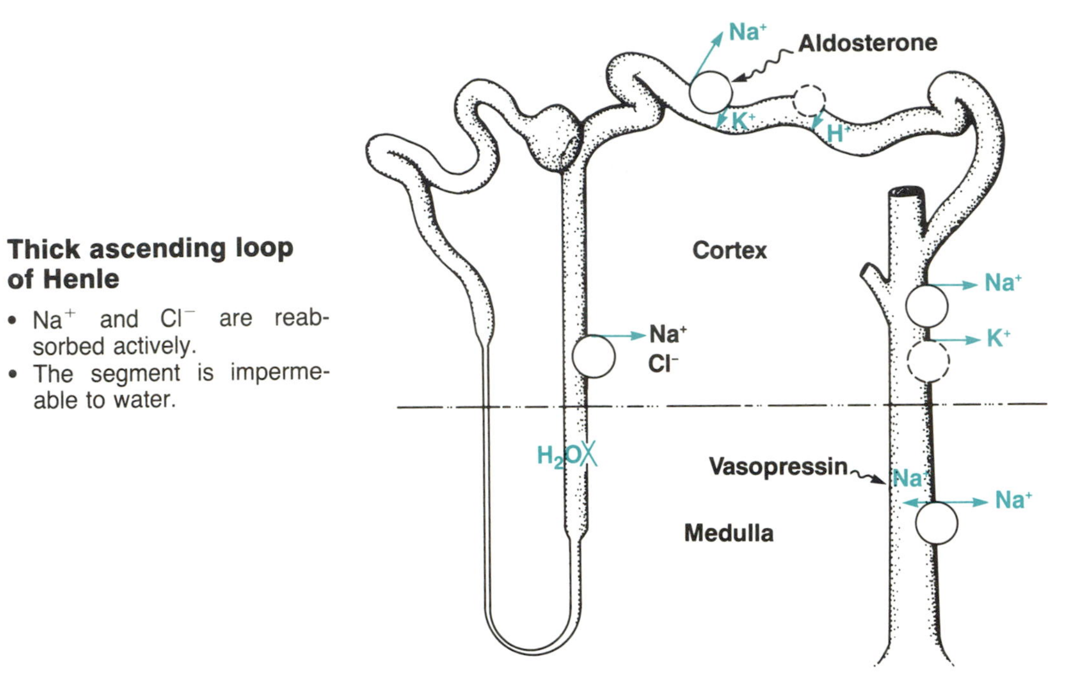

Thick ascending loop of Henle

- Na^+ and Cl^- are reabsorbed actively.
- The segment is impermeable to water.

Cortical collecting tubule

Na^+ and K^+ are reabsorbed actively, and water follows passively.

Medullary collecting duct

- Na^+ can be reabsorbed at varying rates, depending on the status of body electrolytes.
- Water follows passively. Its rate of transport is determined
  —by the osmotic gradient between lumen and medullary interstitium and
  —by epithelial water permeability (which is directly related to the plasma concentration of vasopressin (antidiuretic hormone, ADH).

URINARY CONCENTRATION AND DILUTION

OVERVIEW

While the osmolality of plasma and of glomerular filtrate is remarkably constant at about 300 mOsm/kg, **urine can be as dilute as 50 mOsm/kg in conditions of excess water intake or as concentrated as 1,200 mOsm/kg in severe dehydration.**

Three factors are vital to the production of urine with such a range of osmolality:

- The presence of a very **high osmolality in the medullary interstitium**
- The anatomical **routing of the water-permeable collecting duct through the region of high medullary interstitial osmolality**
- The modulation of water permeability in the collecting duct by the hormone **vasopressin** (also called antidiuretic hormone, ADH)

Medullary Interstitial Osmolality

Medullary interstitial osmolality increases progressively from renal cortex to renal medulla. Creation of highly concentrated fluid in this region requires

- transport of osmotically active material (mostly Na^+, Cl^-, and urea) and
- exclusion of water.

NaCl accumulation

The progressive corticomedullary osmotic gradient is in part established by a **countercurrent multiplier mechanism for NaCl.** This mechanism requires

- anatomic juxtaposition of the two limbs of the Loop of Henle,
- differentially selective permeabilities in each limb, and
- active transport mechanisms that will take either Na^+ or Cl^- out of the ascending limb.

Urea accumulation

Urea is formed during protein metabolism and is excreted mainly in urine. Its excretion is determined

- directly by **tubular load** (the product of plasma concentration and glomerular filtration rate) or
- inversely by **tubular flow rate.**

Urea is responsible for approximately one-half of the total urine osmolality. It becomes concentrated along the nephron as a result of differences in urea permeability among tubular cells along the length of the nephron.

URINARY CONCENTRATION AND DILUTION

Overview

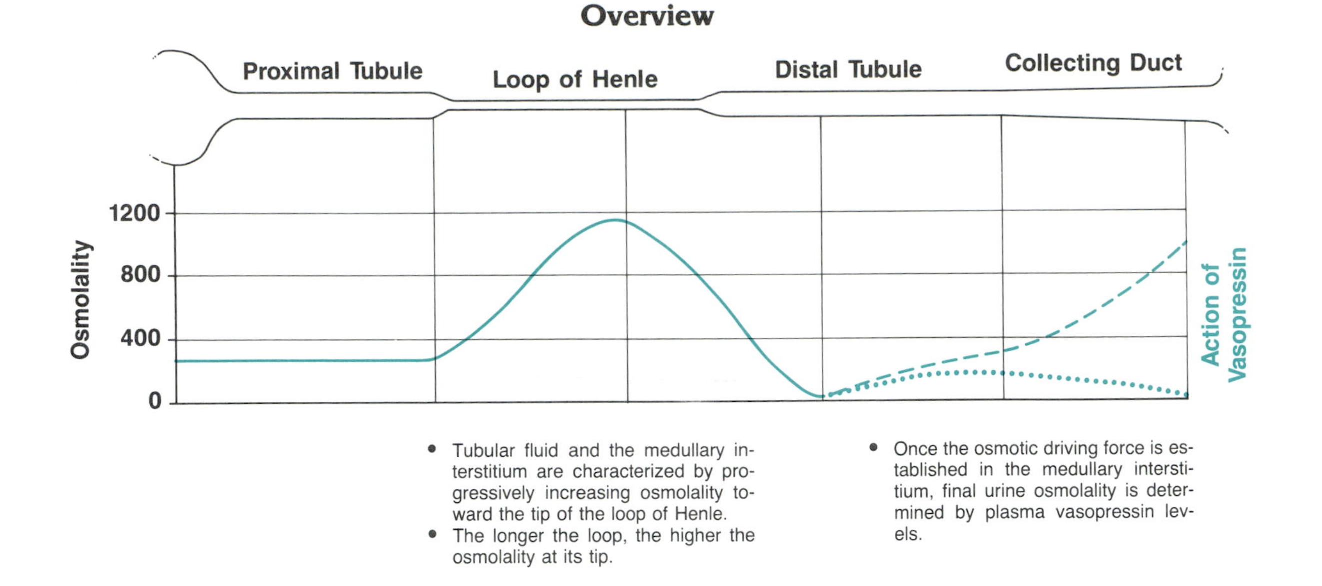

- Tubular fluid and the medullary interstitium are characterized by progressively increasing osmolality toward the tip of the loop of Henle.
- The longer the loop, the higher the osmolality at its tip.

- Once the osmotic driving force is established in the medullary interstitium, final urine osmolality is determined by plasma vasopressin levels.

Medullary Interstitial Osmolality

NaCl accumulation

Countercurrent multiplier for NaCl

- As tubular fluid flows down the descending limb, it enters the region of the active pumps that are located in the ascending limb. These pumps add Na^+ and Cl^- to the descending fluid and to the interstitial space.
- Water does not follow the ions out of the ascending limb because the ascending limb is water-impermeable.
- The net result is that NaCl concentration increases slightly in this portion of the descending limb and at this level in the interstitium.

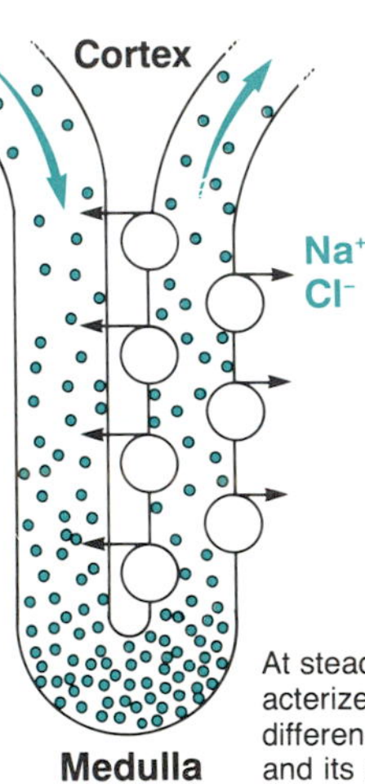

A little deeper into the medulla, more NaCl is added by the active pumps and the osmolality increases further.

At steady state, each level is characterized by a fixed concentration difference between ascending fluid and its surroundings.

Urea accumulation

Urea is a normal constituent of glomerular filtrate. It is present in low concentration up to the thin descending limb of the Loop of Henle because to that point the urea permeability of tubular cells is high and urea can easily move down any concentration gradients that are created by water reabsorption.

- No urea can leave the nephron in the distal convoluted tubule and the early collecting tubule because these segments are urea-impermeable.
- Water reabsorption in these segments causes urea concentration to increase progressively until the tubular fluid reaches the cells of the inner medullary collecting tubules.

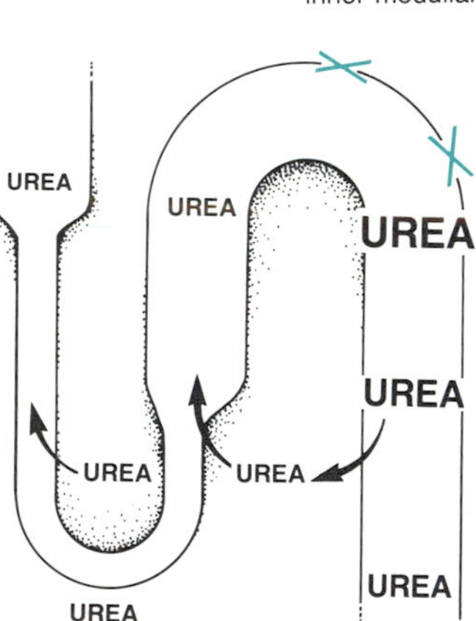

Tubular fluid urea concentration increases slightly between the thin descending limb and the thin ascending limb because
—water reabsorption occurs more rapidly than urea reabsorption (tubular permeability to water exceeds that to urea);
—urea is added to the tubule (driven by a concentration gradient).

Cells of the inner medullary collecting tubule show high urea permeability and, therefore, allow the dumping of urea into the interstitium of the inner medulla.

RENAL REGULATION OF EXTRACELLULAR VOLUME AND OSMOLALITY

Extracellular fluid volume and extracellular osmolality give the appearance of being regulated, and renal excretions of Na^+ and water are the most important effector mechanisms in this regulation. **Physical factors, hormones, and nerves exert significant modulating influences on kidney function.**

Physical Factors

Local **pressures** (both hydrostatic and osmotic) and local rates of **flow** modulate especially proximal tubular mechanisms because they influence

- transmural fluxes and
- the duration of contact between fluid and transporting surfaces.

These factors are critically determined by the **tone in afferent and efferent arterioles.** Both arteriolar resistances are under the control of myogenic, nervous, and hormonal factors and can be altered independently of each other.

Efferent and afferent arteriolar resistances together

The combined settings of these resistances determine renal vascular resistance and, therefore, determine

- **renal blood flow** at a given renal arterial blood pressure;
- hydrostatic **pressure in peritubular capillaries** (relative to a given renal arterial pressure). The significance of this pressure is that it determines reabsorption from the interstitium.

Efferent relative to afferent arteriolar resistance

The relative settings of these resistances determine the partitioning of flow between glomerular filtrate and peritubular capillaries. This partitioning is called the **filtration fraction.** It is normally 20 percent of renal plasma flow.

Hormones

Epinephrine, norepinephrine, angiotensin II, prostaglandins, atrial natriuretic factor, and endothelium-derived factors **influence local blood pressures and flows.** Therefore, they influence renal physical factors. In addition, vasopressin (ADH), aldosterone, atrial natriuretic factor, and kinins influence **tubular reabsorption.**

At present the actions of **vasopressin** and **aldosterone** are understood best. Their importance derives from their ability to **affect salt and water excretion directly,** not via renal blood flow or glomerular filtration rate.

Nerves

Sympathetic constrictor nerves to vascular smooth muscle in afferent and efferent arterioles **alter local blood pressures and flows.**

Sympathetic efferent nervous activity to juxtamedullary cells **increases renin release.**

RENAL REGULATION OF EXTRACELLULAR VOLUME AND OSMOLALITY

The Role of Physical Factors

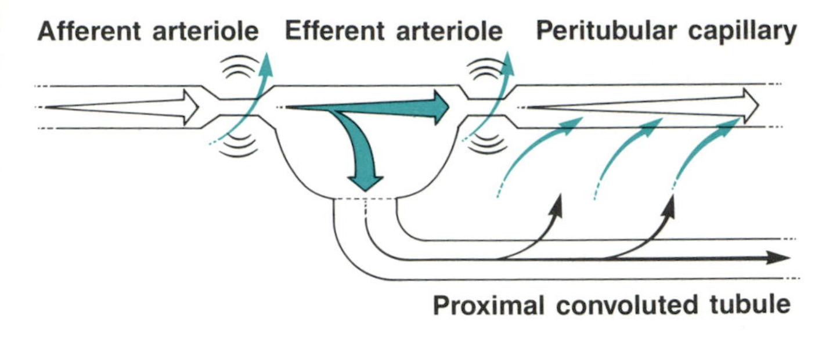

Two variables are determined by the tone in afferent and efferent renal arterioles:

- Renal blood flow
- Filtration fraction

- A relative increase in efferent resistance increases filtration fraction and thereby increases protein osmotic pressure in the peritubular capillary plasma.
- This favors the reabsorption of interstitial fluid and, thereby, promotes reabsorption of fluid from the proximal nephron.
- If peritubular capillary reabsorption were depressed, then the associated increase in interstitial pressure would depress further fluid reabsorption from the proximal tubule.

Hormones

The role of vasopressin

The plasma concentration of vasopressin is regulated by afferent input from

- hypothalamic osmoreceptors and
- atrial stretch receptors.

Its major renal actions are to increase water permeability in the collecting duct and to increase NaCl pumping in the thick ascending limb. Thereby, the rate of water reabsorption from the collecting duct into the medullary interstitium is increased.

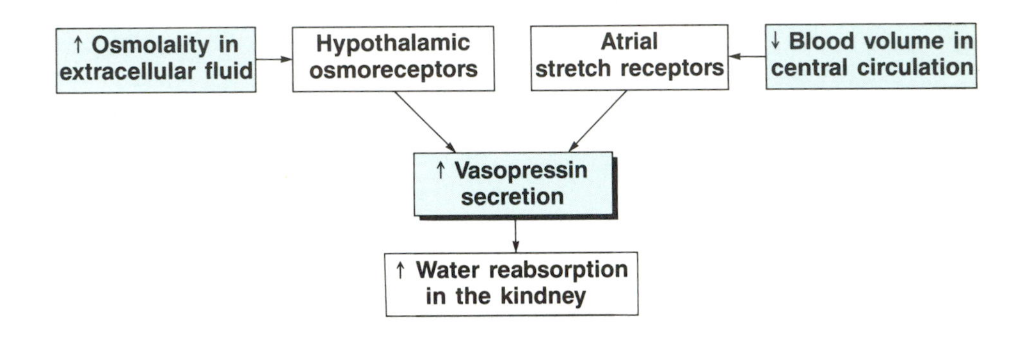

The role of aldosterone

The plasma concentration of aldosterone is altered either

- by a change in the ratio of plasma $[K^+]$ to plasma $[Na^+]$ (acting directly on adrenal cortical cells) or
- by increased cardiovascular hardship.

Aldosterone acts on the Na^+-K^+ transporters in the distal convoluted tubule to increase Na^+ reabsorption.

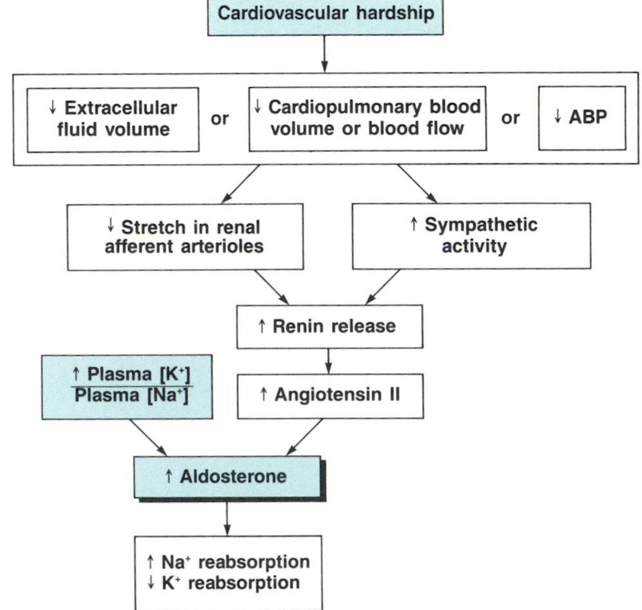

THE ROLE OF THE KIDNEY IN ACID-BASE BALANCE

The kidney plays three parts in maintaining the H^+ concentrations of body fluids within their normal, narrow limits:

- It **reclaims** any $\mathbf{HCO_3^-}$ that enters the nephron in glomerular filtrate.
- It **generates new** $\mathbf{HCO_3^-}$ to replenish body buffer stores.
- It **excretes acids** that cannot be converted to CO_2 and H_2O (fixed acids).

Reclaiming of Filtered HCO_3^-

Bicarbonate ions are the major extracellular buffer. They are filtered freely through the glomerular membrane and would be lost in the urine if they were not reabsorbed.

Reabsorption occurs **mostly in the proximal convoluted tubule** because the **presence of carbonic anhydrase** in the proximal tubular fluid (but not in more distal luminal fluids) allows the required chemical reactions to proceed rapidly.

The essence of the reabsorption mechanism is the formation, diffusion, and hydrolysis of carbon dioxide.

Although a great deal of H^+ is secreted in the process of HCO_3^- reabsorption, this process does *not* eliminate H^+ from the body.

Generation of New HCO_3^-

New HCO_3^- is generated in the proximal and distal tubules as well as in the collecting duct.

◆ The mechanism depends on **intracellular hydrolysis of body** $\mathbf{CO_2}$ **and subsequent removal of** $\mathbf{H^+}$ **from the cell.**

Depending on the destination of the secreted H^+ in the luminal fluid, new HCO_3^- can be generated by the formation of either **titratable acid** or **ammonium** ($\mathbf{NH_4^+}$)**.**

Titratable acid excretion

The kidney is capable of excreting buffered H^+, either as NH_4^+ or as $H_2PO_4^-$.

- $H_2PO_4^-$ is termed "titratable acid" because it will liberate its H^+ if the urine is titrated back to plasma pH.

◆ (NH_4^+ would give up little of its H^+ during titration to plasma pH because of the high pK of the ammonia-ammonium system.)

Ammonium excretion

Renal cortical cells increase **glutamine metabolism in response to low extracellular pH.** This yields ammonia (NH_3), a diffusible gas.

Excretion of Acid

Although the kidney does excrete some acid in the form of free H^+, **the properties of the distal nephron are such that it cannot maintain, across the tubular cell, a gradient in** $\mathbf{H^+}$ **concentration of sufficient magnitude to meet the body's needs for** $\mathbf{H^+}$ **excretion.** As a result,

◆ most acid excretion takes place in buffered form, the H^+ appearing in urine either as **titratable acid** or as $\mathbf{NH_4^+}$**.**

THE ROLE OF THE KIDNEY IN ACID-BASE BALANCE

Reclaiming of Filtered HCO_3^-

- A fraction of proximal Na^+ reabsorption takes place in exchange for H^+ secretion via a specific Na^+-H^+ transporter.
- Within the tubular lumen, secreted H^+ combines with filtered HCO_3^- to form H_2CO_3 (carbonic acid).
- Carbonic acid is, in the presence of carbonic anhydrase, quickly dissociated to CO_2 and H_2O.
- CO_2 diffuses easily into the proximal tubular cell and there combines with water to form H_2CO_3 within the cell.

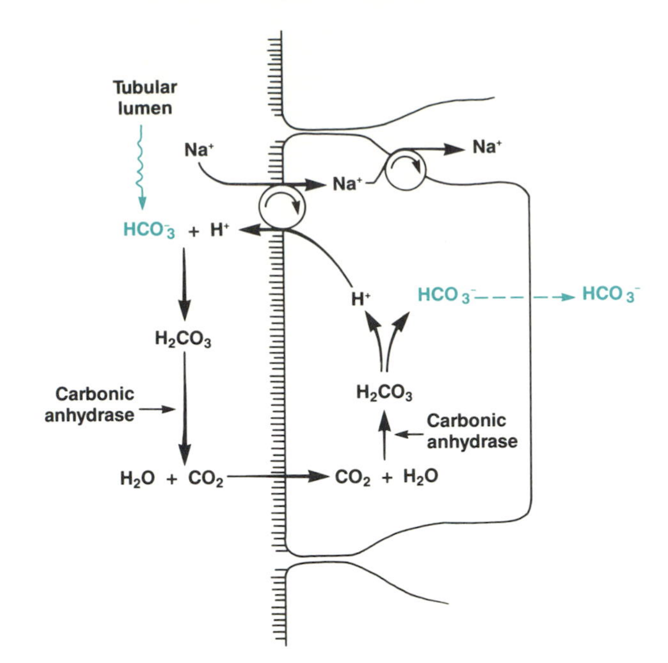

- Cellular H_2CO_3 also dissociates, forming H^+ and HCO_3^-.
- The H^+ thus formed is used to capture the next HCO_3^- in the lumen.
- The HCO_3^- formed within the cell diffuses down a concentration gradient into the peritubular space and from there into peritubular capillaries.

Generation of New HCO_3^-

Titratable acid excretion

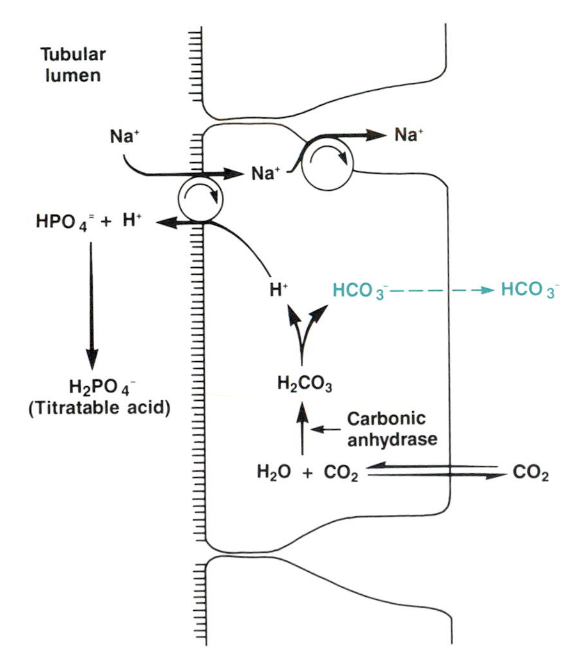

- Titratable acid ($H_2PO_4^-$) is formed from filtered phosphate and from H^+ that entered the tubular fluid in exchange for Na^+.
- Most H^+ that are secreted from a tubular cell leave behind an HCO_3^- ion because H_2CO_3, formed from metabolically produced CO_2, is a major proton source.

Ammonium excretion

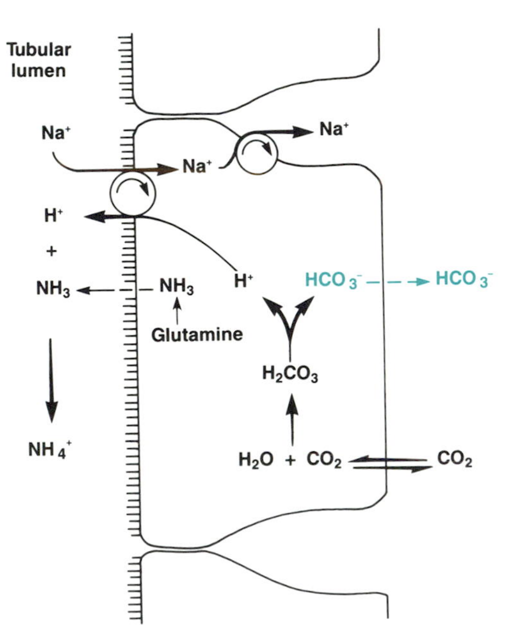

- In the tubular lumen, NH_3 combines with H^+, and the resulting cation, NH_4^+, is trapped within the electronegative lumen.
- NH_4^+ is excreted mainly as the neutral chloride salt.
- As in the case of titratable acid, most H^+ leaving the tubular cell generate an HCO_3^- for the buffer pool of the body.

7

GASTROINTESTINAL PHYSIOLOGY

GI MOTILITY

OVERVIEW

Motility is determined by the **mechanical activity of intestinal smooth muscle.** It has three functions:

- To **transport** food
- To **mix** ingested food with internal secretions that aid digestion
- To **remove** undigestibles and waste to the rectum

Contractions are evident in the muscle units that surround the lumen of the GI tract as well as in those that form sphincters within it.

◆ **The initiating action potentials are generated spontaneously, but the consequent mechanical activity is coordinated by central nervous influences and modulated by local influences (mechanical, chemical, neural, and hormonal).**

Different regions along the length of the tract show different rates of basic, slow-wave electrical rhythm (3 per minute in the stomach to 12 per minute in the duodenum).

Vomiting and Retching

Vomiting and retching are complex reflexes of motility that are coordinated by the medulla. They are initiated by stimuli that include

- chemical input from the duodenum,
- vestibular sensory input from the middle ear, and
- emotional input from cerebrocortical centers.

Vomiting involves three components:

1. GI smooth muscle
   - Spastic contractions of the gastric antrum, but
   - complete relaxation of all structures from the gastric body to the upper esophagus
2. The respiratory system
   - Slow, deep inspirations decrease intraesophageal pressure.
3. Skeletal muscle
   - Contraction of abdominal striated muscle raises intra-abdominal pressure.

As a result of the gradient between intra-abdominal pressure and esophageal pressure, gastric contents are forced upward into the relaxed esophagus.

- If the upper esophageal sphincter is opened, then evacuation into the mouth (i.e., vomiting) takes place.
- If the upper esophageal sphincter remains closed, then distension of the esophagus initiates a wave of secondary peristalsis that sweeps the gastric contents back from the esophagus into the stomach. This is perceived as retching.

GI MOTILITY

OVERVIEW

Muscle

Almost all of the GI muscle coat is smooth muscle, arranged in two layers around the lumen.

- The inner layer encircles the lumen.
- The outer layer is arranged longitudinally.

The two layers are separated by an extensive network of neurons, called the **intrinsic plexus** (myenteric plexus; Auerbach's plexus).

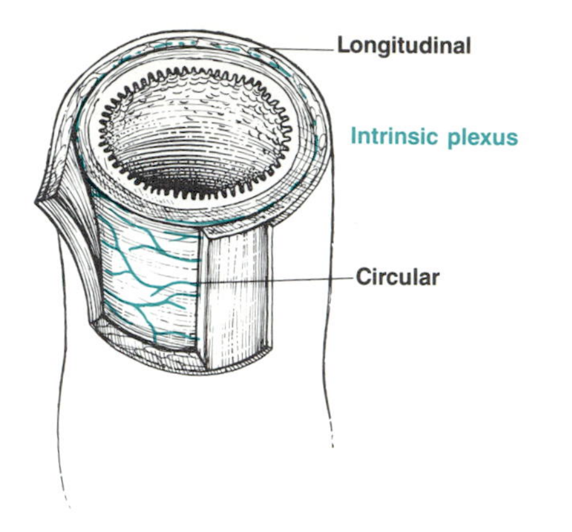

Innervation

This is supplied mostly by the intrinsic plexus. It modulates the activity of smooth muscle cells, secretory cells, and sensory receptors.

- Some sympathetic postganglionic fibers supply the smooth muscle cells directly.

Intrinsic plexus activity is itself modulated
—by parasympathetic fibers from the vagus or pelvic nerves;
—by splanchnic sympathetic fibers.

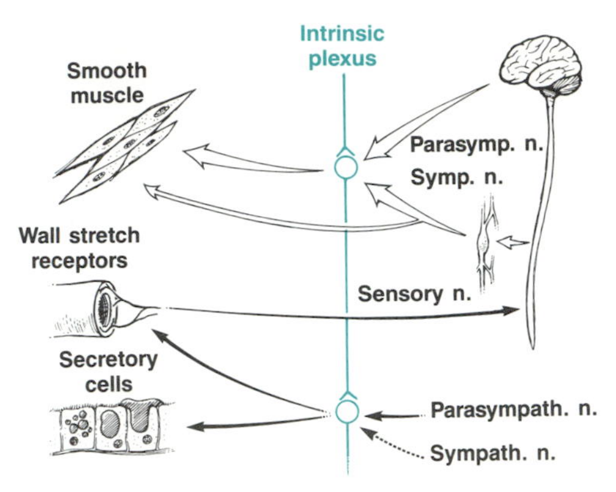

Motility in Different Regions

- A basic rhythm is set by spontaneous intrinsic muscle activity.
- Extrinsic nervous or hormonal influences modulate the basic rhythm.

Mouth and esophagus

Food transport is accomplished by **chewing, swallowing,** and **peristalsis.**

Chewing

- can be voluntary, but is mostly a rhythmic reflex.

Swallowing

- Placement of food at the entrance to the esophagus results from a hypothalamically coordinated program of sequential activation of vagal fibers that supply muscles of the tongue, pharynx, glottis, and epiglottis.
- Transport of food to the lower esophageal sphincter is due to sweeping waves of peristalsis:
  —A **primary peristaltic wave** is initiated and coordinated by the hypothalamus.
  —A **secondary peristaltic wave** occurs in response to esophageal stretch receptors, activated by food that was not propelled by the primary wave.

Stomach

Reception and temporary storage of ingested food:

Food reception involves relaxation of the fundus in synchrony with relaxation of the lower esophageal sphincter. This is part of the swallowing reflex.

Production and controlled release of chyme:

Gastric pacemaker cells originate a peristaltic wave that is directed toward the pyloric sphincter. Neurogenic and hormonal reflexes, responding to osmolality, [H^+], or fatty acid concentration, modulate gastric emptying.

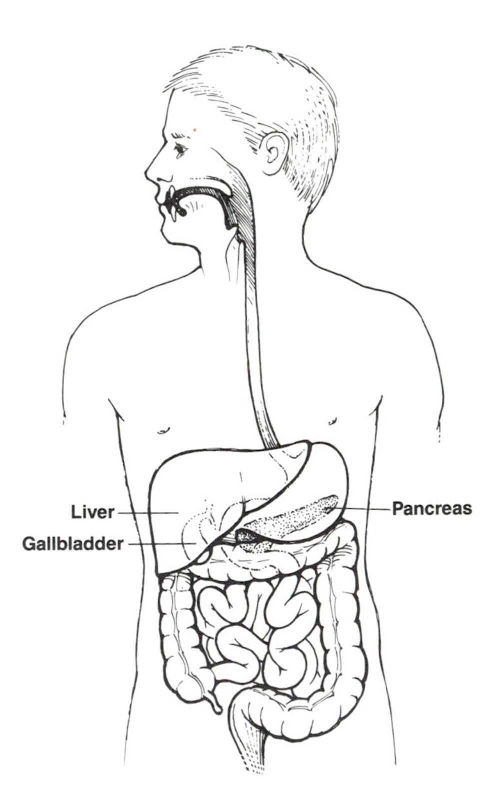

Small intestine (Duodenum, Jejunum, Ileum)

- Semiliquid chyme is received from the stomach, is mixed with bile and digestive secretions, and is moved longitudinally by spontaneous, rhythmic smooth muscle activity (modulated by nerves and hormones).
- Two kinds of contractions occur:

Localized circumferential contractions: These predominate. They cause mixing and short-distance propulsion. Hence, they are also called segmenting contractions.

Longitudinal (peristaltic) contractions: These also occur. They cause long-distance propulsion.

Colon and rectum

- Segmental (nonpropulsive) and peristaltic (propulsive) activities occur similarly to those seen in the small intestine.
- Peristalsis moves feces into the rectum.
- Rectal distension is perceived as an urge to defecate. Those above diaper age can, under most circumstances, voluntarily suppress or facilitate this urge by volitional control over two sets of striated muscle:
  —The external anal sphincter "minds the gate."
  —Muscles of chest and abdomen modulate intra-abdominal pressure.

GI HORMONES

OVERVIEW

The principal hormones controlling GI function are gastrin, cholecystokinin (CCK), secretin, and gastric inhibitory peptide (GIP). Their actions can be viewed as a sequence with a **stomach phase** and a **small intestine phase.**

- **Gastrin** is involved in the **stomach phase;** it promotes events that result in the placement of food and H^+ into the small intestine.
- In the small intestine, both **H^+ and certain constituents in the ingested food activate specialized cells to release secretin, CCK, and GIP.**

GI HORMONES

OVERVIEW

Anticipatory vagal activity and the subsequent arrival of food in the stomach stimulate secretion of gastrin into the blood.

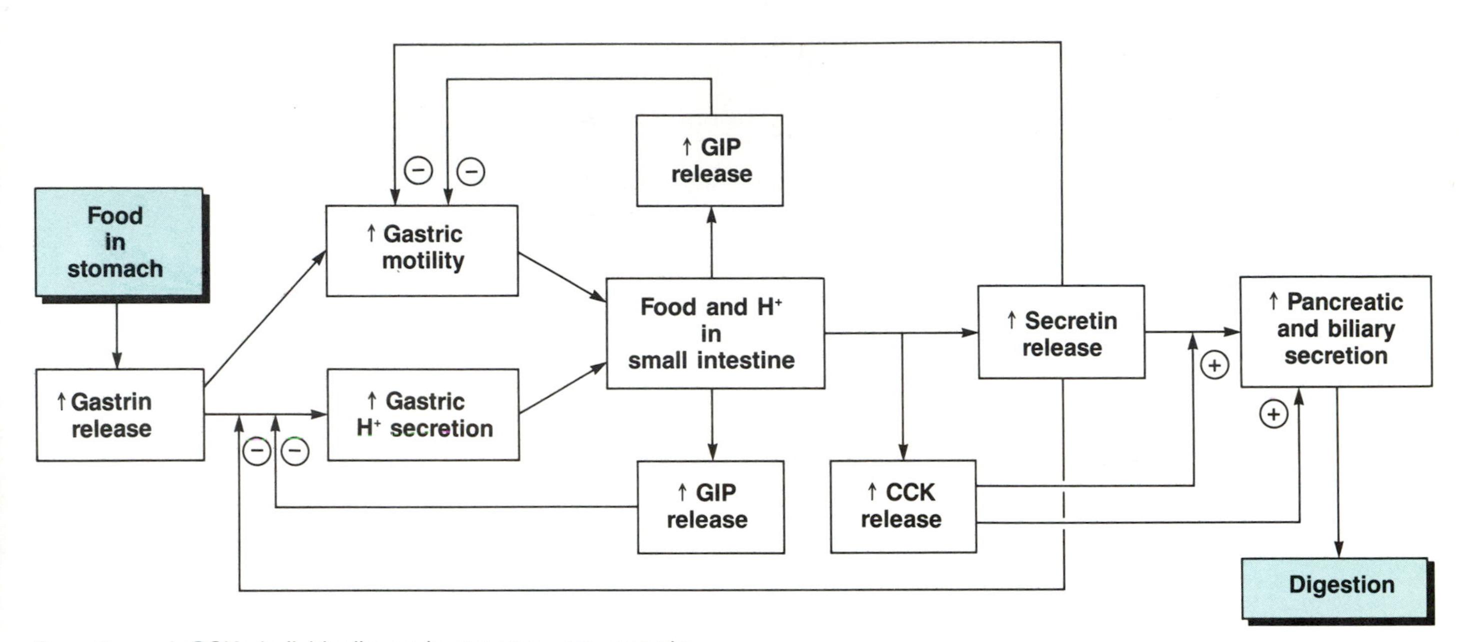

Secretin and CCK, individually or in synergy, are secretagogues for pancreatic and biliary products that are required for digestion and absorption.

Secretin and GIP control the entry of acidic chyme into the duodenum by inhibiting both gastric motility and gastrin-induced H^+ secretion from oxyntic cells.

GI HORMONES—CONT'D.

Gastrin

Its major actions are

- to **stimulate secretion** from oxyntic glands (mucus, H^+, and pepsinogens) and from pancreatic acini (enzymes);
- to **enhance gastric motility;**
- to **maintain gastric mucosal growth.**

Cholecystokinin (CCK)

CCK and gastrin have identical structures with respect to the last five amino acids at the NH_2 terminus and are, therefore, partial agonists of each other.

The major actions of CCK are

- stimulation of **pancreatic secretion** and
- **gallbladder** contraction and relaxation of the sphincter of Oddi.

Secretin

Secretin regulates the concentration of H^+ in the small intestine. It accomplishes this via three paths:

- It promotes elaboration of HCO_3^--rich fluid from cells located in the neck of pancreatic acini and in the bile ducts.
- It inhibits gastrin-mediated H^+ secretion.
- It inhibits gastric motility.

Secretin and CCK are synergistic in many of their actions.

Gastric Inhibitory Peptide (GIP)

The major functions of GIP are in the **glucose-insulin system,** but, by virtue of its inhibitory effect on gastric motility and gastrin-induced H^+ secretion, GIP acts as a **regulator of gastric activity relative to the capacity of the upper small intestine.**

Motilin

This polypeptide is released from jejunal mucosal cells in response to local acid and fat. It enhances phasic contractions of the lower esophageal sphincter and accelerates gastric emptying.

Pancreatic Polypeptide

Pancreatic polypeptide derives from specific cells in the acini of the pancreas and appears to counteract the specifically pancreatic actions of CCK.

Vasoactive Intestinal Peptide (VIP)

VIP is found in endocrine cells of the GI mucosa as well as in the brain and in certain neurons of the autonomic nervous system. Hyperacidity of the small intestine is a potent stimulus for local VIP release. Its prominent GI effect is inhibition of gastric H^+ secretion, but it is not a potent inhibition.

GI HORMONES—CONT'D.

Gastrin

- Gastrin is synthesized and released from mucosal G-cells in the gastric antrum, duodenum, and jejunum.
- Antral cells produce the 17-amino-acid form (G17). The small intestine produces the less potent but longer-lasting G34.
- Release is influenced by a variety of chemicals including gastrin-releasing peptide (GRP), which is secreted from noncholinergic vagal fibers.

Stretch
Peptides
Amino acids
H^+
Somatostatin
GRP
Cholinergic
Vagus
G cell
Gastrin

Cholecystokinin (CCK)

- CCK is synthesized and secreted by specialized cells in the mucosa of the duodenum, jejunum, and, to some extent, the ileum.
- The major stimulus for release is fatty acids.

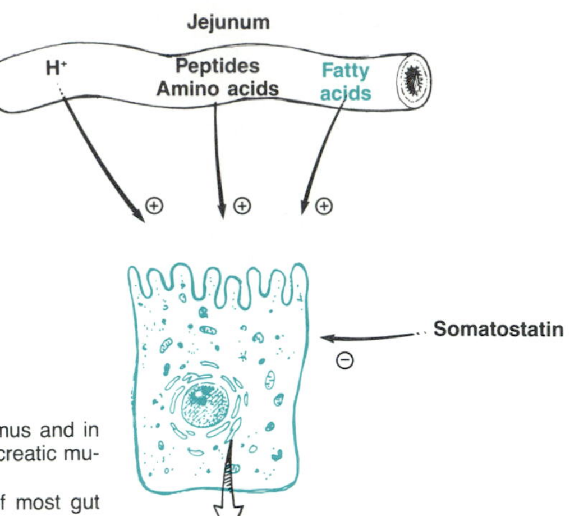

Somatostatin

- Somatostatin is found in the hypothalamus and in D-cells of the antral, duodenal, and pancreatic mucosa.
- It is a strong inhibitor of the release of most gut hormones.
- It inhibits the actions of CCK and inhibits gastric acid secretion.

Secretin

Secretin is synthesized and secreted by the S-cells of duodenal and jejunal mucosa.

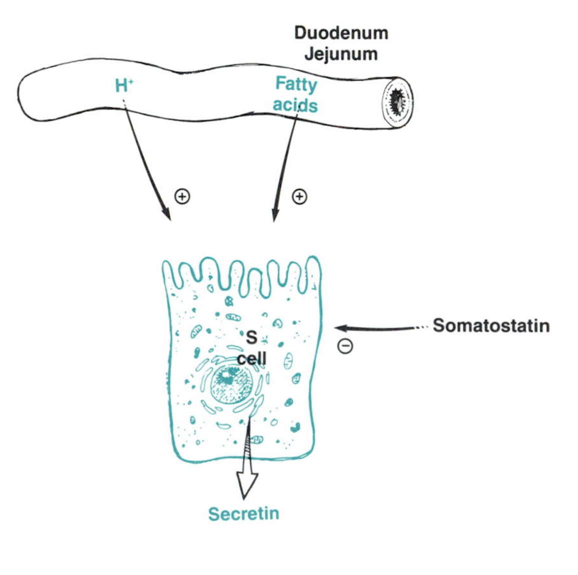

Gastric Inhibitory Peptide (GIP)

- GIP is produced mostly in the jejunum, but also in the duodenum and upper ileum.
- Its rate of release shows two peaks after a meal:
  —The first is stimulated by glucose.
  —The second is stimulated by fatty acids.

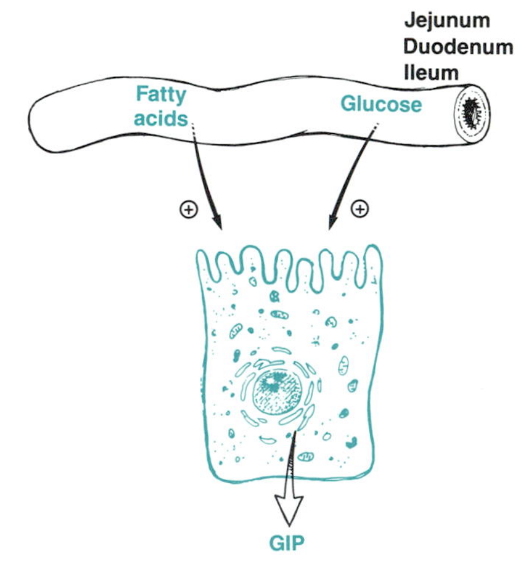

NONHORMONAL SECRETIONS OF THE GI TRACT

Different portions of the GI tract secrete substances that have three basic functions:

- Mechanical modification of ingested food
- Digestion of food
- Protection of the GI tract from its own destructive actions

From the Mouth

Salivary acini, when they are stimulated by the autonomic nervous system, elaborate copious volumes of fluid that is rich in **HCO_3^-** and contains **mucin** as well as the enzymes **α-amylase** and **lipase.**

Bicarbonate ions (HCO_3^-)

These ions have protective functions against the acids that are constantly being produced by oral micro-organisms.

Mucin

Mucin has buffering functions similar to those of HCO_3^-. In addition, it forms a protective, low-friction film in the mouth and around food particles.

α-Amylase

α-amylase begins to break down starch.

Lipase

Lipase begins fat digestion.
Neither α-amylase nor lipase has significant digestive function, because both are inactivated by gastric acidity.

From the Stomach

The secretory unit of the stomach is the **oxyntic gland.** It forms three types of secretions in large volumes of fluid:

- **Pepsinogens** (precursors of the proteolytic enzymes pepsin 1 and 2)
- **H^+** (required to activate the pepsinogens and to stimulate enzyme secretion from later portions of the GI tract)
- **Mucus** (agent to protect the GI mucosa from being digested)

NONHORMONAL SECRETIONS OF THE GI TRACT

From the Mouth

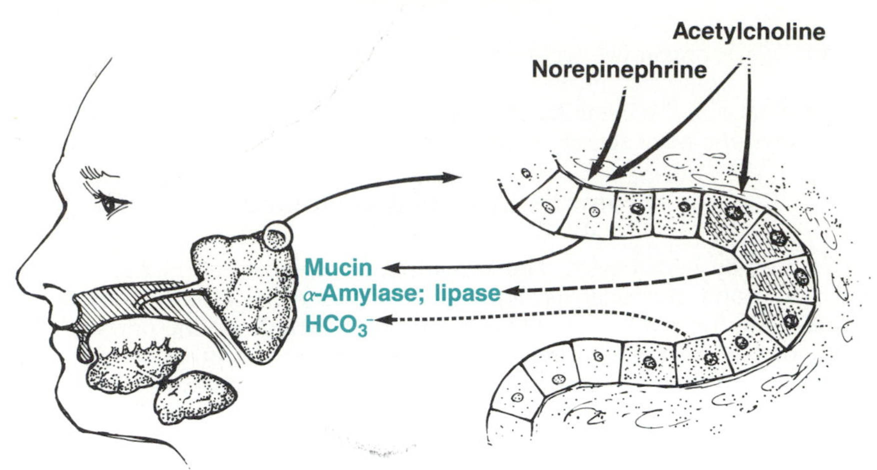

From the Stomach

Different cells in the oxyntic gland produce **mucus, H^+,** and **pepsinogens.**

- Partial stimulation of the three types of secretions is achieved by vagal cholinergic fibers.
- Full secretory activity is achieved by gastrin.
- Histamine is also a potent stimulus.
- Secretin is a synergist of gastrin in its actions on peptic cells, but an antagonist of gastrin in its actions on oxyntic cells (H^+ secretion).

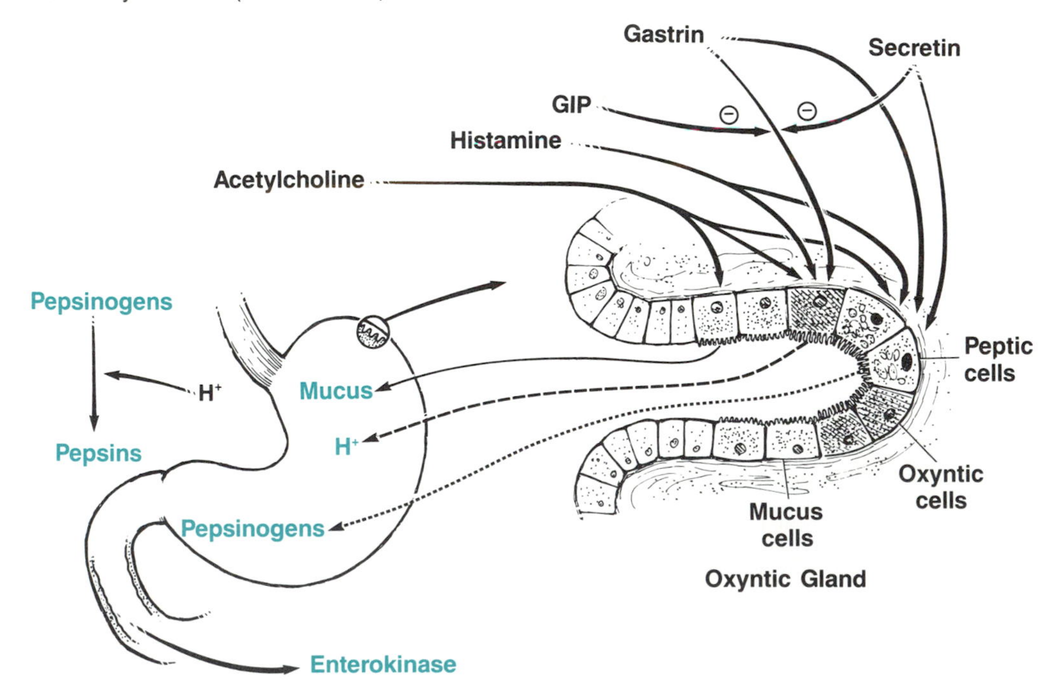

Pepsinogens are inactive. They acquire full proteolytic activity by exposure to high H^+ concentration ($pH < 2$), but even then their contribution to total protein digestion is only 10 to 15 percent.

NONHORMONAL SECRETIONS OF THE GI TRACT—CONT'D.

From the Pancreas

Pancreatic acini contain two types of secreting cells:

- Cells near the neck of the interlobular ducts secrete a fluid that is rich in $\mathbf{HCO_3^-}$.
- Cells deep within the acini secrete **digestive enzymes** or enzyme precursors.

From the Liver and Gallbladder

Bile is formed by hepatocytes. The breakdown of cholesterol produces the primary bile acids—cholic acid and chenodeoxycolic acid. Secondary bile acids—deoxycholic acid and lithocholic acid—are formed by the action of intestinal bacteria on the primary acids.

The rate of bile formation is feedback-controlled by the amount of bile acid that is actively absorbed from the ileal lumen and returned to the liver via portal blood.

NONHORMONAL SECRETIONS OF THE GI TRACT—CONT'D.

From the Pancreas

Almost all of the pancreas is devoted to the elaboration of digestive secretions; only 2 percent of it is concerned with glucose metabolism.

There are four classes of pancreatic enzymes:

1. Amylolytic (starch breakdown)
2. Lipolytic (breakdown of lipids and cholesterol)
3. Proteolytic (protein breakdown)
4. Nucleolytic (breakdown of RNA and DNA)

The significant members of each class are these:
Amylase (class 1)
Lipase; phospholipase A; cholesterol esterase (class 2)
Tripsinogen; chymotrypsinogen; procarboxypeptidase (class 3)
Ribonuclease; deoxyribonuclease (class 4)

Proteolytic enzymes are secreted from acinar cells as inactive precursors (zymogens). Their activation begins with the activation of trypsinogen by the duodenal brush-border enzyme, enterokinase. The active form, trypsin, then acts on the other proteolytic zymogens (chymotrypsinogens, procarboxypeptidases, and proelastases) to form the respective fully active proteolytic enzymes.

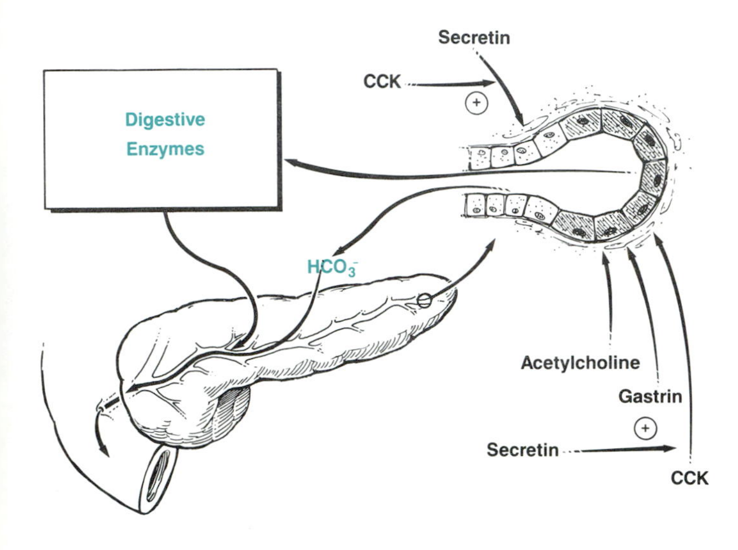

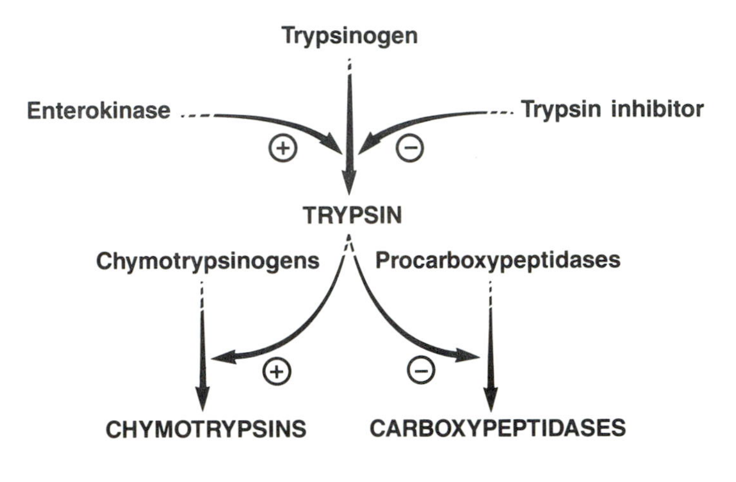

From the Liver and Gallbladder

- The initial step in bile production is formation of two primary free bile acids, cholic acid and chenodeoxycholic acid.
- These are immediately conjugated with the amino acids taurine and glycine, thereby becoming water-soluble and resistant to precipitation in acid media.
- Even then, conjugated bile salts will precipitate out of solution at pH < 4.
- To maintain the pH above that level, cells in the bile ducts secrete an HCO_3^--rich ultrafiltrate of plasma when they are stimulated by secretin.

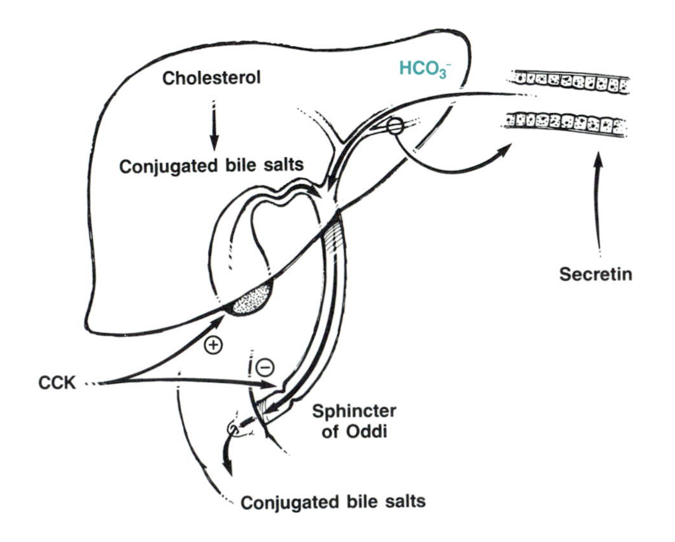

- Hepatic bile is secreted into the bile canaliculi and enters the gallbladder via bile ducts.
- The gallbladder removes up to 90 percent of the water from the fluid reaching it, stores the concentrate, and controls its delivery to the duodenum.

DIGESTION AND ABSORPTION

OVERVIEW

Absorption of nutrients is the primary function of the digestive system. It **requires that ingested food be broken down into smaller and chemically simpler units that will yield absorbable molecules.**

Carbohydrates

The three main carbohydrates in the human diet are sucrose, lactose, and starch. They form two broader classes, **disaccharides** and **starch.**

Disaccharides

Sucrose and lactose are broken down by **disaccharidases** in the **brush border of the small intestine.** Absence of one or more of the disaccharidases causes intolerance to certain dietary sugars. For example,

- Congenital lack of β-galactosidase causes lactose intolerance. In this common disorder, dietary lactose (contained in milk products) is not broken down to reabsorbable monosaccharides. As a result, it holds water in the intestine by osmotic forces and, thus, causes diarrhea. Some of it is degraded by intestinal bacteria to organic acids and CO_2. These cause bloating, belching, flatulence, and cramping.

Starch

Dietary starch consists largely of two polysaccharides, **amylopectin** (75%) and **amylose.** They are broken down initially by **α-amylase.**

Indigestible carbohydrates such as cellulose are not reabsorbed but are excreted in the feces. They are commonly called **"bulk"** or **"fiber"** and are believed by some to be an indispensable part of digestive happiness.

DIGESTION AND ABSORPTION

OVERVIEW

The normal dietary ingredients are

Carbohydrates Proteins Fats	Undergo enzymatic degradation followed by absorption of the resulting subunits
Electrolytes Trace metals Vitamins	Absorbed by passive or specific active transport mechanisms

Carbohydrates

Disaccharides (e.g., sucrose)

The action of disaccharidases yields monosaccharides, which are absorbed from the intestinal lumen by passive transport down a concentration gradient and by specific active membrane transport systems that couple the monosaccharide with Na^+.

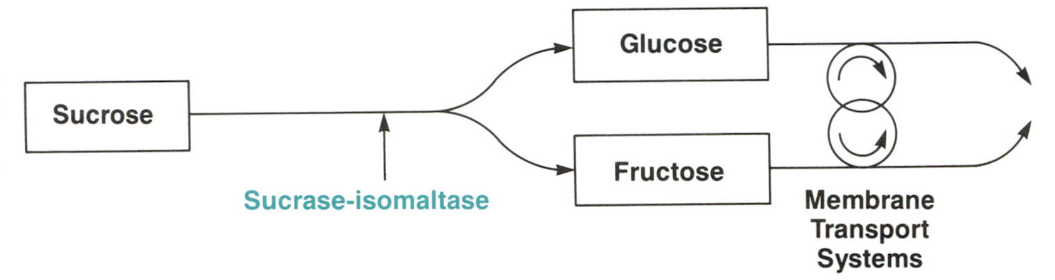

Starch

- The large polysaccharide constituents of starch are broken down to smaller polysaccharides mostly by pancreatic α-amylase.
- They are quickly digested by brush-border enzymes.
- The resulting glucose is absorbed by the time the chyme has travelled 20 cm into the jejunum.

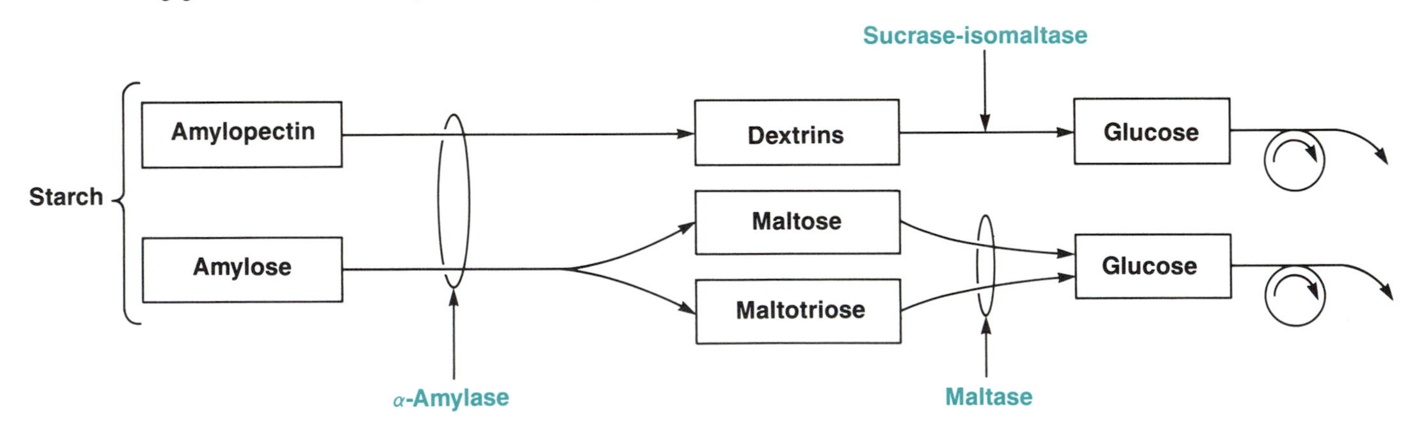

Proteins

Protein in the intestinal lumen derives from three sources:

- The diet
- Sloughed mucosal cells
- Secreted enzymes

Some digestion of protein does occur in the stomach (as a result of the enzymes pepsin 1 and 2), but it accounts for no more than 15 percent of total protein digestion.

Most digestion of protein occurs via **pancreatic peptidases** that work **in sequence with duodenal-jejunal brush-border surface enzymes.**

Fats

A small portion of fat in the diet is in the form of lecithin, cholesterol esters, and medium-chain triglycerides (6 to 12 carbon atoms), but the majority is **long-chain triglycerides** (esters of glycerol and long-chain fatty acids of 16 to 18 carbon atoms).

Long-chain triglycerides

Digestion and absorption of these fats involve

- emulsification (conversion of fat drops to small droplets) and
- **micelle** formation in the lumen of the duodenum.

These are followed by

- formation of **chylomicra** within mucosal epithelial cells.

Chylomicra enter the intestinal lymphatic system and are delivered to the blood for subsequent fat metabolism in tissues.

Medium-chain triglycerides

These are digested and absorbed by a much simpler process because they are rapidly and completely hydrolyzed by lipases and isomerases in the intestinal lumen.

The resulting fatty acids are more water-soluble than long-chain fatty acids and require no micelle formation for absorption into epithelial cells. In addition, from the epithelial cells they are not reabsorbed as chylomicra into the lymph, but instead are absorbed directly into the portal blood without being reconstituted as triglycerides.

DIGESTION AND ABSORPTION—CONT'D.

Proteins

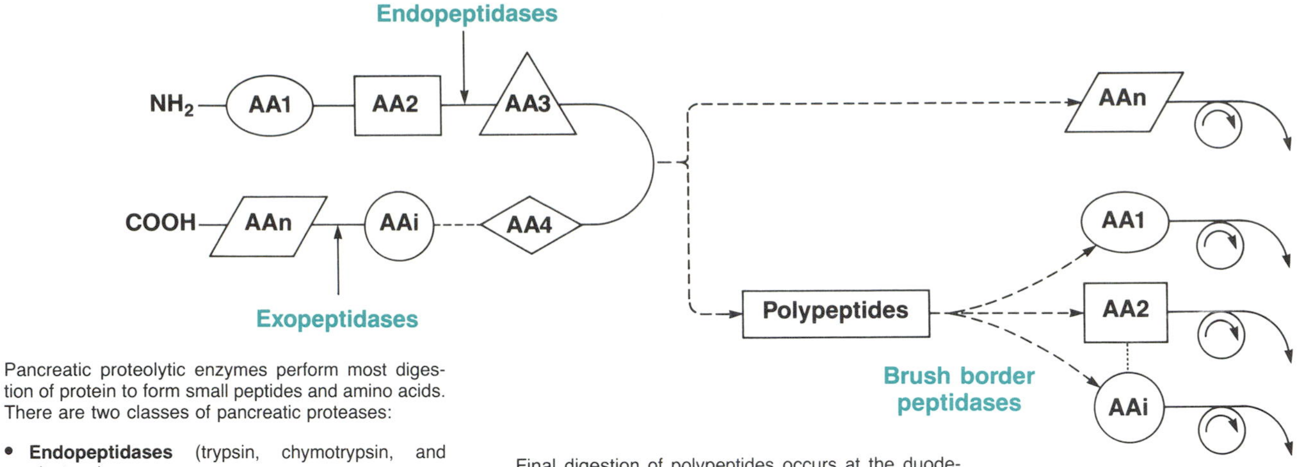

Pancreatic proteolytic enzymes perform most digestion of protein to form small peptides and amino acids. There are two classes of pancreatic proteases:

- **Endopeptidases** (trypsin, chymotrypsin, and elastase)
  —attack the interior peptide bond between neighboring amino acids;
  —differ from one another in their specificity for amino acids adjacent to a cleavage site.
- **Exopeptidases** act on the last peptide bond on the COOH (carboxyl) end of protein molecules.

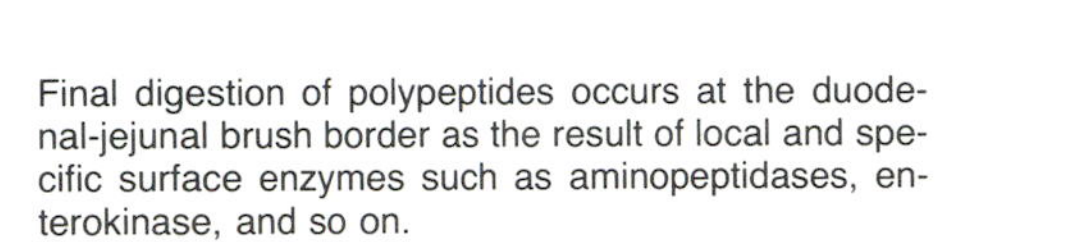

Final digestion of polypeptides occurs at the duodenal-jejunal brush border as the result of local and specific surface enzymes such as aminopeptidases, enterokinase, and so on.

The small peptide or amino acid end products are reabsorbed by specific transport mechanisms.

Fats

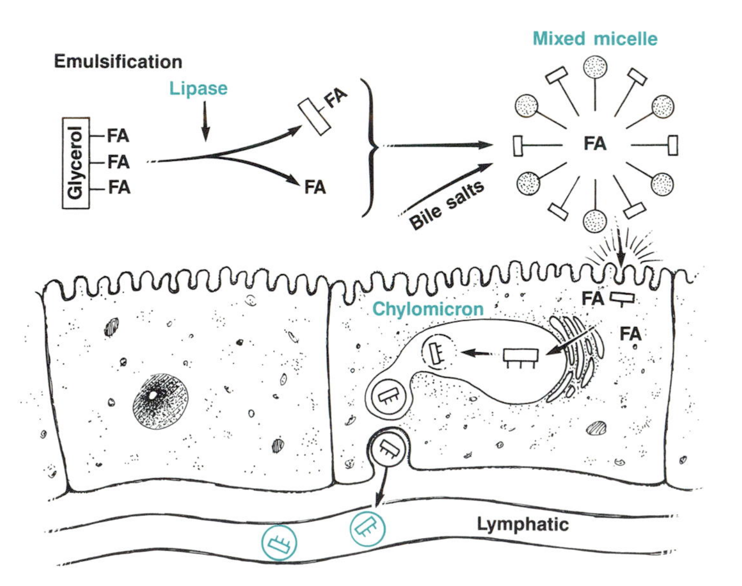

The churning action of the stomach creates large fat drops within the aqueous chyme. These drops are emulsified (converted to stable small droplets) in the duodenum by the combined action of bile salts, monoglycerides, fatty acids, and lecithin.

The purpose of emulsification is to increase the surface area over which pancreatic lipase can attack the ingested triglycerides.

Pancreatic lipase hydrolyzes triglycerides within the emulsion, forming mostly β-monoglycerides and free fatty acids in the process. This action requires the presence of co-lipase (also secreted by the pancreas) because lipase itself is strongly inhibited by conjugated bile salts.

Long-chain fatty acids and β-monoglycerides are insoluble in water and, therefore, difficult to absorb. They are brought into solution by a special property of bile salts:

Bile salts resemble detergent molecules in that they have a water-soluble portion and a fat-soluble portion. When such molecules are present in sufficient concentration, they form clusters, known as micelles, in which the water-soluble portion of each molecule faces outward while the lipid-soluble portion faces inward.

Long-chain fatty acids and β-monoglycerides dissolve into the structure of the bile acid micelles to form mixed micelles and are transported in that form to the brush border of the small intestine.

Mixed micelles disintegrate upon collision with the brush border. Their contents are released and enter the mucosal epithelial cell, possibly aided by specialized transport proteins.

Within the endoplasmic reticulum of the epithelial cell, absorbed fatty acids and β-monoglycerides are reassembled with endogenously synthesized glycerol to form triglycerides again. These and dietary cholesterol are formed into chylomicra (droplets coated with protein and phospholipid) and are delivered in that form to the intestinal lymphatic system, and from there to the blood.

DIGESTION AND ABSORPTION—CONT'D.

Cholesterol

Dietary cholesterol is digested by cholesterol esterase (secreted by the pancreas), incorporated into micelles, and absorbed from the small intestine in chylomicra. It eventually emerges from the liver, incorporated along with triglycerides into **very-low-density lipoprotein** (VLDL), a particle that is even smaller than a chylomicron.

There is a strong positive correlation between serum LDL levels and ischemic heart disease, and a strong negative correlation between serum HDL levels and ischemic heart disease. The processes by which changes in serum lipoprotein levels lead to the formation of atheromatous plaque in the intima of blood vessels are not yet known with certainty.

DIGESTION AND ABSORPTION—CONT'D.

Transport Systems for Fats and Cholesterol

Fats and cholesterol are used by cells as energy sources and as structural components of membranes. They are insoluble in water and, therefore, require a special transport system. This is provided by a family of particles that differ in size, but agree in structure:

- Each has a lipid core surrounded by apolipoproteins.

The major classes of particles are chylomicra, very-low-density lipoproteins (VLDL), low-density lipoproteins (LDL), and high-density lipoproteins (HDL).

At the capillary wall the triglycerides are removed from the chylomicra by the action of lipoprotein lipase and are deposited in adipose tissue and in skeletal muscle.

The chylomicron remnant, containing mostly cholesterol, is removed by the liver.

Hepatocyte action has one of two results: The remnant is

- either broken down and secreted as bile acids or
- packaged as particles of VLDL and then secreted into the circulation.

In fat and in skeletal muscle the triglycerides are removed from the VLDL by lipoprotein lipase, and cholesterol remains behind, quickly forming LDL packages.

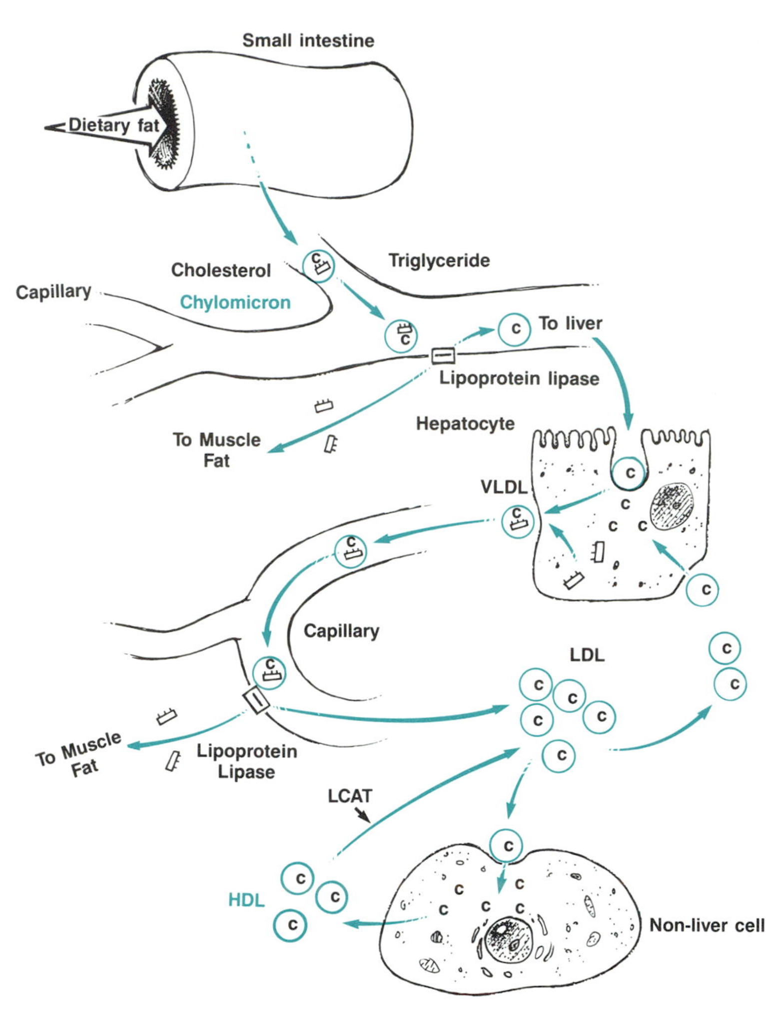

Dietary cholesterol eventually emerges from the liver, incorporated along with triglycerides into VLDL, a particle that is even smaller than a chylomicron.

Most LDL is removed from the circulation by nonliver cells. Their cholesterol content, together with cholesterol that was produced by the cell, leaves the cell via HDL that are broken down by the enzyme lecithin-cholesterol acetyl transferase (LCAT) to form LDL.

THE LIVER

Most substances that enter the body by mouth reach the liver via portal venous blood after they have been absorbed by the intestinal mucosa. While the liver's metabolic functions are probably its most important, it does perform several other vital tasks.

Metabolic Functions

Gluconeogenesis and glycolysis

When plasma glucose concentration is low, the liver **forms glucose**

- from amino acids (gluconeogenesis),
- from glycogen (glycolysis),
- from fatty acids, or
- by converting fructose or galactose to glucose.

When plasma glucose concentration is high, the liver

- stores glucose as glycogen,
- breaks down glucose to amino acids, and
- converts glucose to fatty acids.

Protein synthesis

Plasma albumin, fibrinogen, and most globulins are synthesized by the liver from amino acids that are delivered to it via the portal blood.

Fat metabolism

Phospholipids and lipoproteins are synthesized from fatty acids.

Circulatory Functions

Energy dissipation

The structure of the liver sinusoids allows dissipation of hepatic arterial blood pressure before mixing with portal venous blood occurs at a considerably lower pressure.

Clotting factors

The liver synthesizes most of the clotting factors and, thereby, ensures a functioning hemostatic system.

Secretory Functions

Cholesterol

This is synthesized from acetate portions and is either secreted within VLDL or broken down to form bile salts.

Bilirubin

This end product of heme metabolism derives from the hemoglobin of degraded erythrocytes and is carried in plasma as a bilirubin-albumin complex. It reaches the hepatocytes via large fenestrations in the liver sinusoids and is transported into the cytosol. There a fraction of it is returned to the plasma. The remainder is conjugated mostly to glucuronate and secreted into the bile canaliculi.

THE LIVER

Functions

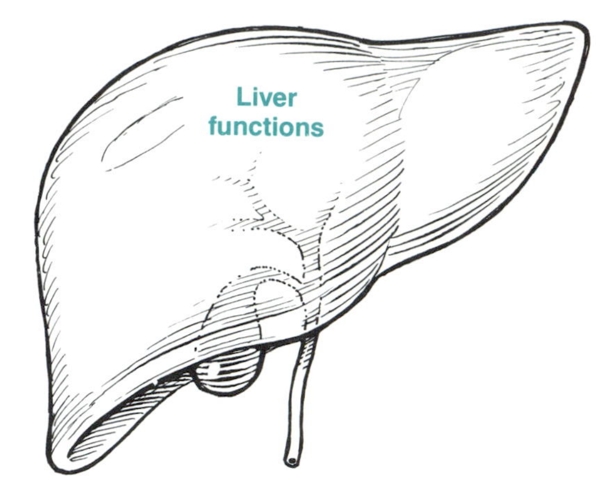

Metabolic

- Gluconeogenesis and glycolysis
- Synthesis of proteins and lipoproteins
- Metabolism of fats and vitamins

Circulatory

- Mixing of low-pressure portal blood with high-pressure hepatic blood
- Synthesis of clotting proteins
- Volume storage organ?

Secretory

- Cholesterol, lecithin, bile salts, and bilirubin

Conservatory

- Bile salts and amino acids

Excretory

- Drugs
- Metabolites
- Urea

Conservatory Functions

Bile salts

These are reclaimed from portal blood and are reused.

Excretory Functions

Metabolic filter

The liver acts as a filter for a variety of substances and uses conjugation, oxidation, reduction, methylation, and other processes to detoxify and excrete potentially harmful agents.

Urea

The liver is the major site of urea formation, removing in the process the ammonia that results from catabolism of amino acids.

8

NEUROPHYSIOLOGY

COMPONENTS OF NERVOUS SYSTEM FUNCTION

Cells of the Nervous System

The nervous system contains **glial cells** and **neurons.**

Glial cells

Glial cells have a supportive function.

- They do not generate action potentials when they are depolarized.
- They are classified according to location and cell type.
  - In the central nervous system:
    - Oligodendrocytes
    - Astrocytes (protoplasmic or fibrous)
    - Microglia
    - Ependymal/cells
  - In the peripheral nervous system:
    - Schwann cells

Neurons

Neurons **generate action potentials and conduct them, via their axons, to synapses with other cells.**

- Axons may be surrounded by **myelin,** a good electrical insulator that prevents current leakage. As a result,
  - an action potential generated at one point is conducted rapidly to the nearest **node of Ranvier** and there generates another action potential that is conducted to the next node.
    - —This arrangement (termed **saltatory conduction**) allows high transmission speed.
    - —Close spacing of nodes is necessary because neurons are incapable of generating sufficient current to overcome leakage and longitudinal resistance and then depolarize a region at the end of a long axon.

COMPONENTS OF NERVOUS SYSTEM FUNCTION

Cells of the Nervous System

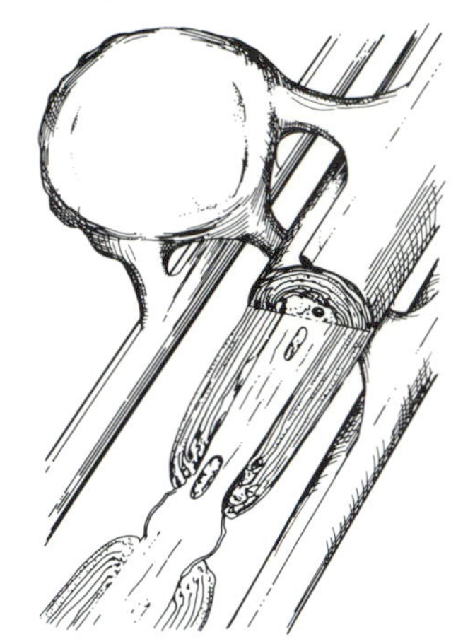

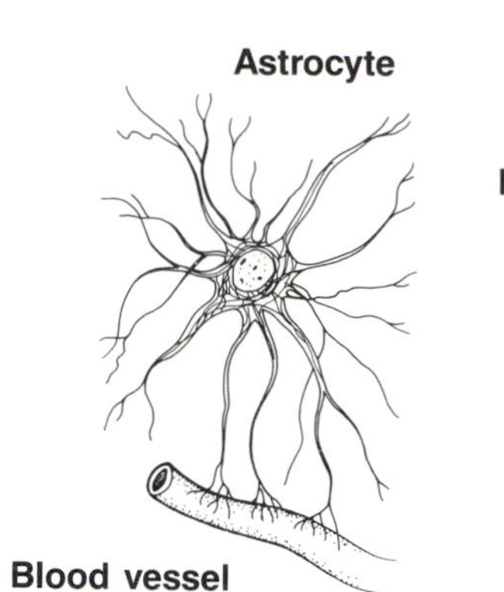

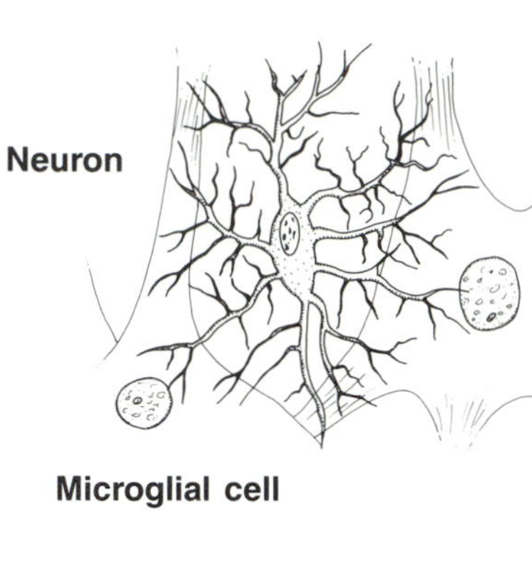

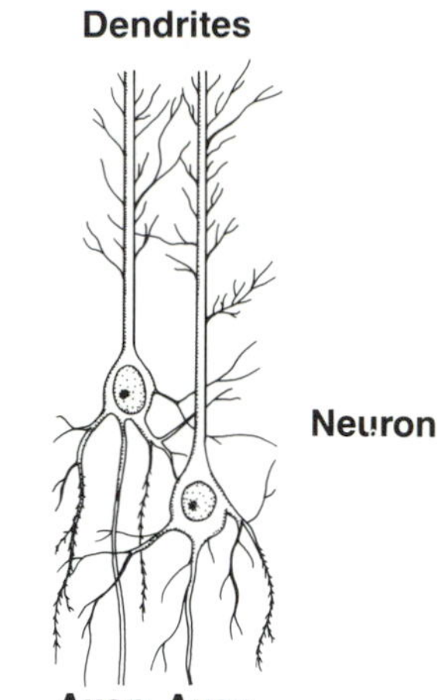

Adapted from Bunge et al. J. Biophys. Biochem. Cytol. 1961; 10: 67.

Oligodendrocytes

- produce, maintain, and repair the myelin sheath that surrounds central nervous system axons.

Schwann Cells

- are the peripheral equivalent of oligodendrocytes;
- surround axons with a myelin sheath. The sheath is interrupted at intervals along the axon (at Nodes of Ranvier) where myelin from one glial cell meets myelin from a neighboring glial cell.

Astrocytes

- surround cerebral blood vessels and contact them with extended foot processes.
  —Brain areas rich in neuronal cell bodies and synapses contain protoplasmic astrocytes.
  —Brain areas rich in axons contain fibrous astrocytes.
- Astrocytes may
  —act as physical dams between adjacent neurons;
  —participate in regulating cerebral interstitial [K^+];
  —participate in synthesis, release, and uptake of transmitter substances;
  —repair wounds and provide scar formation.

Microglia

- may be of intravascular origin;
- are found at sites of injury or inflammation, where they perform phagocytic functions.

Neurons

have elongated processes (dendrites and axons).

Ependymal Cells

make up the lining of the system of intracerebral ventricles and canals.

Classification of Nerve Fibers

Nerve fibers are classified according to their conduction velocities. Two classification systems are in use:

The Gasser System

- Rapidly conducting, large myelinated fibers are designated as Group A.
  —Within the A group, subgroupings α, β, γ, and δ exist according to axon diameter. (Aα fibers have the largest diameter and therefore the highest conduction velocity.)
- Autonomic myelinated fibers are designated as Group B.
- Unmyelinated fibers are designated as Group C.

The Lloyd System

This system ranges

- from Group I (a and b; large, fast, myelinated fibers)
- to Group IV (unmyelinated, slow fibers).

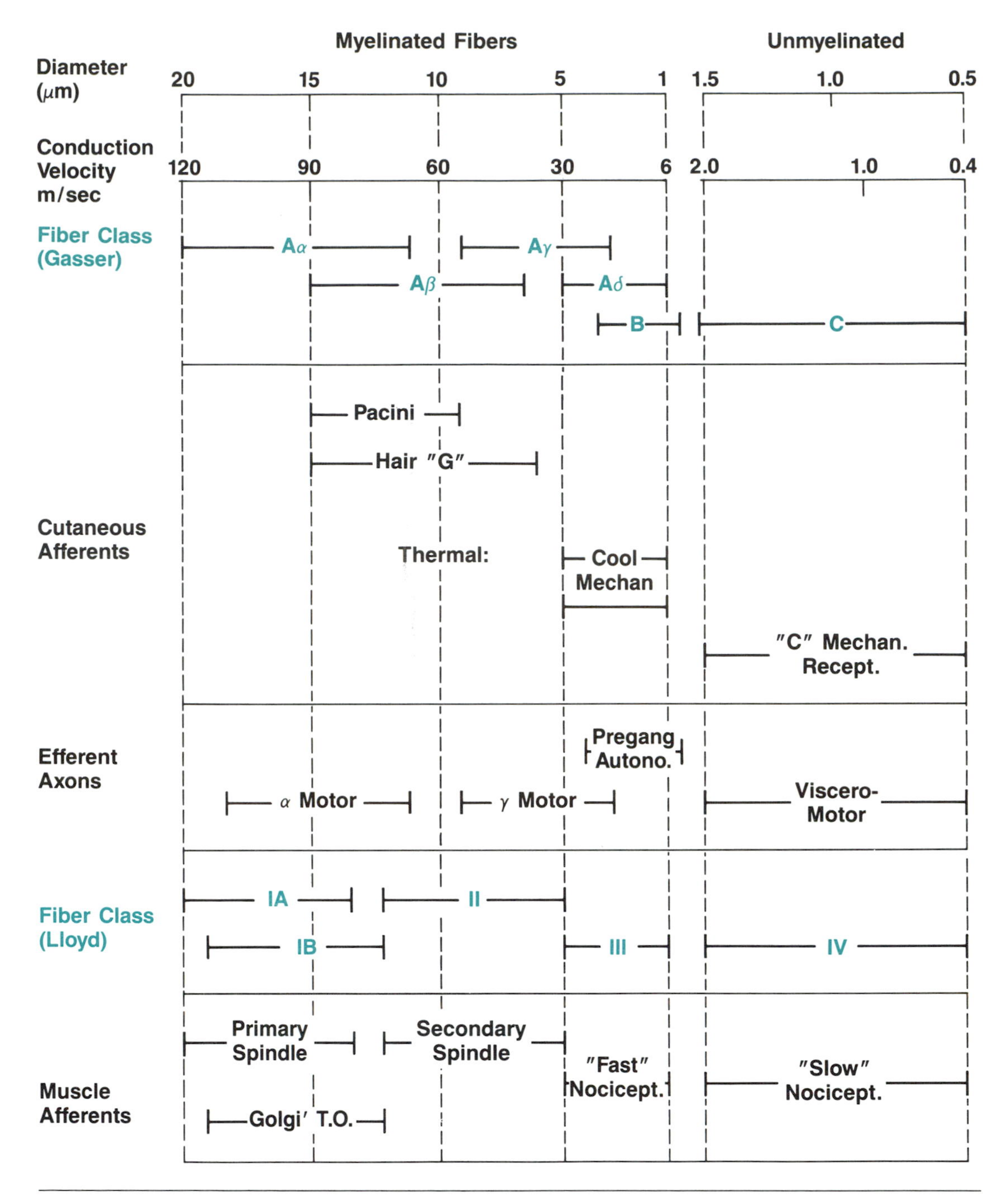

Adapted from Somjen G. Neurophysiology: The essentials. Baltimore: Williams and Wilkins, 1983.

Synapses

Most synapses are chemical. The presynaptic side, but not the postsynaptic side, of each synapse contains many intracellular vesicles (20 to 80 nm in diameter) filled with **neurotransmitter.**

Neurotransmitters

These chemicals are synthesized

- either locally (e.g., the motor nerve of a neuromuscular junction)
- or at a distal site within the nerve (e.g., hypothalamic releasing hormones acting on the anterior pituitary).

- Their interaction with a **postsynaptic receptor**
  - causes changes in transmembrane ion flux and
  - generates
    - —either an **excitatory postsynaptic potential** (EPSP)
    - —or an **inhibitory postsynaptic potential** (IPSP).
- Transmitters that lead to EPSPs appear to open large, nonspecific membrane channels, permitting simultaneous movement of Na^+, K^+, and Cl^-.
- IPSPs are caused by Cl^- flux only.

Convergence

A postsynaptic action potential is generated in a neuron if and only if the algebraic sum of all concurrent EPSPs and IPSPs brings its postsynaptic membrane potential to threshold.

- This **allows neurons to make decisions based on dominant input.** For example, neurons in the central nervous system may have several dozen excitatory and inhibitory synapses converge on them.
- The neuron of convergence is a nodal decision point because it generates an action potential only if the sum of all its synaptic potentials is adequate to achieve threshold.

Divergence

In nerve networks, a single presynaptic neuron may synapse with several postsynaptic neurons.

- This **causes wide distribution of a stimulus.**

Synapses

A synapse is a point of "contact" for the purpose of information transfer between a nerve and another cell. The two synapsing cells do not touch physically, but are separated by a narrow cleft.

Chemical

In chemical synapses, information is transmitted by diffusion of neurotransmitter.

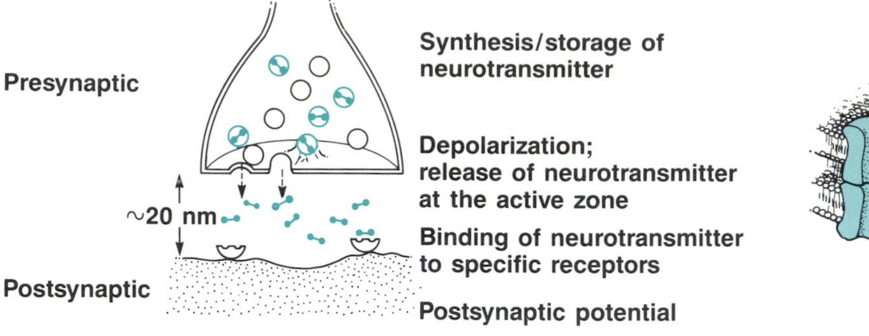

Electrical

In electrical synapses, the pre- and postsynaptic cells share ion channels across a gap junction, and information is transmitted by current (ion) flow through the channels.

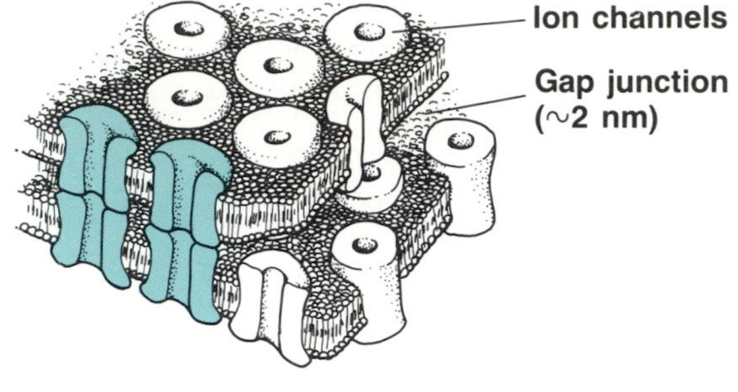

Adapted from Stein JF. An introduction to neurophysiology. London: Blackwell, 1982.

Events During Chemical Synaptic Transmission

1. An action potential in the presynaptic nerve reaches the terminal bouton.*
2. Ca^{++} enters the bouton via a voltage-gated channel.†
3. Synaptic vesicles migrate to the active zone of the presynaptic membrane (mechanisms are not known).
4. Vesicles fuse with the membrane, open, and release their contents into the synaptic cleft.
5. The vesicle contents (transmitter) reach the postsynaptic membrane.
6. Interaction of the transmitter with a postsynaptic receptor alters the permeability of the postsynaptic membrane to one or more ions.
7. Postsynaptic ion flux changes and, depending on the direction of change as well as on the charge of the ion(s), the postsynaptic membrane potential may become less negative (excitatory postsynaptic potential) or more negative (inhibitory postsynaptic potential).
8. If the sum of all postsynaptic potentials is enough to raise the postsynaptic potential to threshold, then a postsynaptic action potential will be generated at that point in the membrane where the threshold is lowest. (In a nerve this is typically the axon hillock.)

*Some synapses function without an action potential.
†Some synapses function without Ca^{++}.

Postsynaptic Potentials vs. Action Potentials

Postsynaptic Potentials	Action Potentials
• Make the membrane potential either more negative (an inhibitory postsynaptic potential, IPSP) or less negative (an excitatory postsynaptic potential, EPSP)	• Always make the membrane potential less negative
• Are graded corresponding to the magnitude of presynaptic stimulation	• Are an all-or-none phenomenon
• Are not propagated along the axon	• Are propagated
• Last about 20 ms, making temporal summation possible	• Last about 1 ms (in nerve)

THE NERVOUS SYSTEM AND REGULATION OF PHYSIOLOGIC FUNCTIONS

Categories of Nervous Function

There are three categories of physiologic function:

- Mental processes
- Actions on the external environment (sensory-motor integration)
- Actions on the internal environment (autonomic reflexes)

Mental processes guide voluntary actions on the external environment but are not normally involved in the maintenance of the internal environment or in involuntary reactions to external stimuli.

Patterns of Physiologic Activity Involving the Nervous System

The common pattern of physiologic activity is that

- central nervous command patterns of action potentials to effector mechanisms lead to a mechanical effect;
- the mechanical effect is sensed in the periphery, transformed, and transmitted to the central nervous system;
- the success of the mechanical effect is evaluated relative to a goal;
- a modified central nervous command pattern is issued to effectors.

Functional Components of the Nervous System

The typical pattern of activity described above requires five functional components:

1. A peripheral nervous sensing mechanism that transforms a specific stimulus modality (touch, smell, light, sound, taste, temperature, stretch, or pain) into trains of action potentials
2. Afferent nerves and interneurons that conduct the action potentials toward the central nervous system
3. A center that evaluates all incoming information, determines whether the stimulus represents a deviation from the normal set point, and generates an appropriate outgoing program
4. Efferent paths (which may be nerves) that convey the outgoing program to the periphery
5. Effector mechanisms that are capable of converting the outgoing motor program into an appropriate somatic or autonomic response

THE NERVOUS SYSTEM AND REGULATION OF PHYSIOLOGIC FUNCTIONS

Categories of Nervous Function

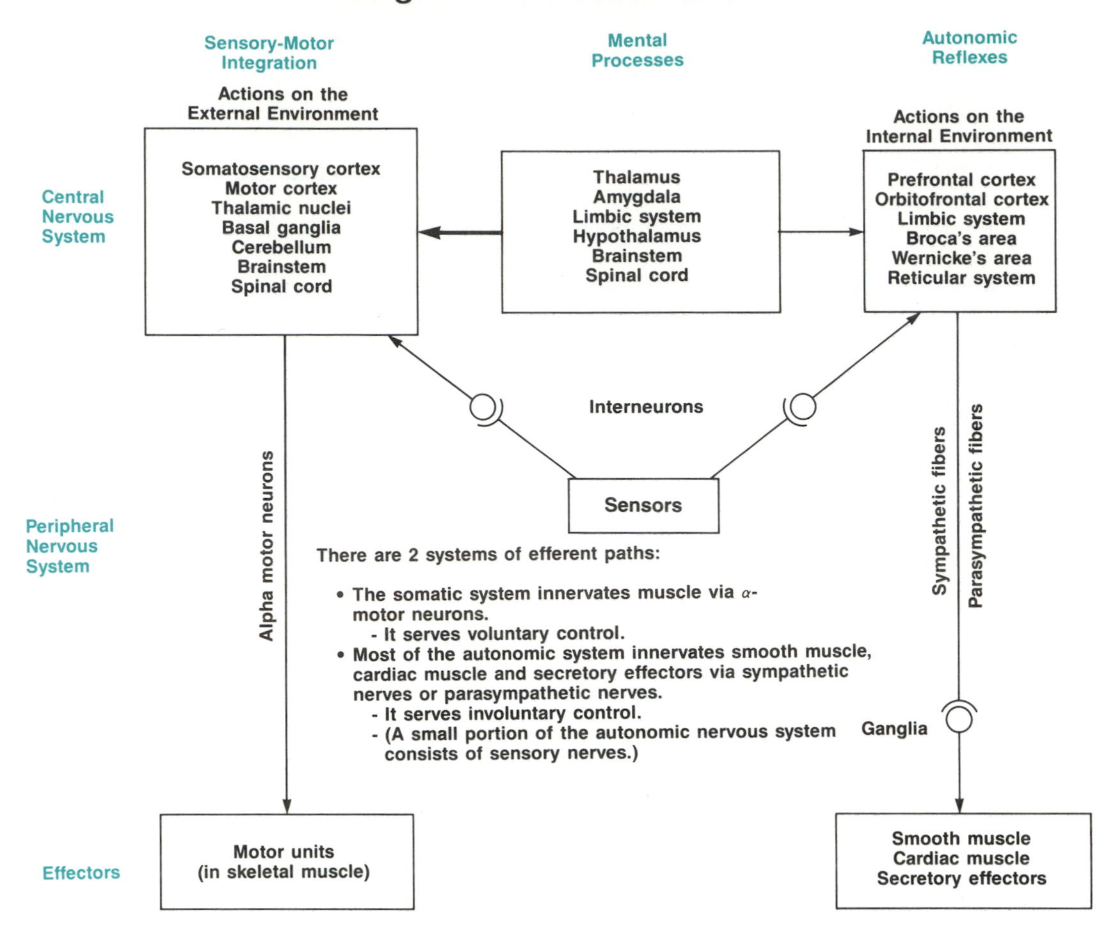

Patterns of Physiologic Activity Involving the Nervous System

A common pattern of physiologic activity is that

1. A group of effectors causes a change in a physiologic variable.
2. Peripheral nervous sensors relay to the central nervous system (CNS) the current status of the variable.
3. The CNS compares current status with desired status, and if there is a discrepancy between them, then
4. The CNS issues a program of action potentials to appropriate effectors.

- The starting point for voluntary action is 3, whereas 2 is normally the starting point for involuntary actions.
- Involuntary actions are automatic responses to a stimulus.
  —They are called reflex responses.

Sensory Receptors

These specialized neuronal structures transform the energy of a specific, local stimulus into a **frequency-encoded train of action potentials.**

Receptor activation

The sensing element is generally a portion of the membrane of the afferent nerve.

- It may be surrounded by accessory structures such as pacinian corpuscles or the optical system of the eye.

- **Activation results from stimulus-induced changes in membrane conductance, ion flux, and membrane potential.**
  - A sufficiently strong excitatory stimulus will raise the membrane potential of the sensing neuron to threshold and generate an action potential.
  - The rate at which subsequent action potentials are generated depends on stimulus strength and **receptor adaptation.**
- **Most sensors function best as detectors of change because they adapt to continuous stimulation with a decrease in action potential frequency.**
  - Some receptors adapt quickly (e.g., the pacinian corpuscle).
  - Others adapt slowly (e.g., the muscle spindle receptor).

Receptive fields

Each receptor is surrounded by a region of sensitivity within which a stimulus will activate the receptor. This region is called the receptive field. Its size determines the discriminatory power of the receptor.

- **The smaller the receptive field of a receptor, the more discriminating its response.**
  - Receptive fields can be **excitatory or inhibitory,** depending on the effect of impulse activity from their respective receptors on subsequent neurons.
  - Inhibitory fields may be located in the immediate vicinity of an excitatory field (surround inhibition) or in another part of the body.

Sensory Receptors

These are specialized structures that transform the energy of a specific, local stimulus into a frequency-encoded train of action potentials.

There are four basic types of sensory receptors:

- Mechanoreceptors (sensing touch, stretch, distending pressure, acoustic vibration, and gravitational acceleration)
- Thermoreceptors (sensing changes in temperature)
- Chemoreceptors (sensing taste, smell, and specific chemicals such as O_2 or H^+)
- Photoreceptors (sensing light)

Sensors convey environmental information to the CNS along specific nerves (labeled lines) at varying impulse frequencies.

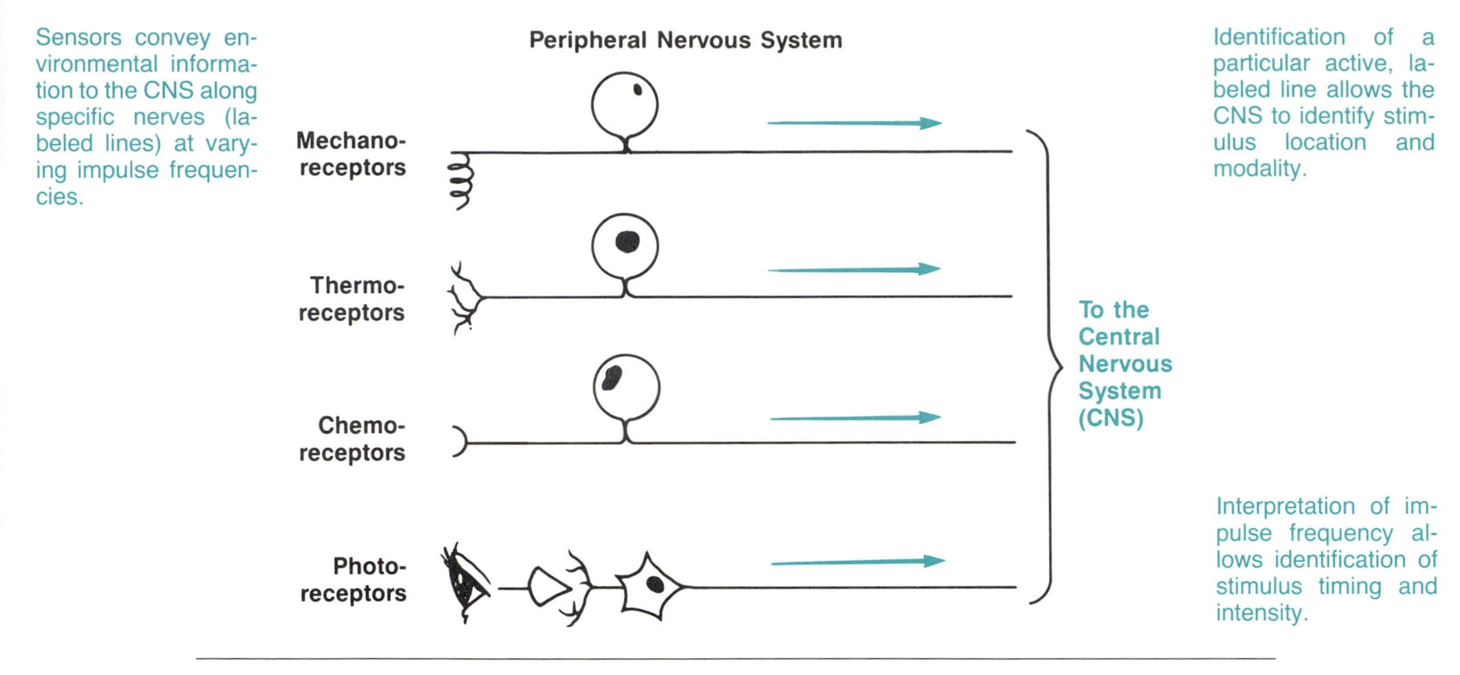

Identification of a particular active, labeled line allows the CNS to identify stimulus location and modality.

Interpretation of impulse frequency allows identification of stimulus timing and intensity.

Central Nervous Connections

- For most receptor types, the neuronal cell body is located at some distance along the afferent nerve, in the dorsal root ganglion.
  —Near it, the apparently single afferent axon divides in the shape of a T, one arm going to the cell body, the other toward the central nervous system.
- Other receptors (e.g., photo receptors) do not have their cell bodies in a dorsal root ganglion. They synapse a short distance away with an afferent neuron.

Sensory Receptors—Cont'd.

Mechanoreceptors

These structures respond to changes in physical deformation of their environment, such as

- touch,
- vibration,
- acceleration, and
- stretch.

They are found in

- skin,
- the inner ear,
- skeletal muscle, and
- blood vessel walls.

Within each of these tissues special accessory structures may influence the quality of the transduction process.

In the skin

In the skin are found the following:

- Extensively branching **free nerve endings**
  - They respond to a variety of stimuli (pressure, temperature, or chemical irritation).
- **Hair follicle endings**
- **Pacinian corpuscles and Meissner's corpuscles**
  - These corpuscles act as mechanical filters for the nerve terminal (which is the primary transducer element).
  - They interpose a mechanically stiff microregion of extracellular fluid between the axon and its surroundings. As a result,
  - these receptors are able to **detect rapidly fluctuating pressures (vibrations).**
- **Merkel's discs**
  - They are disc-shaped attachments of terminal nerve twigs to epithelial cells.
  - They also have a mechanical filtering function for the nerve terminal.
  - They adapt slowly and are, therefore, appropriate for the detection of **continuous touch or pressure.**

Mechanoreceptors

In the Skin

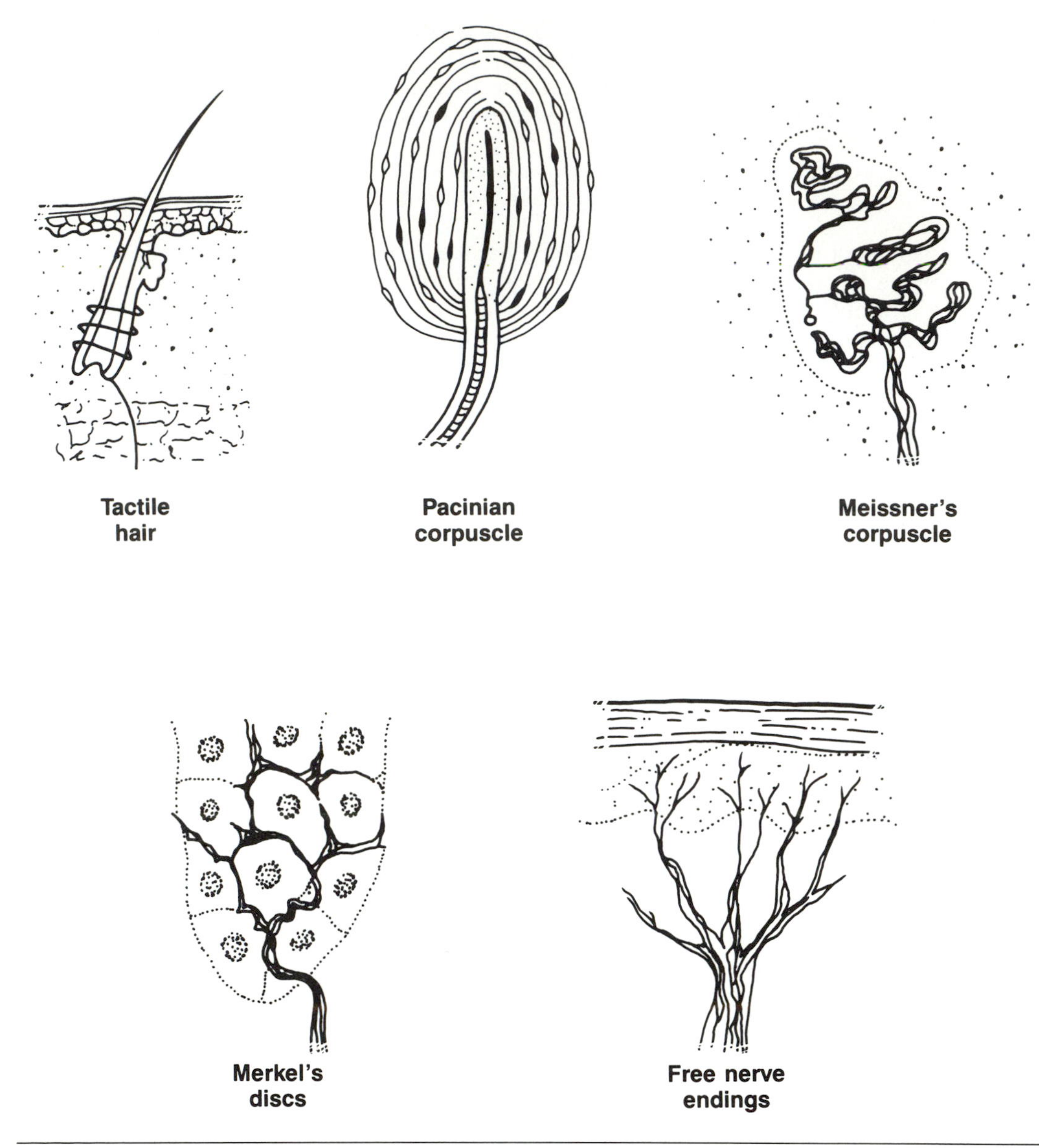

Adapted from Stein JF. An introduction to neurophysiology. London: Blackwell, 1982.

Mechanoreceptors (Cont'd.)

In the vestibular organs

- The relative densities of endolymph and of the gelatinous mass covering the cilia of hair cells determine responses to spatial orientation.
- In the otoliths (utricle and saccule), the gelatinous mass is called the **otolith membrane.**
  - It contains enough calcium carbonate deposits to make its **density higher than that of the surrounding endolymph.**
- In the semicircular canals, the gelatinous mass covering the cilia is called the **cupula.**
  - The cupula contains no mineral deposits, so its **density is equal to that of the endolymph.**
- The otoliths, but not the canals, respond to changes in spatial orientation because
  - **a change in orientation with respect to gravity moves, relative to the surrounding endolymph, only the denser gelatinous mass around the hair cells of utricle and saccule.** This bends their cilia and generates receptor potentials.
- In the semicircular canals, the gelatinous mass of the cupula has the same density as the surrounding endolymph. As a result,
  - The two always move at the same rate under the influence of gravity;
  - The canals are able to detect angular acceleration because **the cupula extends across the semicircular canal and attaches to the wall on the other side.**
  - During angular acceleration, the canal walls, hair cells, and attached cupula move in the direction of the rotation. The inertia of the endolymph keeps it stationary for an instant. This causes bending of the cupula in its mid-region and excites the hair cells.

In the cochlea

- Pressure fluctuations arising from sound waves are transmitted to the **organ of Corti on the basilar membrane.** It holds the **hair cells.**
  - The free ends of their stereocilia are attached to the **tectorial membrane** that lies over the organ of Corti.
  - Hair cell receptor potentials are produced by relative displacement of the basilar and tectorial membranes.
- Cochlear hair cells have directional sensitivity:
  - Bending in one direction yields a positive-going receptor potential.
  - Bending in the other direction yields the opposite. (Molecular details of the transduction process are not yet known, but integrity of the ion pumps is known to be important.)
- **Sound intensity is encoded in the firing frequency** of axons.
- **Sound frequency is encoded principally in the location of the excited hair cells** on the basilar membrane.

Mechanoreceptors (Cont'd.)

Hair Cells

Excitation
Inhibition
Kinocilium
Stereocilia
Afferent nerve fiber
Efferent

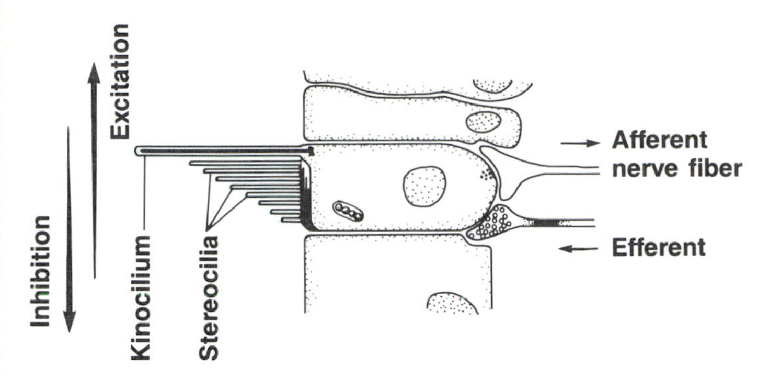

In the inner ear, hair cells convert vibrations or steady displacement into transmitter release.

Their transducer portion is a group of several dozen cilia that project from the free surface.

—Bending of cilia causes receptor potentials and release of transmitter from each activated hair cell into the synaptic clefts of the nerve(s) supplying it.

In the Vestibular Organs

Hair cells are embedded in a supporting endothelial layer within a bulge that is incorporated into each of the three subunits of the vestibular organs—utricle, saccule, and semicircular canals. Several stereocilia and one kinocilium project from each hair cell into the endolymph that fills the vestibular organs. However, the **the cilia do not contact endolymph directly.**

—They are covered by a gelatinous mass whose composition differs between the otoliths (utricle and saccule) and the semicircular canals.

In the Cochlea

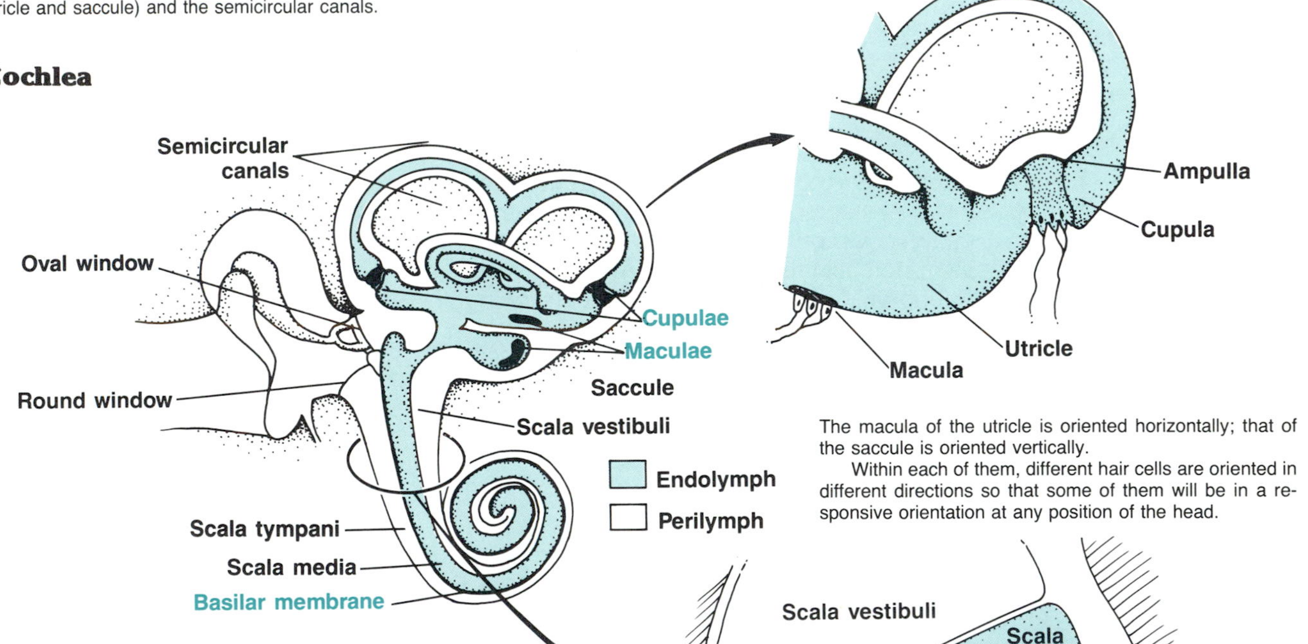

Semicircular canals
Oval window
Round window
Cupulae
Maculae
Saccule
Scala vestibuli
Scala tympani
Scala media
Basilar membrane
Endolymph
Perilymph
Ampulla
Cupula
Macula
Utricle

The macula of the utricle is oriented horizontally; that of the saccule is oriented vertically.

Within each of them, different hair cells are oriented in different directions so that some of them will be in a responsive orientation at any position of the head.

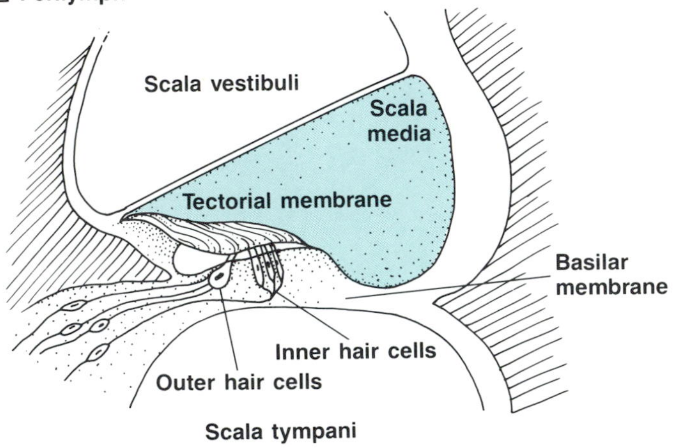

Scala vestibuli
Scala media
Tectorial membrane
Basilar membrane
Inner hair cells
Outer hair cells
Scala tympani

Cochlear hair cells have no kinocilium. They are arranged in two rows:

- Inner hair cells lie near the central axis of the cochlear turns.
- Outer hair cells lie near the outer wall.

About 90 percent of the cochlear afferent nerves derive from the inner hair cells because each cell is innervated by 10 to 12 fibers, each fiber contacting only one inner hair cell. The remaining cochlear afferent nerves originate from the outer hair cells, and each of these fibers innervates many outer hair cells.

Mechanoreceptors (Cont'd.)

In skeletal muscle

Stretch receptors are located in tendons **(Golgi tendon organ)** and in **muscle spindles** (the sense organs of skeletal muscle).

- **Golgi tendon organs** are the extensively branched terminations of a large Group Ib nerve fiber.
  - They are woven into the tendons of a bundle of extrafusal fibers.
  - Their **firing patterns mirror muscle tension.**
- **Muscle spindles** contain their own specialized fibers, named **intrafusal fibers** to distinguish them from the larger, working extrafusal fibers that make up the bulk of the muscle.
  - They are innervated by primary and secondary afferent fibers.
  - **Secondary fibers signal** only the **static length of muscle.**
  - **Primary fibers signal** predominantly the **velocity of contraction.**
    —They also change their firing rate in proportion to stretch.
    —They are quiescent while stretch is being decreased.

In blood vessels

The walls of the aortic arch, the carotid sinus, the cardiac atria, and the ventricles contain extensively branched, unencapsulated nerve endings that generate action potentials in response to wall stretch.

Thermoreceptors

Although all receptors change firing rate with temperature, thermoreceptors are nerve endings, located in the hypothalamus, skin, and mucous membranes, that display far greater temperature sensitivity than do any other receptors.

- **Thermosensitive cells of the hypothalamus are most important for the regulation of body temperature.**
  - Relatively few details are known about them.
- Peripheral thermoreceptors respond either to cold or to warm.
  - Cold-sensitive receptors are branched, slightly enlarged terminals of Aδ fibers.
  - Warm-sensitive receptors are the terminations of C fibers.
  - Their action-potential response patterns are unique among receptors in that they do not show a monotonic relationship with stimulus strength.

Mechanoreceptors (Cont'd.)

In Skeletal Muscle

Intrafusal fibers are grouped on the basis of where their nuclei congregate:

- **Nuclear bag fibers** show concentrations of nuclei in a centrally located bulge.
- **Nuclear chain fibers** have nuclei aligned along the length of the fiber and have no central bulge.

Each spindle has at least two nuclear bag fibers and up to 10 nuclear chain fibers.

Only the ends of intrafusal fibers contract. They are innervated by gamma motor nerves.

The central, noncontracting portion contains sensory nerve endings of two types:

- Branches of a single **Group Ia primary afferent** axon wrap around the middle portions of both types of intrafusal fiber.
- **Group II secondary afferent** fibers arise from extensive branchings on both sides of the Group I primary wrap.

α-Motor neuron
Extrafusal fibers
Golgi, Tendon Organ
Intrafusal fibers
II
Ia
γ-Motor neuron
α-Motor neuron
Intrafusal fibers
γ Efferents
Motor unit
Golgi Tendon Organ
Group Ia Primary
Group II Secondary
Afferents

Thermoreceptors

The action-potential response patterns of peripheral thermoreceptors to their specific stimuli are unique among receptors in that they do not show a monotonic relationship with stimulus strength.

Nonmonotonic Response and Hysteresis

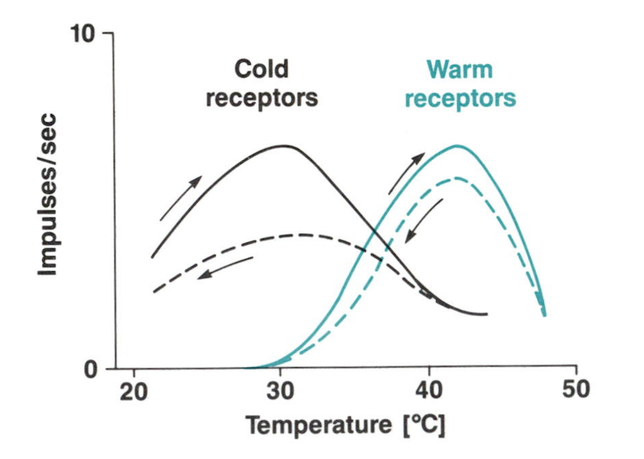

Some receptors show a *decrease* in steady-state firing rate after the temperature has exceeded a certain level.

From Guyton AC. Textbook of medical physiology. 7th ed. Philadelphia: WB Saunders 1986:604.

Evolution to Burst Pattern

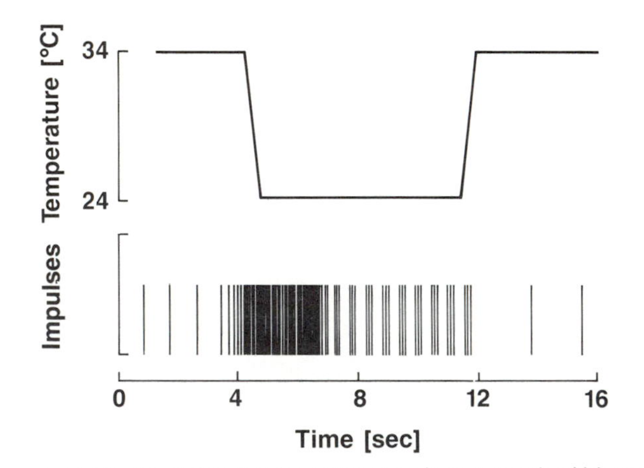

Some receptors change their impulse patterns to a burst pattern in which groups of action potentials are separated by quiescent pauses.

From Darian-Smith et al. J Neurophysiol 1973; 36:325.

Thermoreceptors show directional sensitivity (hysteresis) in that their impulse rate at a given temperature depends on whether that temperature was reached by cooling from a higher temperature or by warming from a lower temperature. It is not yet known whether these changes in impulse pattern carry significant information that governs higher CNS responses.

Chemoreceptors

Chemosensitive cells are found

- in the **taste buds** of the tongue,
- in the **olfactory receptors** of the nose,
- in the central nervous system, and
- in the cardiovascular system (carotid body, aortic body, and cardiac ventricle).

Taste buds

- These special receptor cells **respond to four primary stimuli:**
  —**Sweet**
  —**Sour**
  —**Salty**
  —**Bitter**
- It is not yet known whether each taste receptor cell responds to only one of the four primary stimuli or responds to each with a different sensitivity.

Olfactory receptors

- These are primary afferent neurons whose specialized receptor cilia project into the nasal mucosa.
- Only seven pure chemical elements possess the attributes that are necessary for a substance to have an odor:
  —Soluble in water or lipid
  —Volatile
  —Molecular weight between 17 and 300
- Nevertheless, humans can distinguish a few thousand odors.
- There is no agreement on the underlying receptor physiology that would allow such a discriminatory range.

Cardiovascular and central nervous chemoreceptors

- These specialized neurons produce generator potentials in response to internal H^+ concentration.
- Some also have the ability to convert ambient changes in the partial pressure of CO_2 or of O_2 into changes of intracellular $[H^+]$ (e.g., receptors in the aortic and carotid bodies).
  —These units feature prominently in the regulation of respiration.

Photoreceptors

These are specialized receptor cells (**rods** and **cones**) in the **retina of the eye.**

- Their sensitive portion is a highly specialized single projection that differs from the rest of the cell in histology and in function.

Histology

- In the projection are many membrane-enclosed discs and folds that are filled with photopigment.

Function

- In darkness the Na^+ permeability of the projection membrane is greater than that of the remaining cell membrane.

Electrophysiologic behavior

- is determined by local Na^+ permeability and its changes.

Chemoreceptors

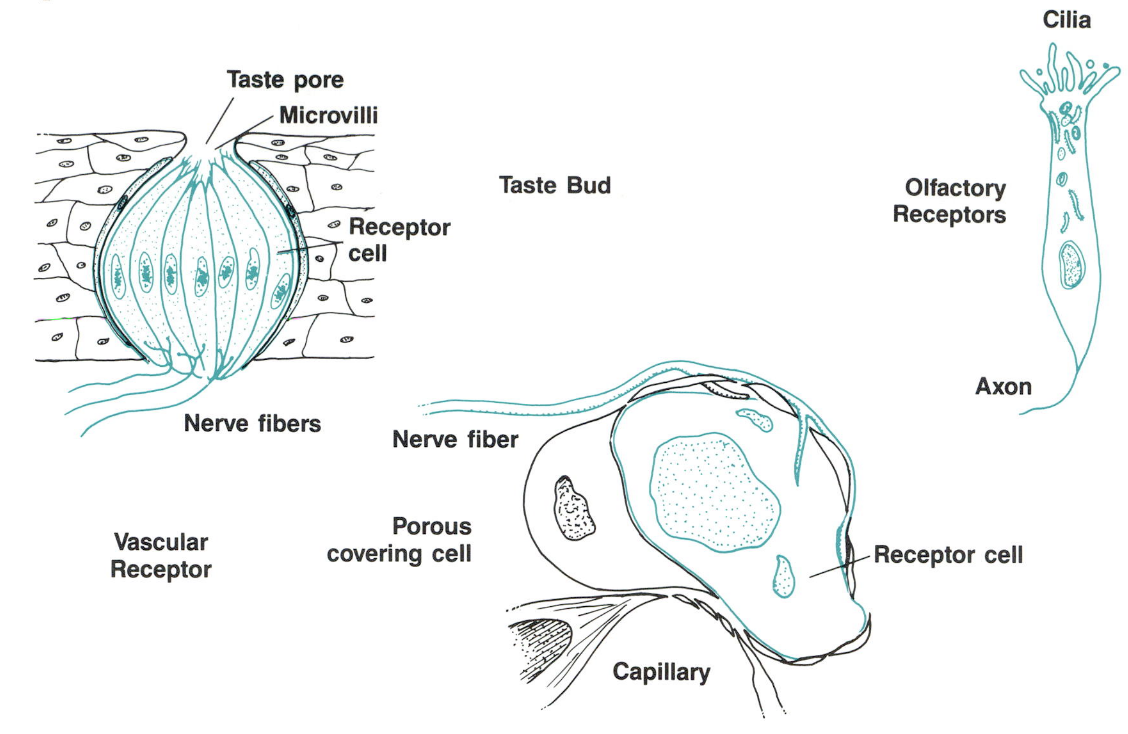

Photoreceptors

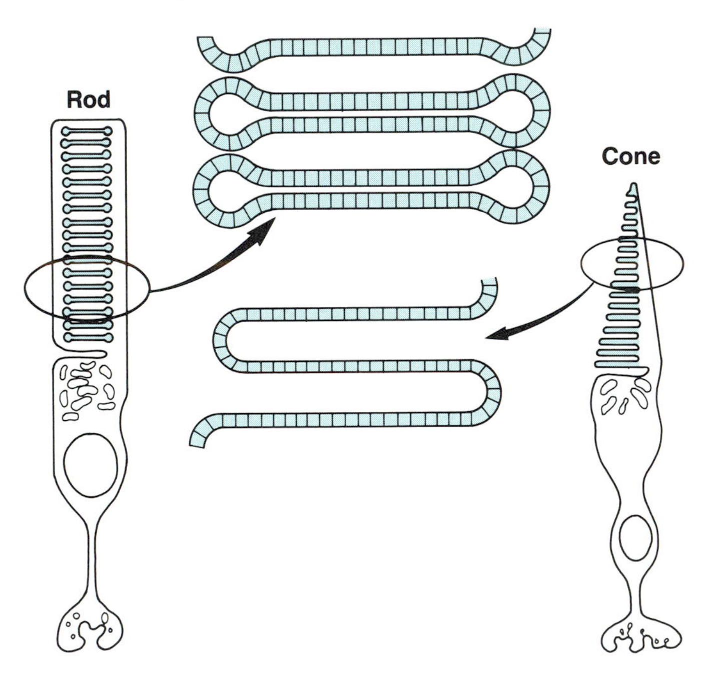

- In the human eye, the ratio of rods to cones is 20:1. They differ from each other
  - —in the structure of the outer, projecting segment of the cell;
  - —in their distribution over the retina;
  - —in the wavelengths to which they are maximally sensitive;
  - —in the light intensity to which they will respond.
- In bright light, rod pigment becomes bleached and only the cones are used.
- In dim light, only the rods are sufficiently sensitive to give sight.
- Rods cannot discriminate between colors because they contain only one photo pigment (rhodopsin).
- Color vision arises from cones because cones contain three pigments, each with a distinct light absorption spectrum.

At rest (i.e., in darkness)

- Na^+ channels in the projection are actively held open by cGMP. This depolarizes the cell partially and causes continuous release of neurotransmitter (glutamate).
- K^+ leaves from the cell body. The Na^+/K^+ flux creates a constant current (the dark current).

During activity (i.e., in light)

- The main action of light on the photo pigments is to activate a cGMP phosphodiesterase.
  The consequent breakdown of cGMP
  —closes membrane Na^+ channels in the cell projection;
  —hyperpolarizes the region;
  —reduces the dark current;
  —decreases neurotransmitter release.

Central Nervous System

Cerebrum

The cerebrum is divided into left and right lateral hemispheres. It includes structures such as the following:

The cerebral cortex

- This consists of a superficial layer 1- to 4-mm thick.
  —Most areas of the cerebral cortex are subdivided into six **layers** and organized into **functional columns** from outside (layer I) to inside (layer VI).
- **All the neurons within each layer serve a characteristic function such as receiving input, sending output,** and so on.

Claustrum

- A bilateral band of gray matter under the lateral cortex

Deep nuclei

- Basal ganglia and amygdala

Thalamus

Its many nuclei serve as interfaces at two levels:

- Upward they act as the **gateway to the cerebral cortex.**
- Downward they **modulate, coordinate, and integrate all body functions.**

Hypothalamus

This region contains nuclei that provide

- an interface between higher centers and the **pituitary and thyroid,** and
- **coordination of whole-body autonomic responses** to behavioral drives (e.g., fear) and to input from autonomic sensors (e.g., baroreceptors).

Cerebellum

The cerebellum is a main locus for guidance and coordination of **skeletal muscle actions.**

Brainstem

This area consists of three anatomically distinct regions: **midbrain, pons,** and **medulla.** They are linked at the core by the **reticular formation.**

Midbrain

- Acts as a conduit for ascending and descending fibers
- Contains nuclei associated with complex neurologic patterns

Pons

- Contains **reflex centers for cardiovascular and respiratory control**

Medulla

- Contains nuclei that act as **reflex centers for autonomic control** or as **intermediaries in somatic control**

Reticular formation

- A collection of both ascending and descending neurons whose major functions are
  — the **determination of the state of consciousness** and
  — **balancing autonomic and somatic activity with the level of consciousness**

Central Nervous System

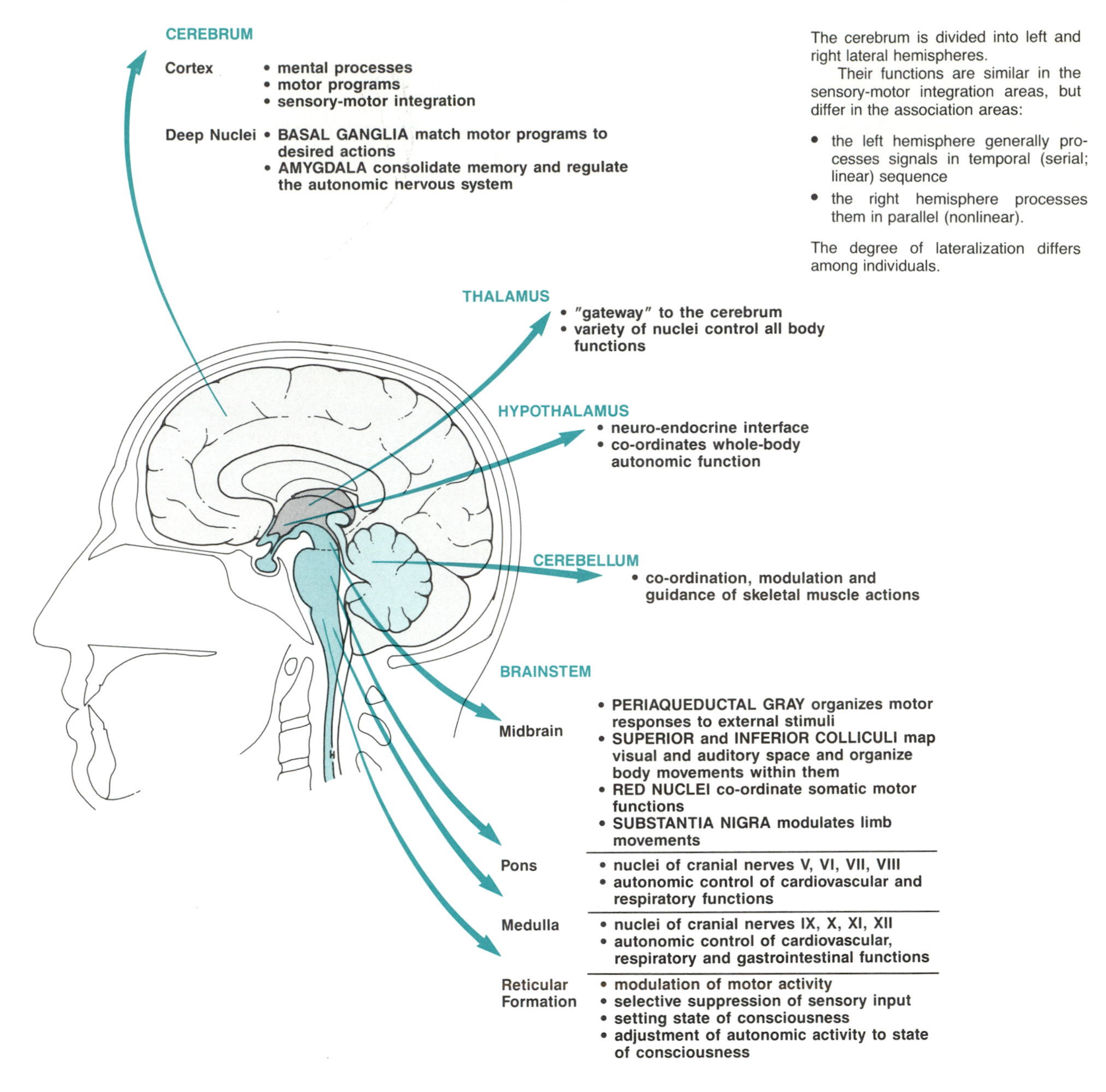

CEREBRUM
Cortex
• mental processes
• motor programs
• sensory-motor integration
Deep Nuclei
• BASAL GANGLIA match motor programs to desired actions
• AMYGDALA consolidate memory and regulate the autonomic nervous system
The cerebrum is divided into left and right lateral hemispheres.
Their functions are similar in the sensory-motor integration areas, but differ in the association areas:
• the left hemisphere generally processes signals in temporal (serial; linear) sequence
• the right hemisphere processes them in parallel (nonlinear).
The degree of lateralization differs among individuals.
THALAMUS
• "gateway" to the cerebrum
• variety of nuclei control all body functions
HYPOTHALAMUS
• neuro-endocrine interface
• co-ordinates whole-body autonomic function
CEREBELLUM
• co-ordination, modulation and guidance of skeletal muscle actions
BRAINSTEM
Midbrain
• PERIAQUEDUCTAL GRAY organizes motor responses to external stimuli
• SUPERIOR and INFERIOR COLLICULI map visual and auditory space and organize body movements within them
• RED NUCLEI co-ordinate somatic motor functions
• SUBSTANTIA NIGRA modulates limb movements
Pons
• nuclei of cranial nerves V, VI, VII, VIII
• autonomic control of cardiovascular and respiratory functions
Medulla
• nuclei of cranial nerves IX, X, XI, XII
• autonomic control of cardiovascular, respiratory and gastrointestinal functions
Reticular Formation
• modulation of motor activity
• selective suppression of sensory input
• setting state of consciousness
• adjustment of autonomic activity to state of consciousness

Cranial nerves

These **12 nerve pairs exit from or enter specific nuclei in the brainstem.**

- Some cranial nerves contain only sensory afferents (I, II, and VIII).
- Some contain only motor efferents (III, IV, VI, XI, and XII).
- The remainder contain both (V, VII, IX, and X).

Spinal cord

This collection of axons and nerve cell bodies is encased within the vertebral column and extends to the level of the first lumbar vertebra (L_1), where it terminates in the cauda equina.

◆ Within each cord segment, two distinct regions can be recognized: **gray matter** and **white matter.**

Gray Matter

- Consists mostly of neuronal cell bodies
- Is subdivided on each side into
  — the **dorsal horn** (synapses of sensory afferents with spinal neurons),
  — the **ventral horn** (groupings of motor neurons supplying skeletal muscle), and
  — the **intermediate zone** (local afferent-efferent interneuron linkages)

White matter

- Surrounds the gray matter
- Consists mostly of axon columns
- Is organized into **fascicles** and **columns:**
  — The dorsal column contains principally ascending fibers.
  — The ventral column contains mainly descending fibers.
  — The lateral column contains a mixture of ascending and descending fibers.

Spinal nerves

Near the spinal cord, each segmental nerve divides into a **dorsal root** and a **ventral root**.

Dorsal root

- The dorsal root contains afferent fibers
- It shows a slight swelling, the **dorsal root ganglion**
  — The cell bodies of primary afferent neurons reside in this ganglion.

Ventral root

- The ventral root contains efferent fibers.
- Efferent **somatic** nerves travel directly to their muscle fibers.
- Efferent **autonomic** nerves synapse in ganglia that may be located near the cord **(sympathetic paravertebral ganglia).**

Spinal Cord

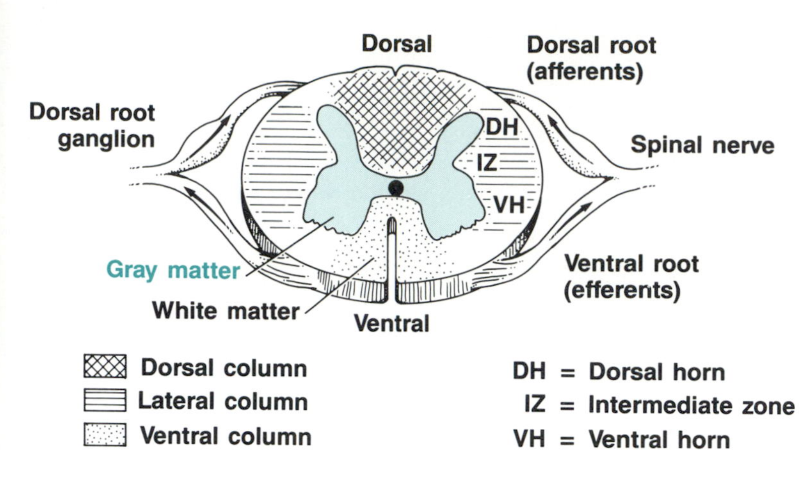

The spinal cord is organized into segments that correspond to the anatomic locations of the vertebrae.

- At each segment, two spinal nerves connect with the cord, one on the left, the other on the right.
  —Each spinal nerve branches into a dorsal root and a ventral root.
- Within each cord segment, two distinct regions can be recognized: gray matter and white matter.

Spinal Nerves

Near the spinal cord, each segmental nerve divides into a dorsal root and a ventral root.

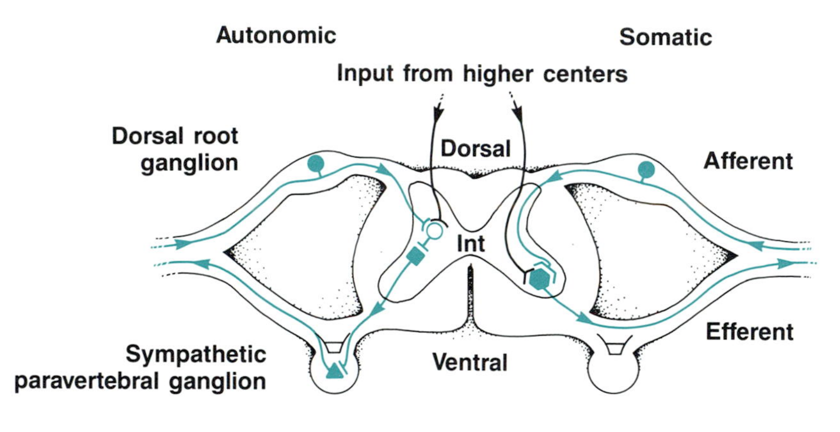

In the autonomic nervous system ventral roots do not innervate peripheral effectors directly. They first form a synapse in a peripheral ganglion.

Each dorsal root contains afferents from a specific region of the body known as a **dermatome.** There is sufficient overlap so that destruction of a given root generally leads to no more than reduced sensibility in the dermatome rather than complete loss of sensation.

Effector Organs

The effector organs of the central nervous system are **skeletal muscle, secretory units, cardiac muscle,** and **smooth muscle.**

Skeletal muscle

This effector is described in more detail in Chapter 2, Muscle Physiology.

- The central nervous control of motor units is described in a later section under "Motor Control."

Secretory effectors

The three types of secretory effectors are **exocrine glands, endocrine glands,** and **neurosecretory cells.**

- Input from the nervous system may be
  — via nerves (e.g., the adrenals) or
  — via secretions that are produced by special **neurosecretory cells** and arrive through the circulatory system (e.g., the hypothalamus-pituitary axis).

Neurosecretory cells

- Are controlled by other neurons
- Look like other neurons, but differ from them in function:
  — They can synthesize and package a secretory product in the cell body.
  — They transport the product along their axons to the capillary bed of a distal release site. (Such **axonal transport** of vesicle-packaged products occurs along axonal microtubules with the help of a special class of enzymes, **force-generating enzymes,** that act by creating a reversible bond between vesicles and microtubules.)
  — Release of secretory product from the terminal end occurs in response to membrane changes that are induced by action potentials.

Cardiac muscle

This tissue differs from skeletal muscle in several functional aspects:

- It is not organized around motor units.
- It requires no nerves for the initiation of contraction.
- It does not use recruitment of muscle fibers as a way of increasing developed tension.
- It cannot use temporal summation as a way of increasing developed tension.

Smooth muscle

Smooth muscle, like cardiac muscle, is **supplied by the autonomic nervous system and usually uses its innervation to modulate contractile behavior.**

Effector Organs

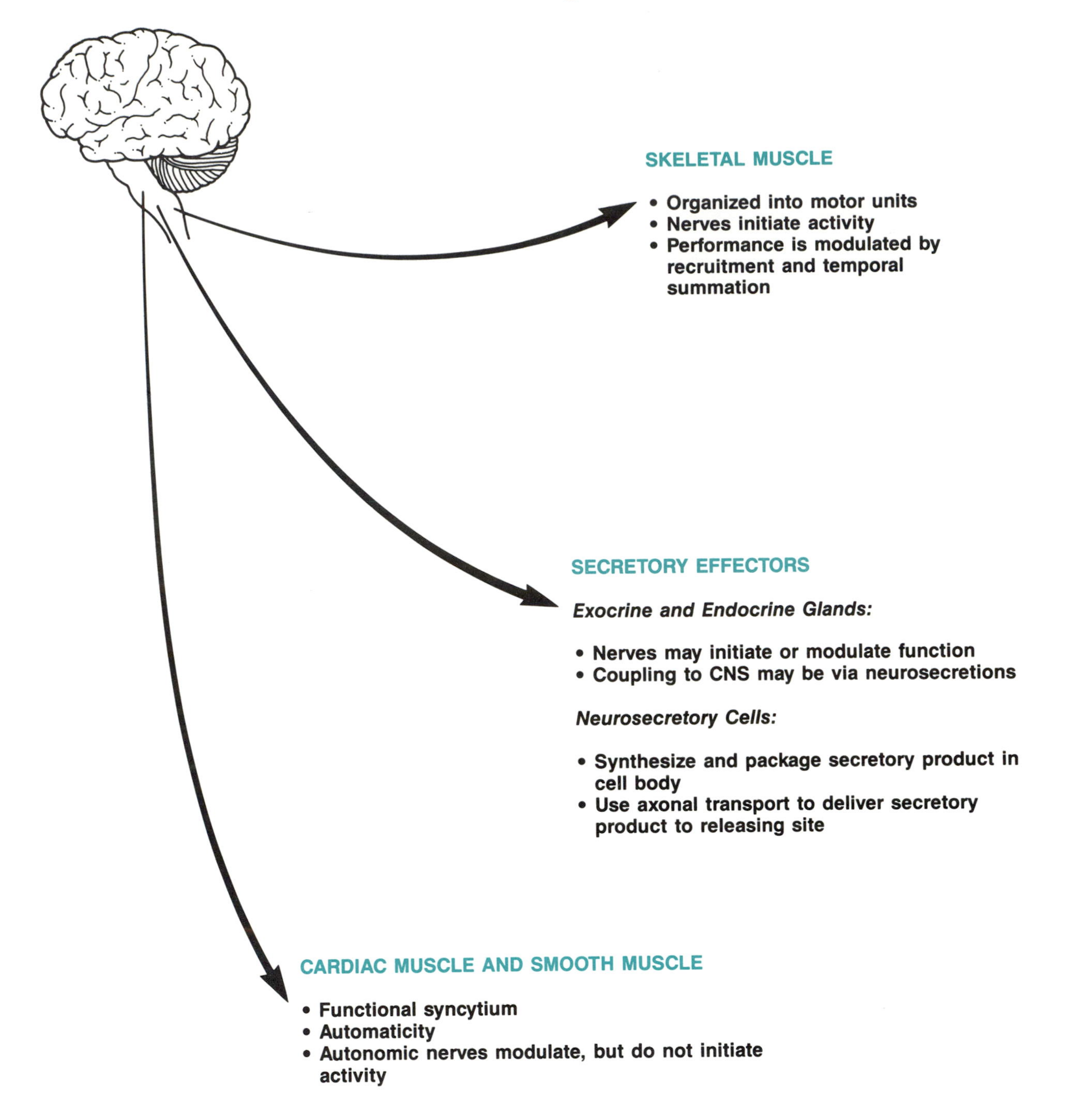

SKELETAL MUSCLE
• Organized into motor units
• Nerves initiate activity
• Performance is modulated by recruitment and temporal summation
SECRETORY EFFECTORS
Exocrine and Endocrine Glands:
• Nerves may initiate or modulate function
• Coupling to CNS may be via neurosecretions
Neurosecretory Cells:
• Synthesize and package secretory product in cell body
• Use axonal transport to deliver secretory product to releasing site
CARDIAC MUSCLE AND SMOOTH MUSCLE
• Functional syncytium
• Automaticity
• Autonomic nerves modulate, but do not initiate activity

SENSORY PHYSIOLOGY

Sensory physiology is classified into

- **somatic senses** (e.g., proprioception, pain, temperature) and
- **special senses** (e.g., taste, hearing, vision).

Somatic Sensation

The somatic senses include

- **proprioception** (as sensed by muscle spindle receptors and Golgi tendon organs),
- **tactile sensations** (pressure or touch as sensed by cutaneous mechanoreceptors),
- **pain,** and
- **temperature.**

Ascending pathways

Afferents from peripheral receptors ascend the spinal cord in three separate paths:

- **Dorsal column**
- **Ventral spinothalamic tract**
- **Lateral spinothalamic tract**
  - Some fibers in these paths terminate in the **reticular formation,** where they provide information relevant to the setting of whole-body behavior states.
  - Most fibers continue toward termination points in intralaminar and posterior **thalamic nuclei.**
  - Ascending paths synapse with other ascending paths at two locations, the medulla and the thalamus.
  - Third-order neurons radiate from the thalamus to the somatosensory cortex.

Descending pathways

Descending paths from the somatosensory cortex synapse with ascending neurons in the medulla and thalamus and can facilitate or inhibit ascending information.

SENSORY PHYSIOLOGY

Somatic Sensation

Ascending Pathways

Afferents from peripheral receptors ascend the spinal cord in three separate paths:

- Dorsal column
- Ventral spinothalamic tract
- Lateral spinothalamic tract

Some fibers in the afferent paths terminate in the reticular formation. Most fibers continue toward termination points in intralaminar and posterior thalamic nuclei.

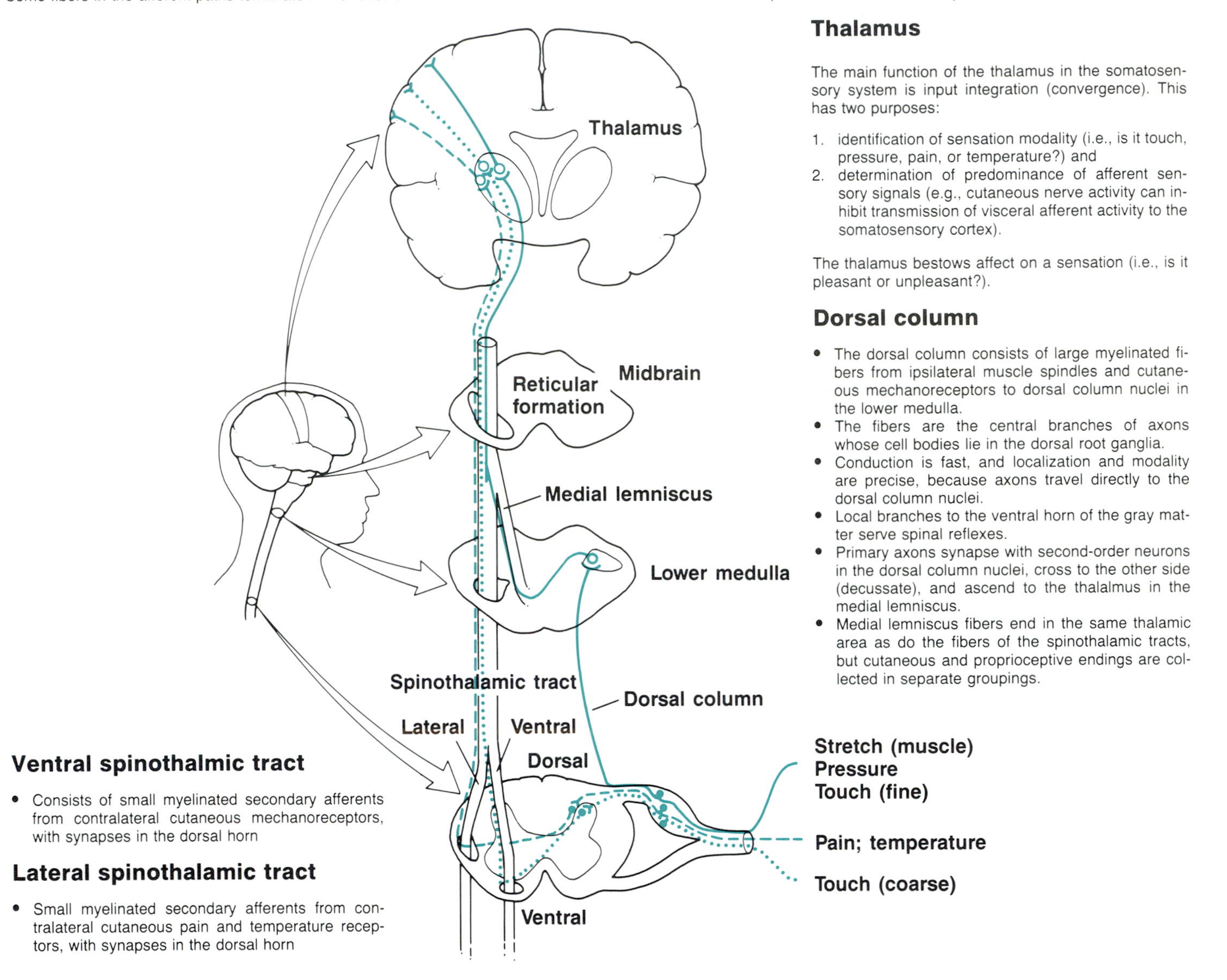

Thalamus

The main function of the thalamus in the somatosensory system is input integration (convergence). This has two purposes:

1. identification of sensation modality (i.e., is it touch, pressure, pain, or temperature?) and
2. determination of predominance of afferent sensory signals (e.g., cutaneous nerve activity can inhibit transmission of visceral afferent activity to the somatosensory cortex).

The thalamus bestows affect on a sensation (i.e., is it pleasant or unpleasant?).

Dorsal column

- The dorsal column consists of large myelinated fibers from ipsilateral muscle spindles and cutaneous mechanoreceptors to dorsal column nuclei in the lower medulla.
- The fibers are the central branches of axons whose cell bodies lie in the dorsal root ganglia.
- Conduction is fast, and localization and modality are precise, because axons travel directly to the dorsal column nuclei.
- Local branches to the ventral horn of the gray matter serve spinal reflexes.
- Primary axons synapse with second-order neurons in the dorsal column nuclei, cross to the other side (decussate), and ascend to the thalalmus in the medial lemniscus.
- Medial lemniscus fibers end in the same thalamic area as do the fibers of the spinothalamic tracts, but cutaneous and proprioceptive endings are collected in separate groupings.

Ventral spinothalmic tract

- Consists of small myelinated secondary afferents from contralateral cutaneous mechanoreceptors, with synapses in the dorsal horn

Lateral spinothalamic tract

- Small myelinated secondary afferents from contralateral cutaneous pain and temperature receptors, with synapses in the dorsal horn

Input from the face ascends separately, via a parallel path that comprises the trigeminal nerve and the trigeminothalamic tract.

Somatosensory cortex

- These areas of the central nervous system **analyze sensation** with respect to
  - intensity,
  - spatial orientation,
  - texture,
  - shape, and
  - weight.
- They are responsible for
  - **pattern recognition,**
  - **determining the significance of the sensation,** and
  - **registering the sensation on consciousness.**

Primary somatosensory cortex (S1)

- Each **cortical column** serves only **one modality** (e.g., touch), and adjacent columns represent the same modality but in another body region.
- The **diameter of a cortical column** representing a body part is **proportional to the number of sensory endings in the part** and not to the physical size of the part.
- **Temperature and pain do not have separate columns.**
  —They are represented in a few cells within mechanoreceptor columns.
- **Area 3b does not receive input from Pacinian corpuscles.**
- Area 3b projects mostly to area 1 of the somatosensory association cortex.

Somatosensory association cortex (areas 1, 2, and 5)

- This cortex receives
  —**direct somatic input from the periphery** and
  —**projections from S1.**
- Area 1
  —receives input from all pacinian corpuscles and
  —**maps the body in mirror image to the mapping in S1.** (The results of dual mapping [in S1 and in the association cortex] are enlargement of receptive fields and integration of information from within the fields.)
- Neurons within areas 1, 2, and 5 are sensitive
  —**to moving stimuli,**
  —to direction of motion,
  —to degree of rotation, and
  —to other indicators of change.

Organization of the second somatosensory area (S2)

- The body is mapped yet again in this area, but with large receptive fields.

Somatosensory Cortex

This is the region posterior to the central sulcus

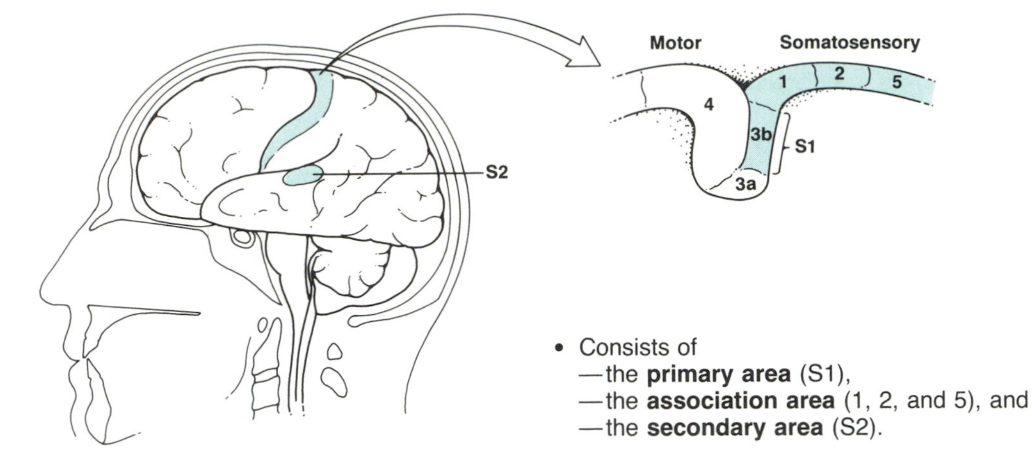

- Consists of
  —the **primary area** (S1),
  —the **association area** (1, 2, and 5), and
  —the **secondary area** (S2).

Organization of the Somatosensory Cortex

The primary somatosensory cortex (S1) consists of area 3b on the postcentral gyrus.

- It is arranged in layers parallel to the surface.
  —They are numbered I to VI (surface to interior).

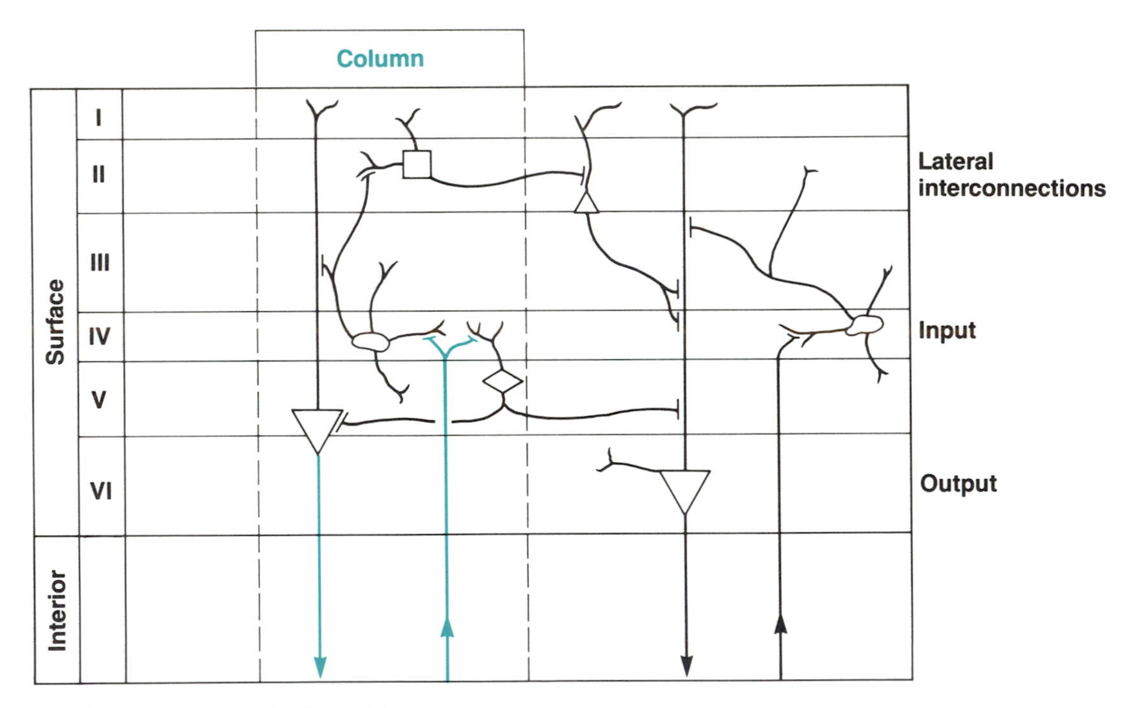

- Afferent neurons enter layer IV.
  —Each point on the skin surface is connected to a specific point in the sensory cortex (somatotopic mapping).
- Neurons directly above the layer IV afferent neuron provide local associations.
- Neurons directly below the layer IV afferent neuron provide output from the somatosensory cortex.

This arrangement causes receptive fields to be mapped in **cortical columns.**

Pain

Pain is detected by **nociceptors.** They are free nerve endings of

- Aδ fibers (**unimodal**—sensitive only to acute mechanical stimuli) or
- C fibers (**polymodal** because they are sensitive to mechanical stimuli, to intense heat, and to chemicals).
  —They have high mechanical and temperature thresholds, so their activation requires intense stimuli.

Afferents

- Extensive interconnections of afferents occur
  — within the substantia gelatinosa of the spinal cord and
  — with other sensory afferents over short distances up and down the cord (via Lissauer's tract).
- Because different afferents can be directly antagonistic to each other, these **extensive interconnections permit suppression of incoming signals (the gate theory of afferent inhibition).** For example, rubbing the skin that overlies a painful area diminishes the pain because, at the level of dorsal horn interneurons, the nociceptive fibers are inhibited by the Aα tactile fibers that are activated by rubbing.

Central projections

These permit

- **localization,**
- **discrimination** (sharp or acute vs. dull or chronic), and
- **modulation** by emotional and behavioral factors.

Efferents

Pain perception is influenced by descending serotonergic neurons.

- They originate in the nucleus raphe magnus.
- They terminate on enkephalin-releasing interneurons within the substantia gelatinosa.

Enkephalin inhibits the afferents from the primary nociceptors.

Pain

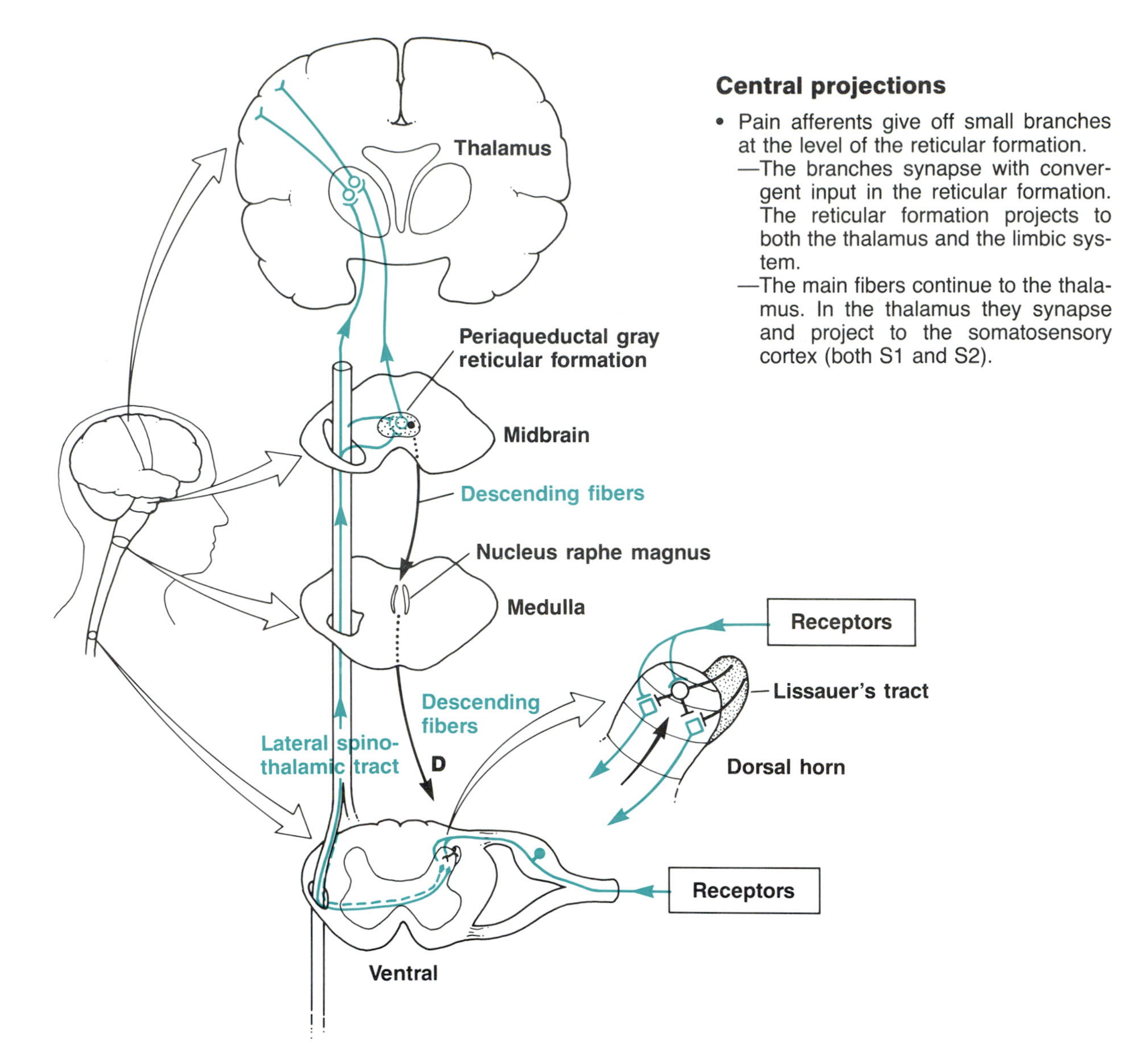

Central projections

- Pain afferents give off small branches at the level of the reticular formation.
  —The branches synapse with convergent input in the reticular formation. The reticular formation projects to both the thalamus and the limbic system.
  —The main fibers continue to the thalamus. In the thalamus they synapse and project to the somatosensory cortex (both S1 and S2).

Efferents and descending fibers

- Beta-endorphins, originating in the hypothalamus and the anterior pituitary, are transported to the periaqueductal gray region of the midbrain via nerve fibers and via blood.
  —There they activate beta-endorphin-sensitive neurons.
- These beta-endorphin-sensitve neurons in periaqueductal gray control serotonin-releasing neurons in raphe magnus.
- Descending serotonergic neurons promote release of enkephalin from interneurons within the substantia gelatinosa.

Afferents and ascending fibers

- Aδ nociceptor afferents enter the dorsal spinal cord at layers I to V of the dorsal horn.
- C fiber afferents enter only in layers II to III, the substantia gelatinosa.
  —The substantia gelatinosa contains many small interneurons whose extensive arborizations provide interconnections within the layer and whose axons pass out into Lissauer's tract, where they travel for short distances up and down the cord, making interconnections with other sensory afferents.
- Output from the substantia gelatinosa is directed ventrally toward neurons of the spinothalamic tract.

Special Senses

The special senses are

- taste
- smell
- hearing
- vestibular reflexes
- vision

Taste

Taste buds respond to four primary stimuli: **sweet, sour, salty,** and **bitter.**

◆ The mechanisms that allow flavor discrimination on the basis of signals received from receptors that respond to only four primary tastes are not fully understood.

It is likely that at least two factors are involved:

- Recognition of certain spacial and electrical patterns of afferent discharge
- Integration of taste sensations with those of smell, touch, and temperature

Smell

Odor discrimination has two aspects:

- **Electrophysiologic patterning of receptor discharges**
- **Spacial patterning of mitral cell activity** within the olfactory bulb

Projection of afferents to the limbic system suggests that

◆ **olfactory sensations are interrelated with such complex phenomena as memory and emotion.**

Special Senses

Taste

- Each taste bud synapses with several secondary afferent nerves, and each afferent nerve synapses with several taste buds.
- The afferent fibers approach the central nervous system via branches of
  - —the facial nerves (VII),
  - —the glossopharyngeal nerves (IX), and
  - —the vagus nerves (X).
- The afferent fibers terminate in the nucleus solitarius of the medulla.
  - —This is a general focus for chemoreceptor afferents not only from the tongue and pharynx, but also from intestinal chemoreceptors as well as from the carotid and aortic bodies.
- Secondary fibers ascend to the thalamus and to the limbic system.
  - —The thalamic connection subserves projections to the tongue-taste area of the sensory cortex.
  - —The projections to the limbic system allow integration of taste with aspects of feeding behavior, with the chemical state of body fluids, and with smell.

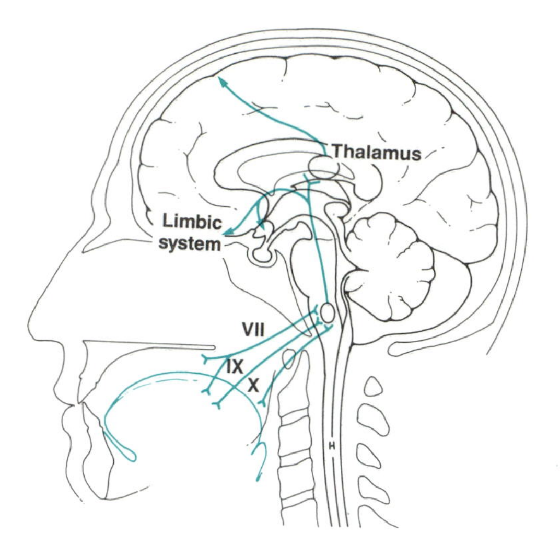

Smell

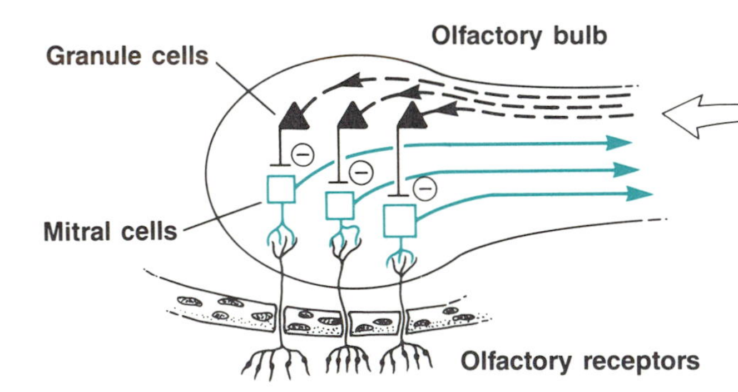

- The unmyelinated fibers of olfactory receptors pass through the floor of the cranium directly into the olfactory bulb.
- There they synapse on the dendrites of mitral cells.
- Mitral cells also synapse with granule cells and are inhibited by them.

The axons of mitral cells project directly to the limbic system and from there

- to the thalamus-frontal cortex as well as
- to the reticular formation.

Hearing

Sounds can be perceived only if they are able to create

- **pressure fluctuations within the fluid that fills the cochlea.**

Such pressure fluctuations can arrive in the inner ear

- directly, through bone conduction, or
- from the outer ear via the mechanical piston-level system that is formed in the middle ear by the eardrum (tympanic membrane) and ossicles.

Hearing involves

- **discrimination of sound frequencies and sound intensities** and
- **localization of sound sources.**

Frequency discrimination

- **Each cochlear afferent fiber and its attached inner hair cell show a highly selective response to sound of a particular frequency.**

This frequency selectivity of inner hair cells arises

- partly from the fact that the basilar membrane is
  —narrower and stiffer at its base (near the stapes) (this makes it more responsive to high frequencies) and
  —wider and more compliant at its apex (this makes it more responsive to low frequencies), and
- mainly from a factor associated with excitation of outer hair cells.
- The absence of synaptic connections between inner and outer hair cell nerve fibers suggests that **central nervous processing is involved** in this cooperative selectivity.

Source localization

The source of a sound is determined on the basis of

- **differences in perceived intensity** of the sound between the left ear and the right ear as well as
- **differences in the timing of perception** between the left ear and the right ear.

Localization analysis begins in the superior olives, the first central auditory nuclei receiving both left and right afferents.

- Incoming cochlear impulses from each side contain information relevant to intensity and timing.
- At each orientation angle, each sound frequency produces a certain difference in the intensities perceived by the ears and a certain delay between left and right excitation.
- **Postsynaptic neurons in the superior olive are selective in that each neuron is maximally excited only when the sound presented to one ear is louder than *and* precedes that at the other ear by certain "memorized" amounts.**

Associations between sound orientation and specific neuronal firing patterns are learned during the postnatal period.

- The resulting **map of the auditory space** around the head is **stored in the inferior and superior colliculi** so that future incoming neuronal activity can be compared with this map for precise auditory localization.

Hearing

The Cochlea

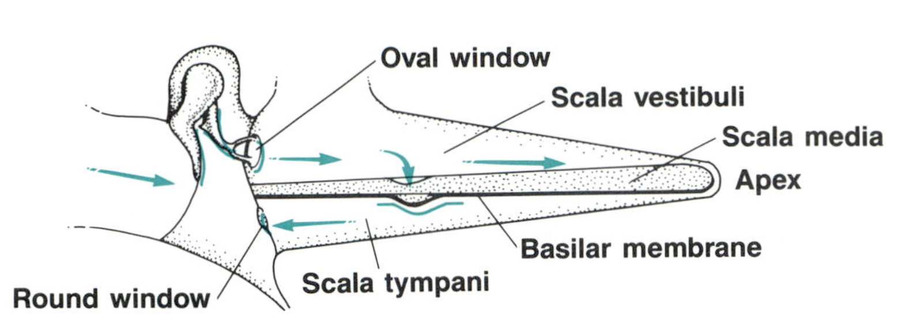

The cochlea is a system of three parallel, lymph-filled, coiled tubes:
—The scala vestibuli
—The scala media
—The scala tympani
Each is separated from its neighbor by a membrane.

- Sound pressure waves, amplified about 20 times by the eardrum-ossicle system, are applied to the scala vestibuli at the oval window and set up vibrations in the perilymph that fills the tube.
- The vibrations are transmitted
  —along the scala vestibuli as well as
  —laterally to the scala media.

Even though the scala vestibuli and the scala tympani are connected at the apex of the cochlea,

- pressure fluctuations are not transmitted instantaneously and equally all the way around from the oval window to the round window.

Therefore,

- local pressure gradients cause bending in the scala media and its membranes.
- The basilar membrane, lying between the scala media and the scala tympani, is of special importance:
  —Its vibrations are transmitted to the hair cells of the organ of Corti.

Central Projections

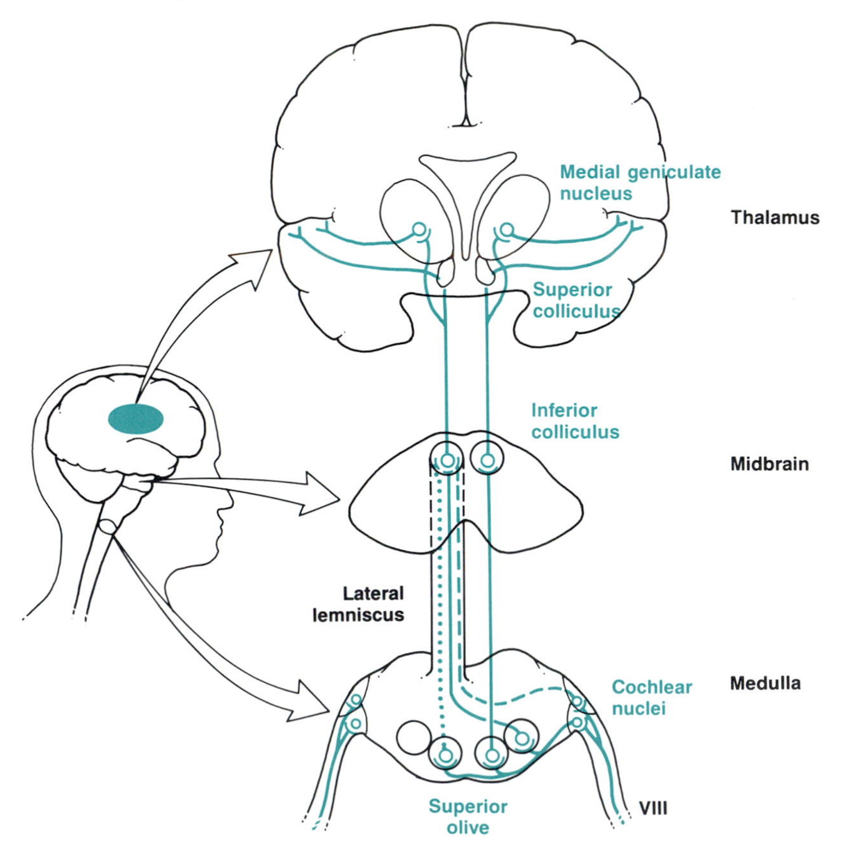

Processing in the auditory cortex is for the purpose of complex analysis. It allows

- recognition of sound patterns and
- tuning in on a sound that is surrounded by other sounds (e.g., a conversation at a cocktail party).

The **inferior and superior colliculi** store a map of the auditory space around the head.
This map associates incoming impulse patterns from the superior olives with a unique sound location relative to the ears.

The **cochlear nuclei** are arranged in lamellae:
Each layer consists of the terminations of VIIIth nerve fibers that respond to a preferred frequency (tonotopic representation).

The **superior olives** are the first level of processes involved in sound localization.

Vestibular reflexes

The vestibular system integrates information about head movements from several peripheral sense organs (otoliths in the middle ear, eyes, and muscle spindles) and from the cerebellum.

- The output of the vestibular nuclei is used to generate two types of reflexes:
  - **Postural reflexes**
    - —Maintenance of balance
    - —Orientation of body relative to gravity
  - **Gaze fixation reflexes**
    - —Stabilization of visual images during body movement

Postural reflexes

- These are patterns of motor programs that simultaneously affect flexors and extensors in the neck, trunk, and limbs.
- There are two classes of postural reflexes:
  - —Those following sudden accelerations and decelerations (sensed by the semicircular canals)
  - —Those following gravitational displacement (sensed by the utricle and the saccule)
- Under **normal circumstances, the reflex responses operate to change the relative positions of the head, trunk, and limbs so as to maintain balance and equilibrium despite body movements.**
- Under **extreme circumstances** (unexpected accelerations or decelerations, or vestibular stimulation in the absence of visual cues), **the reflex responses initiate defensive movements of the head, trunk, and limbs so as to protect the head and ensure appropriate support by the limbs at the point of anticipated surface contact.**

Gaze fixation reflexes

- Are initiated by impulses
  - —from semicircular canals,
  - —from proprioceptors in neck muscles, and
  - —from the retina
- Operate to maintain image stability
  - —when the visual scene is stable, but the head moves, and
  - —when the visual scene moves, but the head is stable
- Ensure eye movements of the same magnitude as, but directed opposite to, the relative motion between head and visual scene

- If the head or the scene is rotated continuously,
  - the eyes maintain a fixed visual image by rotating in the opposite direction until it becomes physically impossible to maintain the fixation point;
  - the eyes then rapidly reverse the direction of their rotation, find a new fixation point, and maintain it as long as possible;
  - the pattern of oppositely directed slow and rapid eye movements is called **nystagmus.**

Vestibular Reflexes

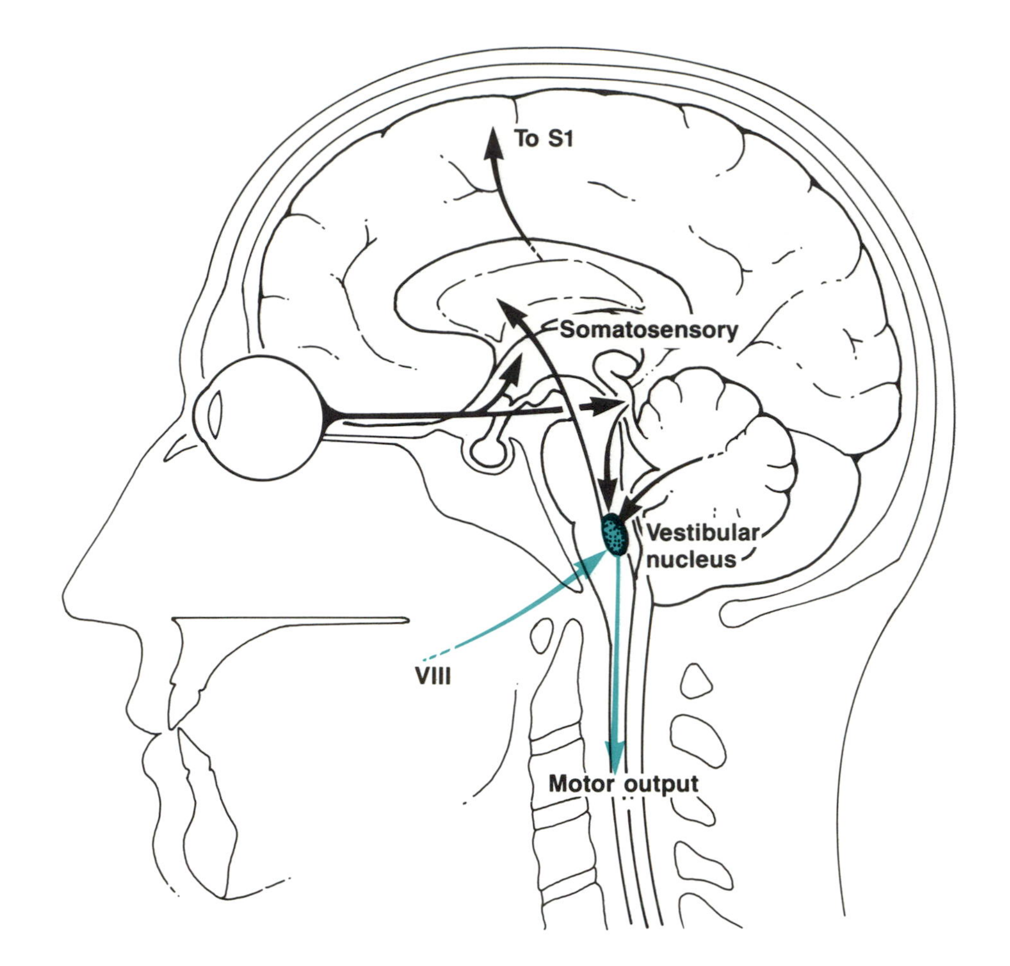

Afferent neurons from the semicircular canals and otoliths enter the brainstem in the VIIIth cranial nerve together with auditory afferents. They terminate in the vestibular nuclei of the pons. These nuclei also receive

- excitatory and inhibitory input from the cerebellum,
- somatosensory input from muscles and joints, and
- visual input via the accessory optic nuclei of the midbrain.

Neurons from the vestibular nuclei project centrally to the thalamus and from there to the primary somatosensory cortex (S1).

Vision

Vision involves

- focusing of electromagnetic radiation by the cornea and the crystalline lens of the eye on the photoreceptors of the retina;
- conversion by the photoreceptors of radiation in the 450-to-700-nm wavelength range into action potentials;
- interpretation of the spacial relationships of incoming light waves.

Optical functions of the eye

The crystalline lens of the eye can change its focal length by changing its curvature:

- The smaller the radius of curvature, the shorter the focal length.
  - Adjustments in lens curvature are called **accommodation.**
  - Accommodation is accomplished by altering the balance between two forces:
    —Natural elastic recoil of the lens tissue
    —Tension in the radially arranged fibers of the suspensory ligament

Focusing on near objects

- An ordinary convex lens brings rays from a near object to a focus at some distance greater than the focal length of the lens.

However,

- **the normal eye can continue to form a sharp image of an object that is being moved closer and closer.** Three processes are involved:

  1. **Accommodation** of the lens is initiated by constriction of the ciliary muscle and allows the lens to relax to its more curved, short-focus shape.
  2. **Convergence** of the eyeballs is initiated by contraction of the medial rectus muscles. It keeps the cornea and lens of each eye pointing at the object.
  3. **Constriction** of the iris narrows the pupil and restricts incident light to the central portion of the lens, where its radius of curvature is smallest. It also increases the depth of field, i.e., the distance range over which an object can be brought to sharp focus.

- When the range of physiologic adjustments is inadequate to form a sharp image on the retina, then external corrective lenses are required.
- A special need for lenses to correct hyperopia occurs with advancing age, as the crystalline lens of the eye loses its elastic recoil and, with that, its ability to accommodate a highly curved shape of short focal length. The condition is called **presbyopia.**

Optical Functions of the Eye

Light enters through the cornea and the crystalline lens. This convex lens concentrates parallel rays into a single point at the principal focal length behind the lens.

Without active tension in the suspensory ligament, the normal eye is focused at infinity. (Given the size of the eye, "infinity" corresponds to distances greater than 6 m.)

When the range of physiologic adjustments is inadequate to form a sharp image on the retina, then external corrective lenses are required. Two basic conditions require such lenses:

1. The curvature of the cornea-crystalline lens unit is not uniform in all directions (astigmatism).
2. The focal length of the cornea–crystalline lens combination is inappropriate for the length of the eye:
   - **Hyperopia** if the eye is too short
   - **Myopia** if the eye is too long

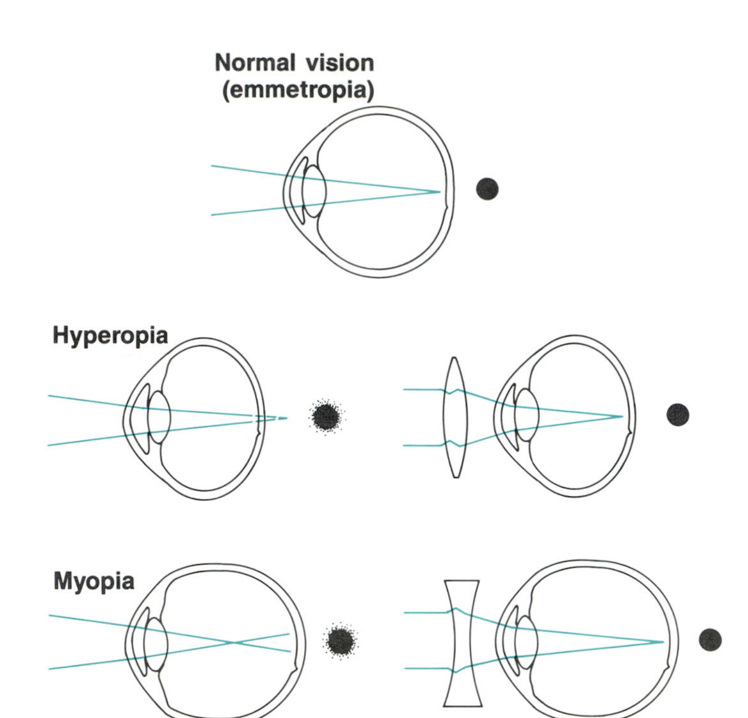

Visual Acuity

Visual acuity is measured with sharp-edged black and white patterns placed at a distance (generally the letters on a Snellen chart).

- The measuring instrument is a pattern that subtends an angle of 1 minute at the optical center of the eye.

Snellen charts are constructed of letters of different standard sizes:

- Large letters are constructed so that the individual strokes making up the letters subtend a 1-minute angle at a large distance. (The strokes of small letters subtend the 1-minute angle at a small distance.)
- Each line of letters is identified by the distance at which its strokes subtend the 1-minute angle (e.g., the 6-m E).

The person taking a visual acuity test sits at a distance of 6 m (20 ft). The Snellen score is expressed as the ratio of testing distance to the distance number of the line of letters that is just legible. For example, if the line containing the 6-m E is just legible, then the Snellen score is 6/6 (20/20 if distances are measured in feet).

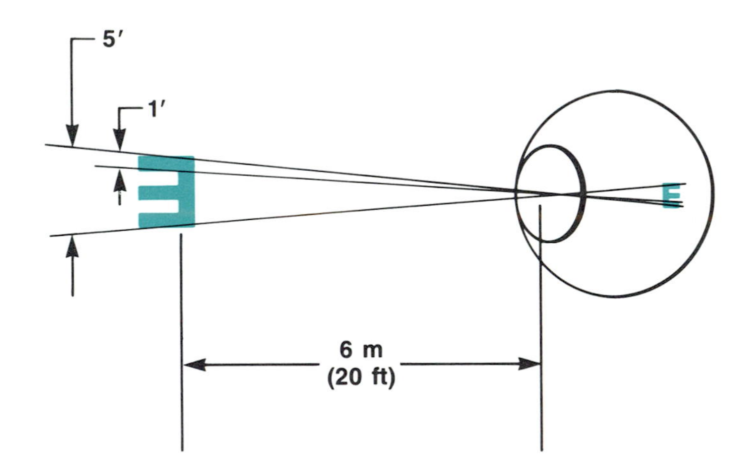

Vision (Cont'd.)

Retinal organization and function

The structure of the retina leads to

- **graded, converging responses from localized and widespread receptive fields.**

The information that is finally transmitted to the central nervous system via the optic nerve is both specific with respect to particular loci and general with respect to the remaining receptor fields.

Light and dark adaptation

- At normal daylight intensity levels,
  —most rod pigment is inactivated (bleached);
  —most visual information derives from cones (**photopic** vision).
- At extremely low levels,
  —only the rods are active;
  —vision is **scotopic.**
- **Rods provide no wavelength discrimination** (hence, all cats are black in the dark), and rods are not sensitive to red light (hence, exposure of the dark-adapted eye to purely red light does not reverse dark adaptation).

- **When illumination is suddenly reduced from daylight levels to darkness, then the minimum intensity that can just be detected decreases exponentially over a period of 30 minutes (dark adaptation).**
  - Two processes are involved:
    — A chemical process involving regeneration of pigment in the rods where it was bleached by daylight intensity
    — Changes in retinal organization that permit increased spacial and temporal summation of incident stimuli
  - When ambient light is made brighter, **light adaptation** occurs; it requires only 3 to 5 minutes.
    — The initial phase of light adaptation is due to neural reorganization within the retina.
    — Later phases are due to more gradual photochemical changes.

Vision (Cont'd.)

Retinal Organization and Function

The retina is unusual among sensory arrays in that the transducer cells (rods and cones) are turned away from the source of their specific stimulus.

- Light must pass through several layers of cells before reaching the photoreceptors
- The electrical activity generated in the photoreceptors must pass back up through the layer of **bipolar cells**, then the layer of **ganglion cells** to reach the optic nerve fibers.
  —Along the way the activity is modulated by a layer of horizontal cells and a layer of **amacrine cells**.

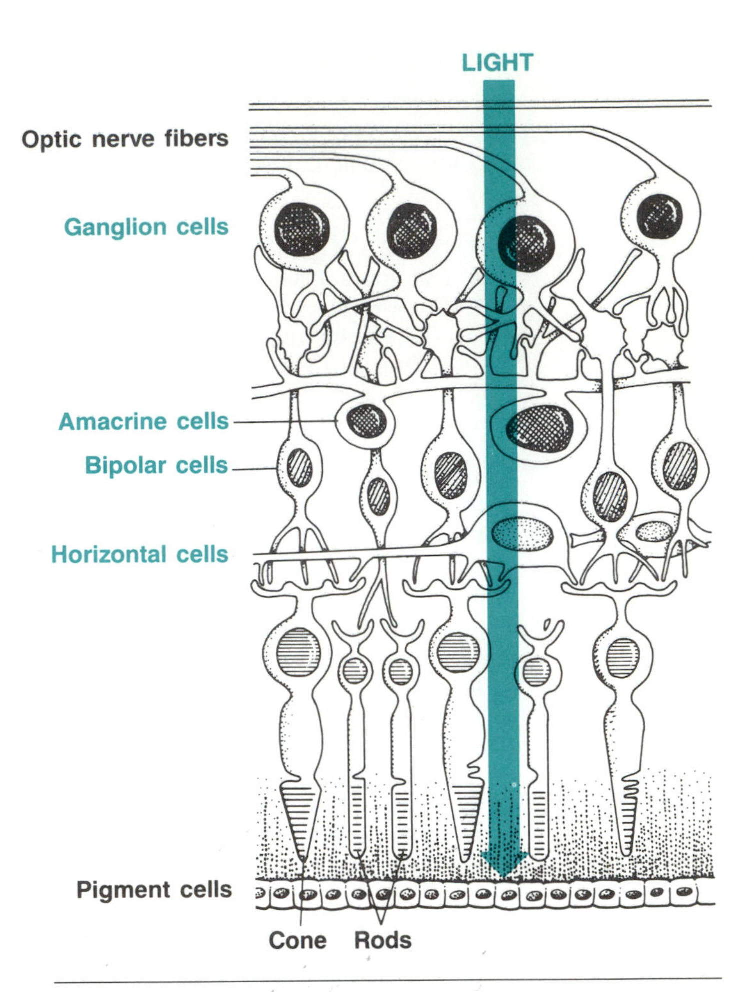

Modified from Dowlins and Boycott. Proc Roy Soc Lond 1966: B166.

Ganglion cells

- Classified by appearance into X, Y or W cells
- X ganglion cells:
  —have small receptive fields
  —receive input mostly from bipolar cells
  —adapt slowly to continuous stimulation
- Y ganglion cells:
  —receive input mostly from amacrine cells
  —adapt rapidly
  —respond best to illumination changes and to moving boundaries
- W ganglion cells:
  —are distinguished by small, slowly conducting axons
  —cannot, because of their heterogeneity, be readily classified by function
- Ganglion cell receptive fields are organized concentrically and surround inhibition is more highly developed in them than in bipolar cells.

Amacrine cells

- Receive input from bipolar cells and are, therefore, of the "on-center" or "off-center" type.
- Transform sustained receptor potentials in bipolar cells into transient amacrine generator potentials that cause amacrine action potentials if threshold is reached

Bipolar cells

- Contact several photoreceptor cells directly and are influenced by more distant receptors via synapses with horizontal cells
- Large bipolar cells synapse with both cones and rods; others synapse with cones only.
- "On-Center" bipolar cells
  —are depolarized by light falling in the center of their receptive field
  —are hyperpolarized by light falling on an annular margin
- "Off-Center" bipolar cells
  —are depolarized by light falling on the margin of the receptive field
  —are hyperpolarized by spot illumination of the receptive field
- By virtue of these 2 types of bipolar cells pattern recognition begins at this level:
  The receptive field of bipolar cells consists of a central area in which light causes a change in membrane potential surrounded by a ring in which light causes a change in the opposite direction (surround inhibition).
- Summated receptor potentials spread passively along bipolar cells and do not generate action potentials, but cause proportional release of transmitter vesicles to amacrine cells and ganglion cells.

Horizontal cells

- Each horizontal cell synapses with a large number of cones as well as with other horizontal cells and with bipolar cells
- Do not generate action potentials, but passively transmit summated generator potentials
- Are hyperpolarized when their input cones are illuminated and use this hyperpolarization to inhibit photoreceptors (lateral inhibition)

Vision (Cont'd.)

Central connections

The visual system performs a variety of physiologic functions in addition to perception and identification of visual patterns. These functions are possible because the optic nerve, after leaving the eyeball, connects to a variety of nuclei in different portions of the central nervous system:

- The **lateral geniculate nuclei** in the thalamus
- The **suprachiasmatic nuclei** in the hypothalamus
- Three midbrain nuclei:
  — **Pretectal nuclei,**
  — **Superior colliculi**
  — **Accessory optic nuclei**

Lateral geniculate nuclei

- These are organized into six anatomic layers that are not interconnected.
- Inputs from the ipsilateral and contralateral retinas are segregated into specific layers, and inputs from X or Y ganglion cells are also segregated into layers.
- Electrical activity resembles that of ganglion cells.

◆ **The primary function of the ganglion cell–lateral geniculate unit is to identify such stimulus attributes as**
— **location,**
— **size,**
— **shape,**
— **rate of movement,**
— **intensity,**
— **steadiness of intensity,**
— **duration,** and
— **color.**

- Organization of receptor fields is concentric.
- Surround inhibition is stronger than in the ganglion cell layer; therefore, areas of activity are more sharply defined.

Suprachiasmatic nuclei

- Synchronize **circadian rhythms** of sleep and wakefulness, adrenal function, eating, and so on with ambient light

Pretectal nuclei

- Integrate reflexes of **ocular fixation**

Superior colliculi

- Integrate visual information with all nonvisual information that identifies objects of visual interest
- Generate motor control patterns for appropriate movement of eyes, head, and trunk

Accessory optic nuclei

- Integrate visual and **vestibular information**
- Generate oculomotor control patterns for gaze stabilization during head movements

Vision (Cont'd.)

Central Connections

Ganglion cell axons travel across the retina, converge at the optic papilla, and leave the eyeball as the optic nerve.

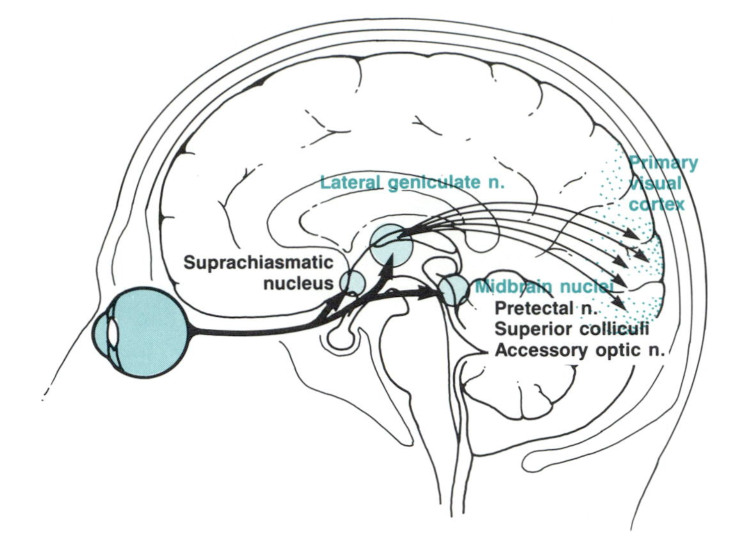

Fibers from the medial side of both retinas cross at the optic **chiasm.** From there they send collaterals

- mainly to the (contra-)**lateral geniculate nucleus** in the thalamus, but
- also to the hypothalamus (suprachiasmatic nuclei) and
- to the midbrain (pretectal nuclei, superior colliculi, and accessory optic nuclei).

Each of the lateral geniculate nuclei projects to the **primary visual cortex** of its own hemisphere and from there to the **visual association cortex.**

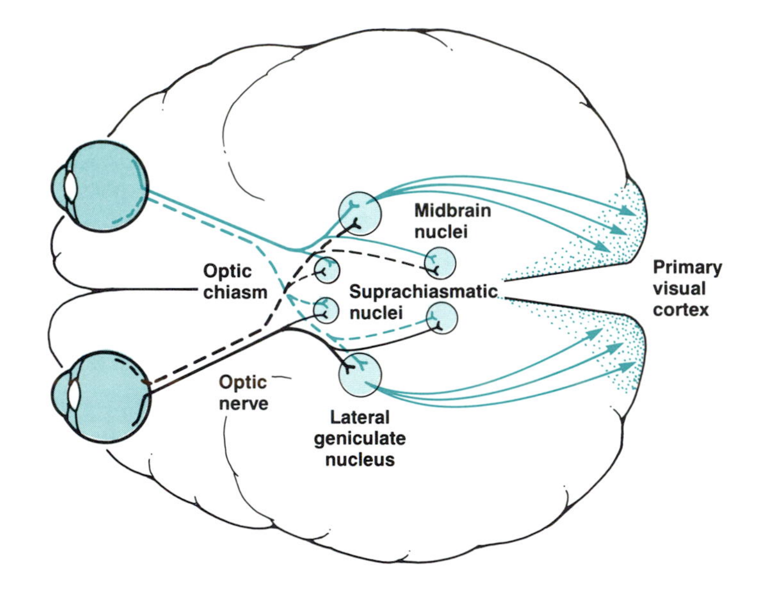

Vision (Cont'd.)

Primary visual cortex

- This is organized into functional units of horizontally layered **hypercolumns.**
- Hypercolumns are divided vertically into right and left **ocular dominance columns** by whether the predominant input is from the right or left eye.
- Clumps of cells rich in **cytochrome oxidase** are found in the center of each ocular dominance column.
  - —These neurons give frequency-specific responses and represent the **first stage of color analysis.**
- The center of the retina is represented in the largest cortical region; the peripheral retina is represented in a small region.
- **Simple cells have elongated receptive fields that are orientation-specific.** This results from convergence and integration, and it means that a given simple cell fires only when a bar of brightness or darkness appears at a specific orientation within the receptive field.
  - — Simple cells of like orientation preference are grouped in vertical slices **(orientation columns)** within each hypercolumn.
  - — Neighboring orientation columns are arranged so that each contains simple cells whose receptive fields are rotated slightly from those of the neighbor.
- **Complex cells** are interspersed among the simple cells.
  - —They have the same orientation-specific receptive fields as their simple-cell neighbors, but they give a response only when their stimulus (a bar of brightness or darkness) is moving at right angles to its long axis within the receptive field.
- **Hypercomplex cells** are also present.
  - —For them the optimal stimulus is also a correctly oriented bar, but its length cannot exceed a certain maximum.

Vision (Cont'd.)

Primary Visual Cortex

The primary visual cortex is organized into six layers, numbered I to VI from the surface inward.

- The functional unit is the vertical **hypercolumn** spanning all six layers.
- Hypercolumns are divided vertically into right and left **ocular dominance columns** by whether the predominant input is from the right or left eye.

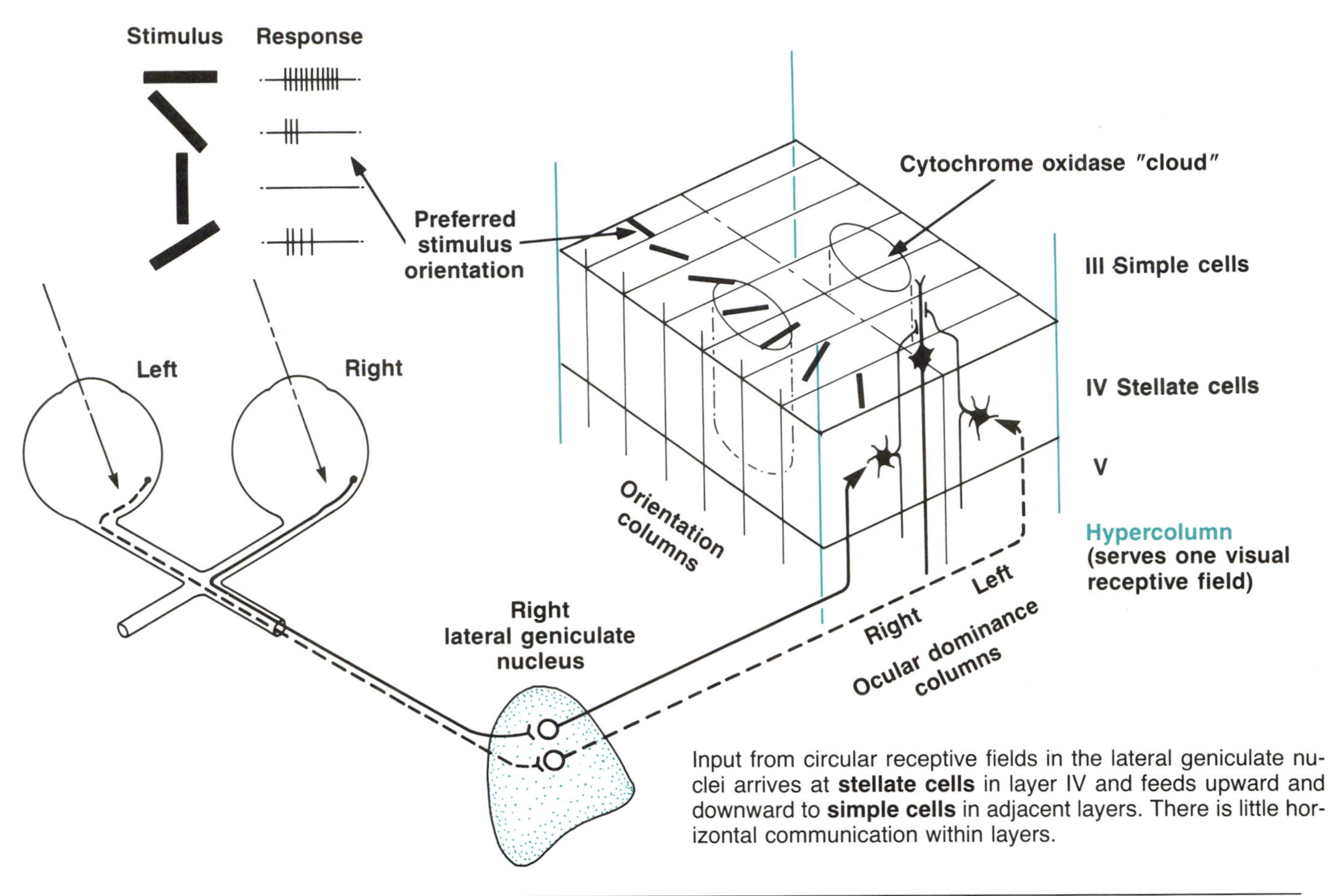

Input from circular receptive fields in the lateral geniculate nuclei arrives at **stellate cells** in layer IV and feeds upward and downward to **simple cells** in adjacent layers. There is little horizontal communication within layers.

Adapted from Somjen G. Neurophysiology: The essentials. Baltimore: Williams and Wilkins, 1983.

Vision (Cont'd.)

Central processing

Different specialized areas of the visual cortex analyze a visual stimulus for a variety of qualities (e.g., location, shape, depth, movement, and color).

- Both the structure and function of these areas and the pathways connecting them depend greatly on the richness of visual experiences during the first few months of life. This can be interpreted as an adaptive process by which

◆ **the visual system develops in accordance with the complexity of the visual environment.**

Location analysis (stereoscopic vision)

- Estimation of object depth and fusion of disparate left and right images occur in specialized cortical neurons **(binocular disparity neurons).**

Movement analysis

- The majority of **neurons in area "MT" of the visual association cortex respond selectively to *moving* retinal stimuli.**
- The response is **direction-specific and highly precise:**
  — Some neurons respond only when the visual stimulus rotates.
  — Others respond only to motion in depth (i.e., motion with a component that points toward or away from the observer).
  — Others respond differently to the image motion produced by motion of the head relative to a stationary object and to that produced by motion of the object relative to the stationary head.

Color analysis

- Retinal **cones** are concentrated in the region of the fovea and contain **three different pigments.**
  — Each pigment has an absorption maximum at a different, narrow wavelength band; therefore,
  — **A given color produces a different impulse frequency in each of the three groups of cones, and color is coded by relative activity in the groups.**
- Bipolar cells and ganglion cells receive input from one of the three cone types, not from all three; therefore,
  — **bipolar cells and ganglion cells are wavelength-specific.**
- Information output from bipolar and ganglion cells is organized in circular center-surround, excitation-inhibition patterns.
  — Center activity and the opposing surround activity originate from cones of different spectral responses **(opponent color coding).**
- Opponent color coding
  — occurs at all levels up to and including the visual cortex;
  — narrows the band width of the spectral response wherever it occurs because the opposing spectra overlap to some extent.

Projections to other cortical areas

- Outputs from the visual cortex and the visual association cortex are directed to other cortical regions for integration of visual information with memory, the limbic system, and the motor system.

Vision (Cont'd.)

Central Processing

Stereoscopic Vision (Stationary Object)

When the gaze is fixed on a stationary object at a given distance, then other objects, placed either in front of or behind the primary object, create retinal images at slightly different locations in the right and left retinas.

Specialized cortical neurons (binocular disparity neurons) are activated only by a given difference in right and left retinal image locations.

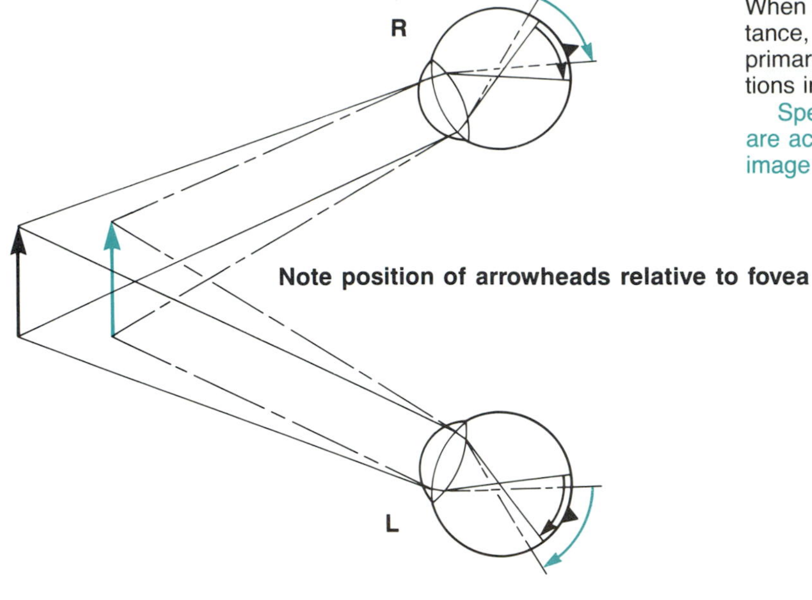

Stereoscopic Vision (Obliquely Moving Object)

Images of obliquely moving objects cast arcs of different lengths on the right and left retinas.

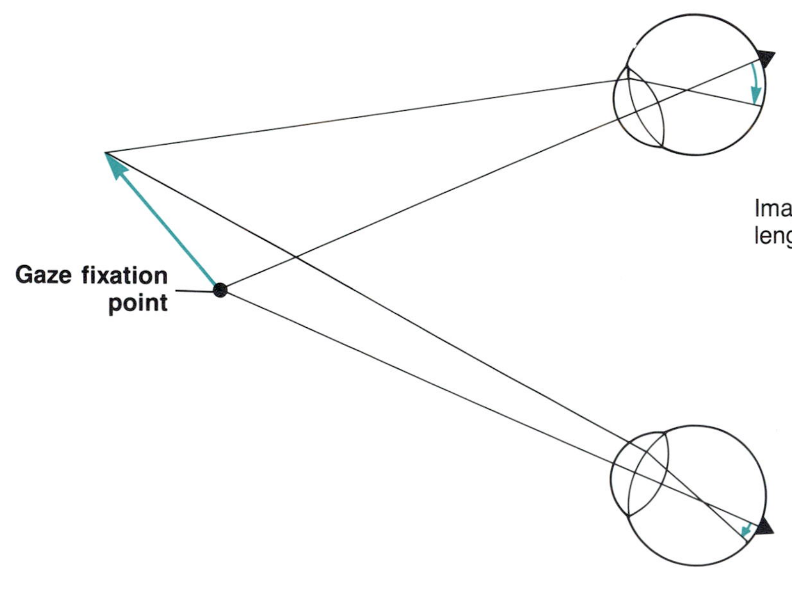

THE AUTONOMIC NERVOUS SYSTEM

OVERVIEW

The autonomic nervous system **governs actions on the internal environment.** It is a system of central nervous nuclei and efferent nerves that

- **integrates and coordinates all reflex mechanisms that maintain a living person in steady state with the changing environment.**

Peripheral Autonomic Nervous System

The two branches of the autonomic nervous system are the **sympathetic** and **parasympathetic** divisions. They differ from each other in the following respects:

- Location of spinal outflow
- Location of ganglia
- Nature of the neurotransmitter released at the effector organ synapse

Ganglia

Sympathetic ganglia are found at two sites:

- In the **sympathetic trunks** alongside the vertebral column and
- In the **viscera** (cervical, stellate, and mesenteric ganglia).
  —Preganglionic fibers that synapse in a visceral ganglion pass through the ganglia of the sympathetic trunk without synapsing there.
- The fibers of the parasympathetic division are not collected in ganglia, but are distributed in the walls of the effector organs.

Receptors

Autonomic receptors are broadly divided into **cholinergic** and **adrenergic** types. Receptor subtypes exist within each division.

Cholinergic receptors

- Postsynaptic receptors in ganglia are nicotinic cholinergic.
- Receptors in effector organs are for the most part muscarinic (M) cholinergic in the parasympathetic system.

Adrenergic receptors

- Receptors in effector organs are **adrenergic** in the sympathetic system.
- There are two types of adrenergic receptors, designated α and β.
- Each has two functional subtypes.
- Within the α grouping:
  —α_1 receptors predominate **postsynaptically** on the effector organ membrane.
  —α_2 receptors dominate **presynaptically** on adrenergic nerve terminals themselves. There they modulate norepinephrine release.
- Within the β grouping:
  —β_1 receptors are found mostly in cardiac myocytes.
  —β_2 receptors are found in smooth muscle and in secretory effectors.

THE AUTONOMIC NERVOUS SYSTEM

Peripheral Autonomic Nervous System

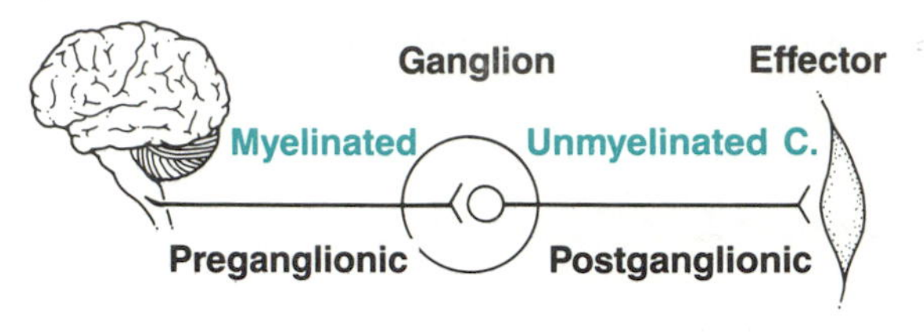

Neurotransmitter: Acetylcholine Acetylcholine or norepinephrine

General Structure

- A neuron in either the brainstem or the spinal cord (in the intermediate zone of the lateral horns) sends a small myelinated cholinergic axon to a peripheral ganglion (the preganglionic fiber).
- There it synapses and activates nicotinic receptors in the postsynaptic neuron.
- The axon of the postsynaptic neuron forms the postganglionic fiber that innervates the effector organ. It is generally an unmyelinated C fiber.

Comparison of Systems

Sympathetic System	Parasympathetic System
• Preganglionic neurons lie in the thoracolumbar region of the spinal cord **(thoracolumbar outflow).**	• Preganglionic neurons lie in the brainstem or the sacral region of the spinal cord **(craniosacral outflow).**
• **Ganglia** form **separate, discrete structures.**	• **Ganglia** are in **the wall of the effector organ.**
• In most cases the **postganglionic** neurotransmitter is **norepinephrine.**	• The **postganglionic** neurotransmitter is **acetylcholine.**

Dominant Effects of Systems

Organ	Regions	Sympathetic		Parasympathetic		For more details, see
		Action	Receptor	Action	Receptor	
Heart	Atria	↑Rate	β_1	↓Rate	M	
		↑Performance	β_1	↓Performance	M	Chapter 5
	Ventricles	↑Performance	β_1	Rhythm change	M	
Blood Vessels	In most organs	Constriction	α_1	No effect		Chapter 5
	Skeletal muscle	Constriction	α_1	No effect		
		Dilatation	β^{*}_2			
		Dilatation?	M†			Chapter 12
	Penis; clitoris; labia; vagina	Dilatation	M	Dilatation	M	Chapter 12
Lungs	Bronchial smooth muscle	Relaxation	β_2	Contraction	M	
GI Tract	Longitudinal and circular muscle	Relaxation	β_2	Contraction	M	Chapter 7
	Sphincters	Contraction	α_1	Relaxation	M	
Kidney	Renin release	Increased	β_2	No effect		Chapter 6
Urinary Bladder	Wall	Relaxation	β_2	Contraction	M	
	Urethral sphincter	Contraction	α_1	Relaxation	M	
Penis	Seminal vesicle; vas deferens; internal urethra	Contraction	M			Chapter 12
Vagina; Labia Minora		Contraction	M			Chapter 12
Skin	Piloerector muscle	Contraction	α_1			
	Sweat glands					
	Epicrine (heat)	Secretion	M			
	Apocrine (stress)	Secretion	α_1			
Glands	Salivary	Secretion	α_1	Secretion	M	Chapter 7
	Stomach			Secretion	M	Chapter 7
	Exocrine pancreas			Secretion	M	Chapter 7
	Endocrine pancreas	↑Insulin	β_2	↑Insulin	M	Chapter 9
		↓Insulin	α_1			Chapter 9
		↑Glucagon	α_1			Chapter 9
		↓Somatostatin	β_2	↓Somatostatin	M	Chapter 9

*This dilator effect is caused by circulating epinephrine.
†This effect is caused by sympathetic cholinergic fibers that innervate large skeletal muscle groups. Their existence in humans is not certain.

Central Autonomic Nervous System and Reflex Centers

Certain spinal segmental circuits allow reflex responses to mild stimuli in patients with high cord transections, but

◆ **true autonomic integration requires the presence of higher nervous centers.**

Brainstem

The pons-medulla region is an **autonomous center** for reflex responses to afferent signals from

- **respiratory,**
- **cardiovascular,** and
- **gastrointestinal receptors.**

The responses are **physically complete and system-specific.**

Hypothalamus

The hypothalamus **coordinates all autonomic systems.** This involves

- integration of responses to desires;
- integration of thermoregulation;
- integration of defense reactions;
- control of several endocrine secretions:
  —Adrenal medulla
  —Posterior pituitary
  —Anterior pituitary

Responses that are coordinated by the hypothalamus **are physically complete** and **involve the whole body.**

Limbic cortex and amygdala

These regions of the central nervous system are the gateway to domains of mental processing.

- Their activity governs the **extent of autonomic correlates that may accompany emotional states such as fear, rage, embarrassment, or sexual desire.**
- They also represent a level of conscious control over some aspects of resting autonomic function. The range of this control is demonstrated during states of **deep meditation.** In this state,
  —metabolic rate,
  —heart rate,
  —arterial blood pressure, and
  —distribution of blood flow
  can all be modified by application of conscious mental effort.
- It has also been suggested that **voodoo death,** in which the victim dies of no apparent external causes after members of the community have stated that death should occur, represents an extreme example of the influence of cortical centers (in the victim) over autonomic functions.

Central Autonomic Nervous System and Reflex Centers

Cortex
Correlation of emotional states with autonomic function

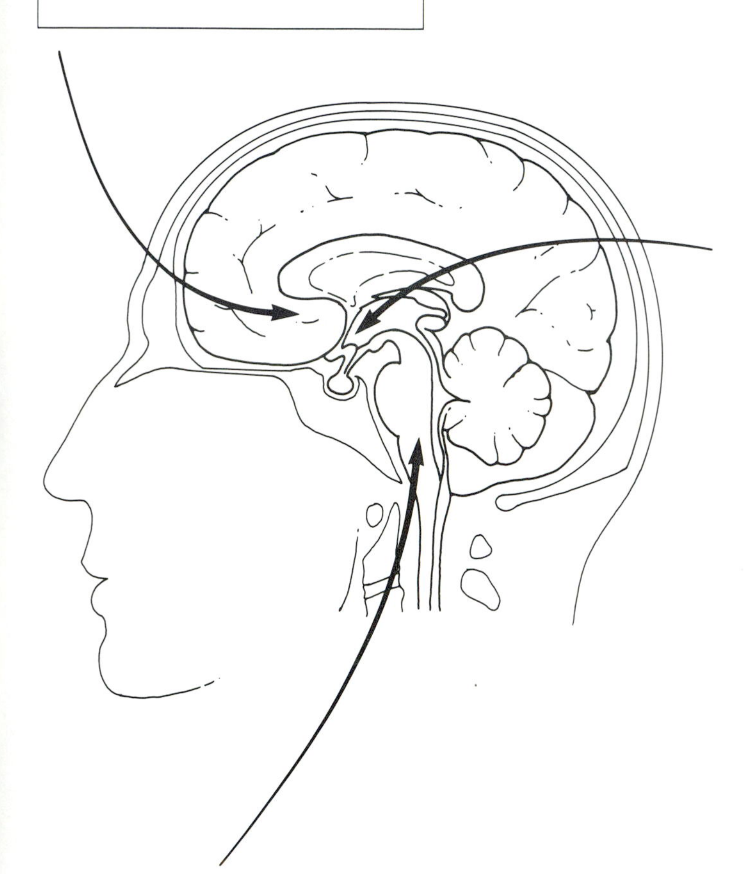

Brainstem
The pons/medulla region is the center for physically complete, system-specific reflex responses to respiratory, cardiovascular or gastrointestinal afferent signals

Hypothalamus

Integrates responses to desires

- Thirst
  dryness of mouth and information from osmoreceptors are translated into decreased urine production and increased drinking
- Hunger
  information about stomach wall stretch and blood glucose initiate feeding behavior
- Sexual drive
  behavior patterns from higher brain centers and information about ambient conditions (hormone levels, sensory input) are integrated to direct physical aspects of sexual activity

Integrates thermoregulation
Information from thermoreceptors is used to direct a wide range of responses:

- thyroid stimulating hormone
- muscular activity (shivering)
- respiration (panting)
- skin blood flow
- sweating

Integrates defense reactions
In response to feelings of fear or rage every system in the body is prepared for emergency action

Controls many endocrine secretions

- Adrenal medulla
  —catecholamines
- Posterior pituitary
  —vasopressin (antidiuretic hormone)
  —oxytocin
- Anterior pituitary
  —thyroid-stimulating hormone
  —adreno-corticotropic hormone
  —follicle-stimulating hormone
  —luteinizing hormone
  —prolactin
  —growth hormone

MOTOR CONTROL

Effector Structures

Skeletal muscle is organized in functional units called **motor units.** Each motor unit consists of

- **one alpha motor neuron** and
- **the single-type (fast, slow or, intermediate) muscle fibers innervated by its branches.**

Motor neurons are located in the ventral portion of the spinal cord gray matter, but they form extensive dendritic synapses into both the intermediate zone of the gray matter and the adjacent white matter.

Sensory Structures

When a muscle contracts, it develops force and it shortens.

- Afferent information regarding muscle tension and length is sensed
  - — By **Golgi tendon organs** and
  - — By **muscle spindles** that are interspersed among the extrafusal fibers throughout the muscle.

Golgi tendon organ

Group Ib afferents from tendons encode muscle force in their firing rate.

- Each tendon organ averages the force of several motor units. (The function of Ib afferents is not yet known. They often excite the same spinal interneurons as do Ia spindle afferents from the same muscle. In other instances they inhibit alpha motor neurons in the homonymous muscle while exciting alpha motor neurons in the antagonist muscle.)

Muscle spindle

Muscle spindles consist of **two nuclear bag fibers** and **several nuclear chain fibers.**

- Bag 1 is a fast fiber.
  - — It is selectively innervated by gamma motor neurons that are active during changes of extrafusal muscle length (dynamic gamma motor neurons).
- Bag 2 is a slow fiber.
  - — Bag 2 and the chain fibers are selectively innervated by gamma neurons that are active during static stretch of the extrafusal fibers (static gamma motor neurons).

◆ **Spindles are transducers of muscle shortening velocity and of muscle length.**

- Some spindles give information that relates primarily to muscle shortening velocity.
- Other spindles give information that relates primarily to muscle length.
- Afferent spindle nerves also respond differently to changes in velocity and changes in length:
  - — Group Ia primary afferents are sensitive to velocity because they show great activity during dynamic phases and little activity during static phases.
  - — Group II secondary afferents show the opposite. Accordingly, they are sensitive to length.
- The central branches of each Group Ia primary afferent axon synapse at several spinal cord segments with **Ia inhibitory interneurons.**
  - — The function of these branches is to activate relevant antagonists.

MOTOR CONTROL

OVERVIEW

Control of skeletal muscle function permits volitional actions on the external environment. It involves

- mental planning,
- composition of appropriate central nervous command patterns,
- sensing of the mechanical effect of the initial command pattern,
- evaluation of the success of the pattern relative to the initial goal, and
- modification of the central nervous command pattern to correct errors between present state and present plan.

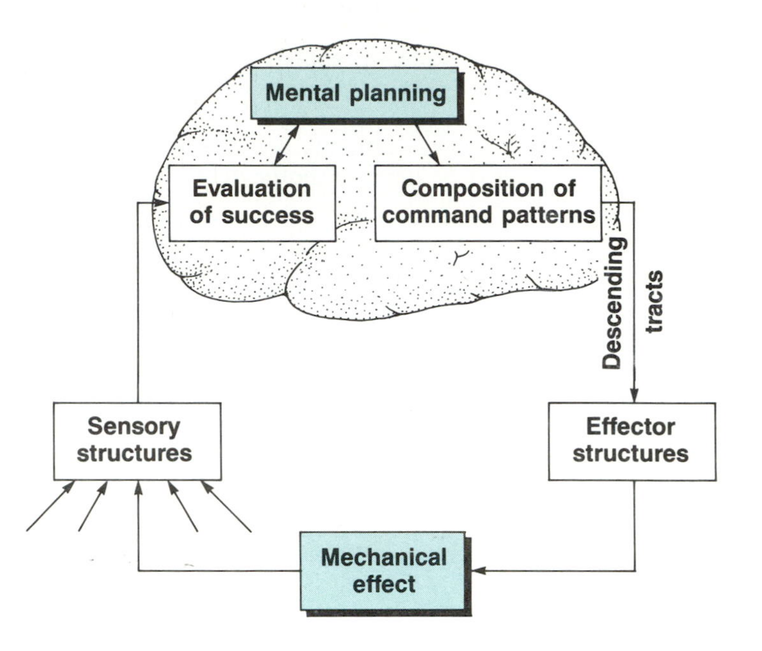

Sensors, Motor Neurons, and Effectors

The firing rate of spindle afferents depends on the length of intrafusal fibers in the spindle. This, in turn, depends on two factors:

- the degree of shortening of the extrafusal fibers and
- the degree of shortening of the intrafusal fibers.
  —Complex motor programs in gamma fibers innervating the contractile portions of each intrafusal fiber determine the relative activity in static or dynamic gamma motor neurons innervating a given spindle.
  This dictates whether the spindle is more sensitive as a transducer of shortening velocity or as a transducer of length.
- The family of motor neurons that innervate all the units contained in a given muscle is aligned in a vertical string in the ventral horn.
- The location of these muscle-specific strings is the same in all individuals.
- The repertoire of types of contraction in a given muscle is accomplished by central nervous programs that activate the appropriate motor units in the appropriate sequence. Examples of types of contraction are
  —repetitive at high load and high velocity;
  —repetitive at low load and high velocity.

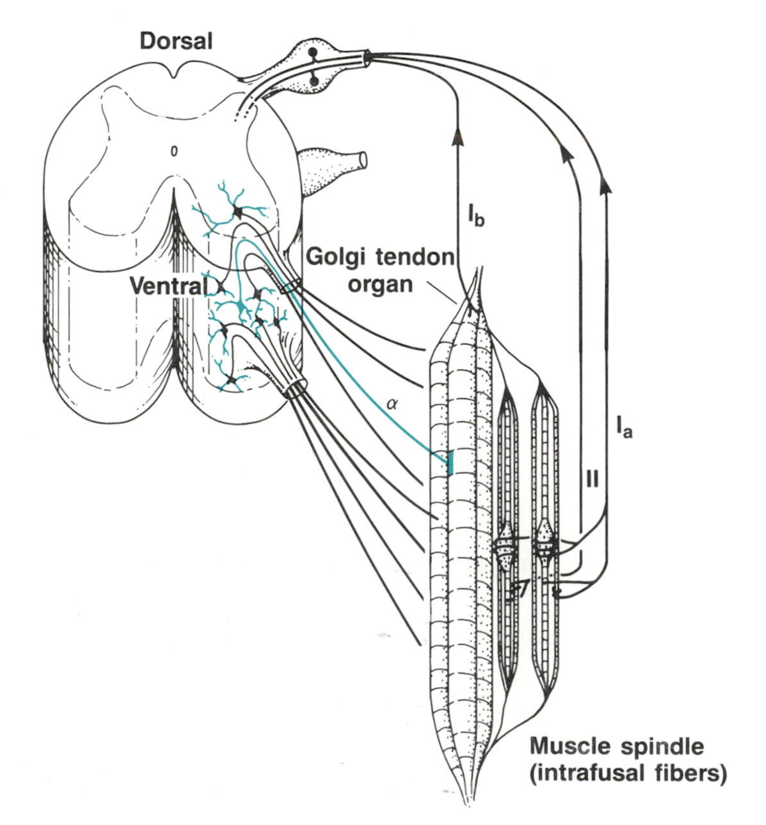

Patterns of Muscle Control

Open loop movements

This pattern applies to movements occurring so rapidly that there is **no time for muscle sensor feedback to influence the movement.**

◆ These movements are **characterized by alpha motor neuron activity without associated drive to gamma motor neurons.**

Feedback-controlled movements

Most muscle movements are of the feedback type,

◆ characterized by concomitant activity in alpha and gamma motor neurons.

Two feedback patterns have been identified:

1. In most cases, **gamma activity to the muscle spindles precedes alpha activity to the extrafusal fibers** (follow-up servo control).
2. In some precisely controlled movements, alpha and gamma activities occur simultaneously (servo-assistance control).

Follow-up servo control

- This begins with spindle contraction in response to gamma motor activity.
- Consequent signals in spindle afferents lead to
  — excitation of the agonist muscle and
  — inhibition of the antagonist muscle.
- **Excitation of agonist alpha neurons stops when agonist shortening has completely unloaded the precontracted spindles.**
- Sensory integration in higher centers evaluates mismatch between intent and present state, leading to generation of corrective signals for gamma motor neurons in agonist and antagonist.
- The sequence then repeats.

Servo-assistance control

- This differs from follow-up servo control in that

◆ **input from higher centers *simultaneously* activates gamma and alpha motor neurons, and both spindle and extrafusal muscle fibers contract simultaneously.**

- The rate of spindle contraction is set by gamma motor action potentials, which are set by what the higher centers "expect."
  — The rate of extrafusal contraction is influenced by both alpha activity and the load against which the muscle is contracting.
- If there is a discrepancy between the expected rate of contraction and the actual rate of contraction, then there will be a change in spinal cord input from spindle afferents and
  — spindle afferent activity will drive an appropriate reflex response in alpha motor neuron firing.

Segmental inhibitory circuits

There are many segmental circuits in which the excitation of a cell serves to inhibit other cells by one of two basic patterns: feedforward inhibition and feedback inhibition.

- **Feedforward inhibition is characterized by inhibition of motor neurons that were not previously excited.**
- **In feedback inhibition, the inhibiting interneuron acts on the spindle afferent that activated the interneuron in the first place.**

Open Loop Movements

These movements occur so rapidly that there is no time for muscle sensor feedback to influence the movement.

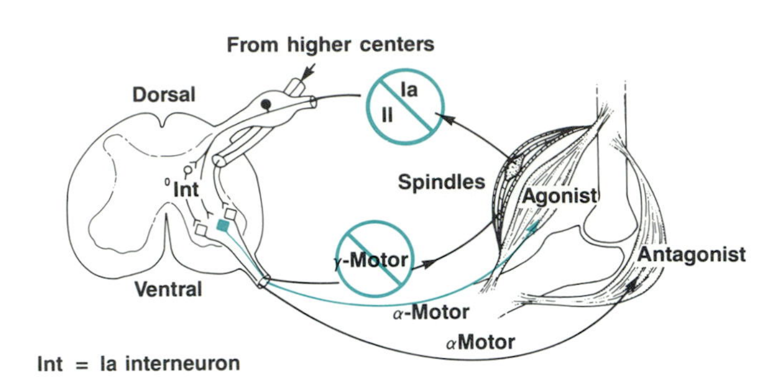

Feedback-Controlled Movements

Follow-Up Servo Control

The sequence of events is this:

1. Input from higher centers activates gamma motor neurons to contract spindles to the desired length.
   —This has a negligible effect on the much bulkier extrafusal fibers, but
2. spindle contraction increases activity in Ia and II afferents.
3. Increased spindle afferent activity
   - excites alpha motor neurons to the agonist muscle and
   - inhibits alpha motor neurons to the antagonist.
4. Agonist extrafusal contraction unloads the spindles and decreases spindle afferent activity.

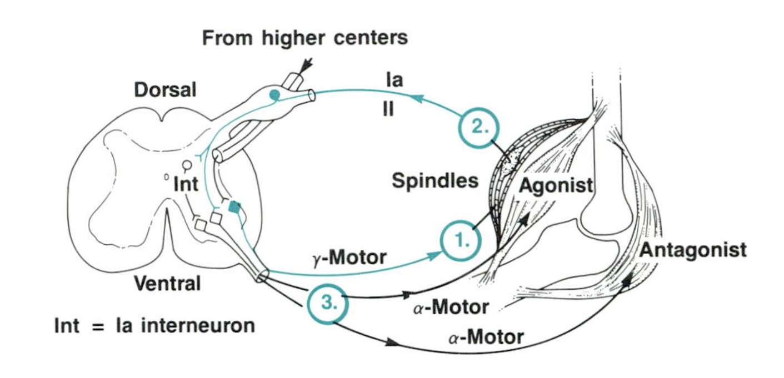

Segmental Inhibition

Feedforward Inhibition

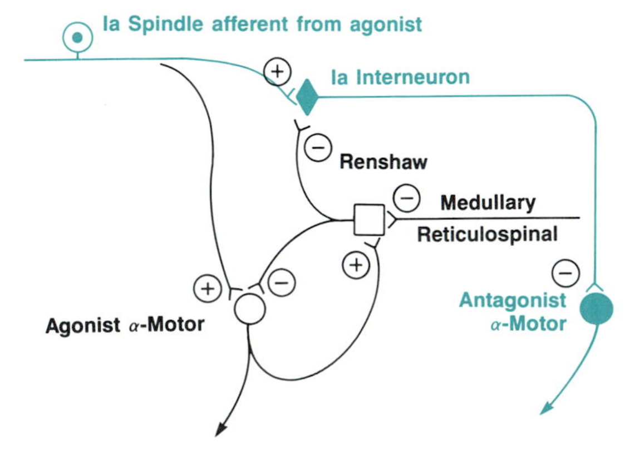

- Activity in the Ia spindle afferent excites the Ia interneuron.
- Motor output to the antagonist muscle is inhibited.

Feedback Inhibition

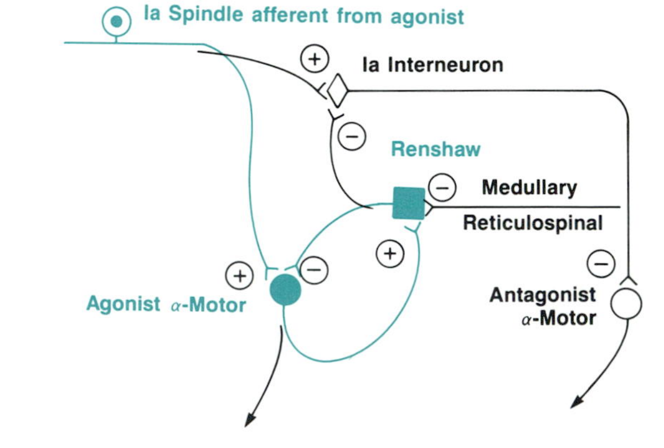

- Activity in the Ia spindle afferent excites an agonist α motor neuron.
- Collaterals from the activated α motor neuron excite Renshaw cells.
- Collateral Renshaw axons feedback-inhibit the activated motor neuron.

Central Nervous Motor Areas

Overview

Normal muscle movement is directed either by reflexes (e.g., posture maintenance) or toward specific, cortically derived goals. It always occurs in the context of the sensory environment, and all movements are designed with special reference to this environment.

Primary motor cortex

The motor cortex **does not originate motor programs.**

- Its **function is to integrate inputs**
  - from the premotor cortex and
  - from two nuclei in the thalamus:
    - — the ventrolateral nucleus, conducting cerebellar instructions, and
    - — parts of the ventroposterior nucleus, conducting proprioceptive information.
- Its **output is directed**
  - partly to the thalamus (ventrolateral nucleus) and
  - partly to the cerebellum, but
  - **mostly to the corticospinal tract.**
- Neurons representing any one body part are arranged in column-shaped "nests":
  - The neurons that drive the most distal portions of a limb (e.g., the finger) are arranged in the center.
  - The neurons driving more proximal portions are arranged in vertical lamellae around the center.
- Individual muscles are represented severalfold by neurons that form columns at several sites.
  - By this arrangement, columns at different sites will be surrounded by columns representing different synergists.
- Those parts of the body with especially good motor abilities (e.g., fingers and tongue) are represented in far larger cortical areas than parts that are capable of only less precise movements.

Supplementary motor area, premotor cortex, and Broca's area

These regions are located in a band adjacent and anterior to the primary motor cortex.

Supplementary motor area (medial area 6)

This area is

- a relay station for somatosensory input and
- **an assembly point for specific motor programs.**

- When movement sequences are rehearsed mentally, but not performed physically, then neurons in the supplementary motor area are active, and the premotor cortex (lateral area 6) is quiescent.

Premotor cortex (lateral area 6)

- When movement actually occurs, then the premotor cortex, under instructions from the supplementary motor area,
  - assembles specific subprograms for components of the overall movement and
  - transmits to the motor areas specific motor unit activation patterns for the intended movement.

Broca's area

- This area receives input from the language area of the temporal lobe.
- Its **function is to organize motor programs of speech and writing into language patterns.**

Central Nervous Motor Areas

Overview

Complex movements, requiring cooperation of many muscles, are carried out by motor programs that contain many subprograms. The selection of appropriate subprograms involves the basal ganglia as well as command interneurons in area 7 of the parietal cortex.

Selected subprograms are then assembled and their execution is directed by neuronal activity in the primary motor cortex.

A feedback path from the motor cortex to the cerebellum and the basal ganglia ensures smooth transitions from one subprogram to the next.

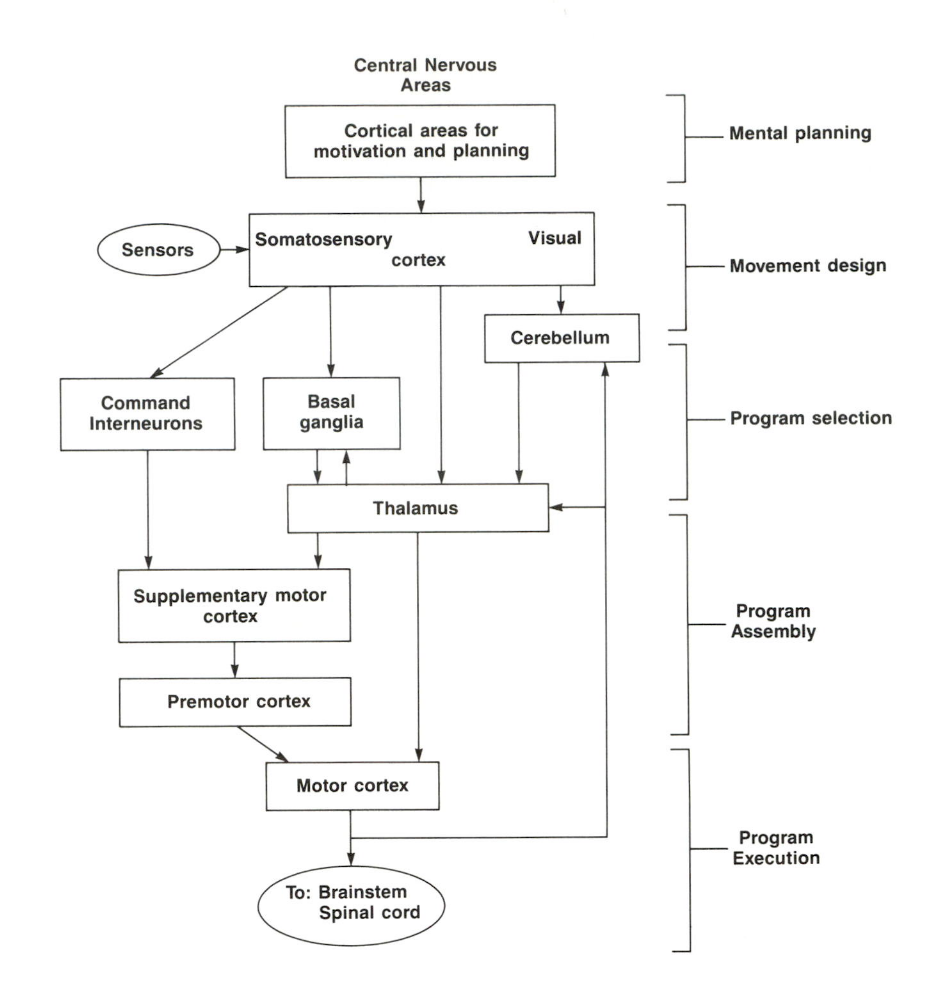

Primary Motor Cortex

The primary motor cortex is located in a narrow lateral band just forward of the central sulcus (the precentral gyrus).

Organization is somatotopic:

- Neurons influencing the lower body are clustered medially.
- Neurons influencing the head region are located laterally.
- Upper body neurons are located in between.

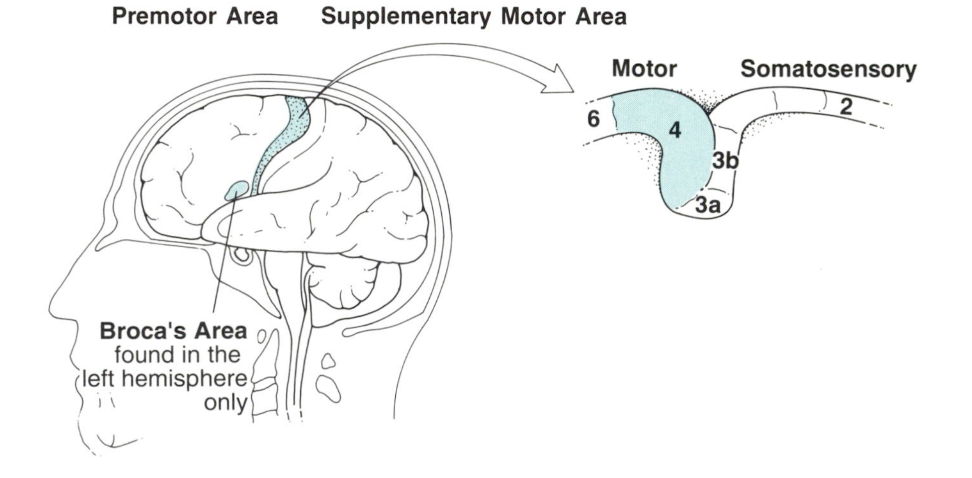

Basal ganglia

The basal ganglia exert their influence over networks that link the motor cortex to other cortical areas. They **behave as a variable filter,** with two primary functions:

- They match the performance specifications of motor programs to the criteria that have been established by the motivational and sensory cues that define a particular circumstance.
- They facilitate the selection of only those motor programs that meet the specific criteria.

- **Basal ganglia participate in motor control only if incoming signals are facilitated by dopaminergic input from the substantia nigra compacta.**
  —Loss of this facilitation leads to Parkinson's disease.
  - In their subsequent function, the basal ganglia facilitate or inhibit the incoming signals and modify them by signals from the subthalamic nucleus.
- **The overall effect of basal ganglia activity on motor activity is inhibition of inappropriate networks that link the motor cortex to the entire nonmotor portion of the cerebral cortex.**
  - They are of particular importance for the selection of bridging (preparatory) subprograms that move a limb or muscle from its initial position to one from which a standard motor program can continue. As a result,
- basal ganglia diseases such as Parkinson's disease are accompanied by impaired ability to perform preparatory movements, and patients appear to "freeze" before they execute major motor tasks.

Basal Ganglia

- The basal ganglia are an extensively interconnected group of nuclei:
  - —Striatum (consisting of the caudate nucleus, putamen, and nucleus accumbens)
  - —Substantia nigra compacta
  - —Substantia nigra reticulata
  - —Pallidum (external and internal)
  - —Subthalamic nucleus
- They receive input from the centromedian nucleus of the thalamus and from several regions of the cortex:
  - —Association cortex projects to caudate nucleus
  - —Sensorimotor cortex projects to putamen
  - —Limbic cortex projects to accumbens

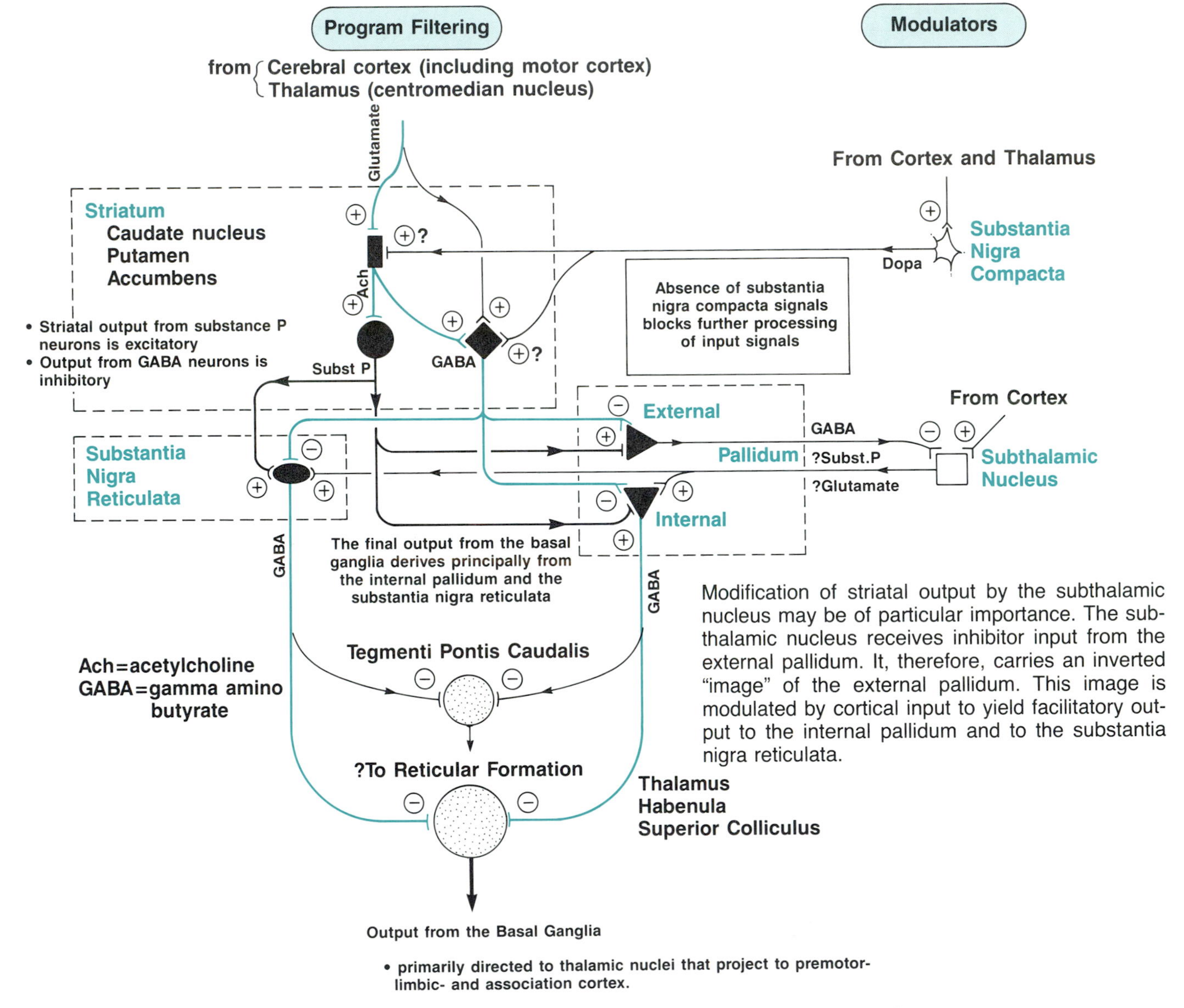

The effect of Basal Ganglia output is inhibition of ongoing activity.

Cerebellum

The cerebellum consists of

- a three-layered **cortex** that is extensively folded from side to side into *folia;*
- pairs of internal nuclei;
- the **white matter,** composed of fibers entering, leaving, and traversing the cerebellum.

It is attached to the brainstem by three distinct fiber bundles.

Internal nuclei

- There are three pairs of major internal cerebellar nuclei:
  - — Fastigial nucleus
  - — Interpositus nucleus (which is subdivided into globose and emboliform nuclei)
  - — Dentate nucleus
- Most Purkinje cells in any one of the three functional cerebellar zones (vermis, intermediate hemisphere, and lateral hemisphere) project to the same internal cerebellar nucleus.
- Projections from the cerebellar nuclei terminate in specific loci and, therefore, modulate specific aspects of motor function:
  - — Fastigial nucleus to the lateral vestibular and reticular nuclei (for balance and posture)
  - — Interpositus nucleus to the red nucleus and the ventrolateral nucleus of the thalamus (for posture, gait, and coarse movements)
  - — Dentate nucleus to the ventrolateral nucleus of the thalamus (for skilled movements of hands and fingers)

◆ **Internal nuclei have an excitatory effect on muscle tone and motor activity.**

Functions of the cerebellum in movement control

- The cerebellum **contributes to voluntary movements.**
- Its general functions are to
  - — **correlate** incoming muscle and other sensory information,
  - — **compute** the most effective deployment of muscular effort necessary to accomplish a required task, and
  - — **compose** the necessary outgoing commands to the spinal motor neurons and the motor cortex.
- Some zones and associated internal nuclei show electrical activity only after the onset of movement (e.g., the interpositus nucleus).
  - — Their purpose is thought to be **compensation on the basis of sensory feedback.**
- Other zones show electrical activity before the onset of movement (e.g., the dentate nucleus).
  - — They probably participate in the generation of motor sequences.
  - — Their particular function is **control of the relative timing of agonist and antagonist alpha motor neuron activity to effect a smooth pattern of limb acceleration, deceleration, stop, and acceleration in the opposite direction.**
- Mild cerebellar dysfunction results in inability to judge the range of limb movements without watching them. Severe cerebellar dysfunction results in inability to perform limb movements smoothly and efficiently even while watching them.

Cerebellum

The building blocks of internal cerebellar organization are

- Incoming tracts
  —Mossy fibers
  —Climbing fibers
- Purkinje cells
- Granule cells
- Internal nuclei along with their outflow tracts

There are also three populations of local inhibitory interneurons that modulate the function of Purkinje and Granule cells:

- Basket cells
- Stellate cells
- Golgi cells

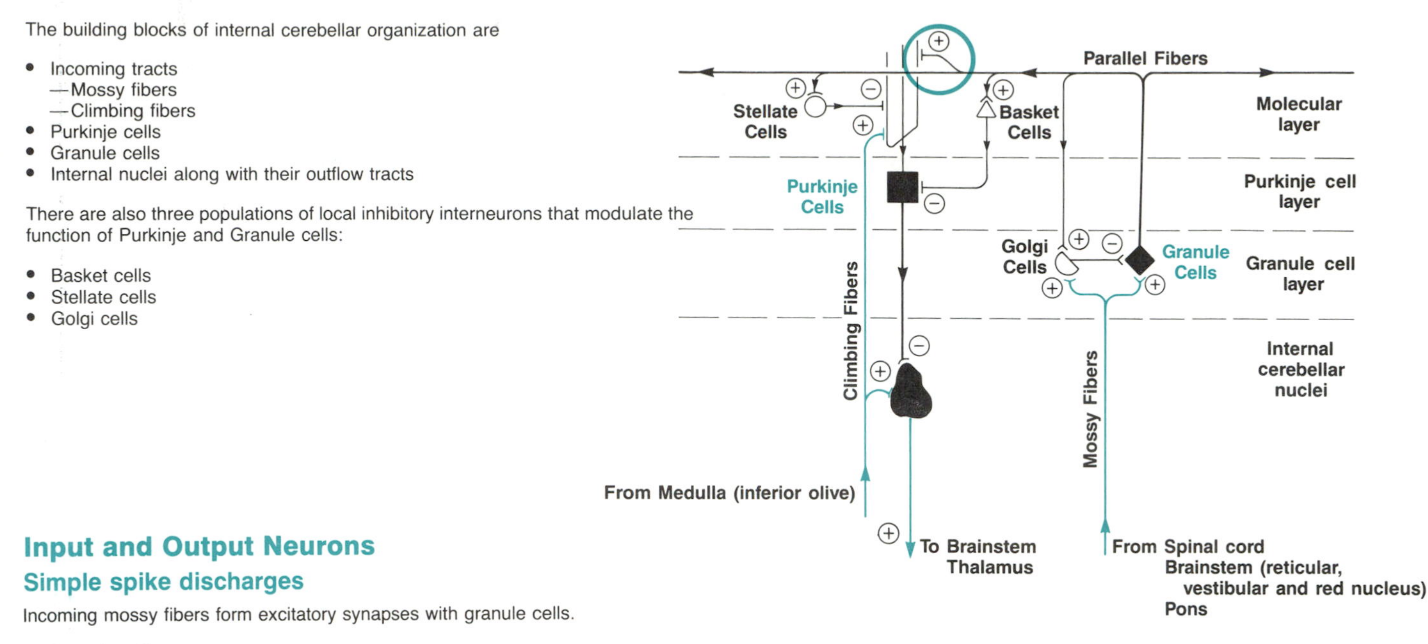

Input and Output Neurons

Simple spike discharges

Incoming mossy fibers form excitatory synapses with granule cells.

- Granule cells
  —are the major input cells.
  —Axons project toward the surface, bifurcate, and form a layer of parallel fibers in the direction of the folia.
- Each parallel fiber forms excitatory synapses with a sequence of several dozen Purkinje cells at their flattened dendritic bushes.
- Purkinje cells are the major cortical output cells.

Excitation of a sufficient number of parallel fibers contacting a given Purkinje cell will cause that cell to discharge trains of simple spikes that inhibit the muscles to which the cell projects.

Complex spike discharges

- Purkinje cells also receive excitatory input from climbing fibers that originate mostly in the opposite inferior olive. (Each region of the inferior olive projects to a separate longitudinal strip of cerebellar cortex.)
- Climbing fibers synapse extensively with the dendrites of Purkinje cells, but not with the cell body.

Climbing fiber input to a Purkinje cell produces in that cell a large, prolonged complex spike discharge.

Inhibitory Interneurons

Stellate cells

—are excited by parallel fibers;

—inhibit Purkinje cells.

Basket cells

—are excited by parallel fibers.
—The axons of each of these form "baskets" of inhibitory synapses around 20 or more Purkinje cells.

Golgi cells

—have dendritic trees that spread in all directions among the parallel fibers (unlike the flattened Purkinje dendrites, which are confined to a narrow longitudinal layer);
—are excited by input from parallel fibers and mossy fibers;
—act to inhibit granule cells.

Internal Nuclei

- In addition to the obvious anatomic division of the cerebellum into horizontal folds, there is a functional subdivision into three vertical strips:
  —Vermis
  —Intermediate hemisphere
  —Lateral hemisphere
- Different parts of the body are topographically mapped in these zones:
  —Trunk and head in the vermis
  —Limbs stretched out toward the lateral hemisphere
- Nearly all Purkinje cells in any one of the three strips project to the same internal cerebellar nucleus:
  —Vermis to the fastigial nucleus
  —Intermediate hemisphere to the interpositus nucleus (globose and emboliform nuclei)
  —Lateral hemisphere to the dentate nucleus
  —(Purkinje cells in the flocculonodular lobe are an exception because they project directly to the lateral vestibular nucleus in the midbrain, where they participate in the control of eye movements.)

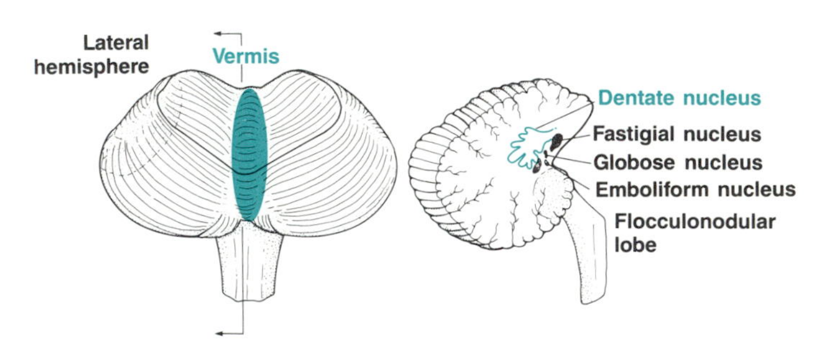

Central Nervous Motor Areas—Cont'd.

Descending Tracts

There are three major descending tracts in humans:

- **Corticospinal**
- **Vestibulospinal**
- **Reticulospinal** (of pontine or medullary origin)

They originate at cortical, midbrain, and brainstem levels, descend to different levels in the spinal cord, and terminate at preferential sites in the spinal gray matter.

Corticospinal (pyramidal) tract

◆ This tract **functions to modulate the activity of alpha or gamma motor neurons as directed by the motor cortex.**

Vestibulospinal tract

This tract, whose nerves run on both sides of the cord, originates in

- the **ipsilateral Lateral Ventricular Nucleus (Deiter's nucleus)** and
- the **contralateral Medial Ventricular Nucleus.**

These nuclei integrate higher-center motor control patterns with sensory information from mechanoreceptors in the vestibular organs and in muscles of the neck region. Therefore, their output, conveyed to the periphery via the vestibulospinal tract,

◆ **modulates activity in muscles that rotate the head and upper torso** and
◆ **modulates adjustments pertinent to limb and body orientation in the gravitational field.**

Reticulospinal tract

The role of reticulospinal neurons in muscle control is poorly understood. In general, they

◆ **organize movements by the shoulders, elbows, and face muscles.**

Two particular reticulospinal circuits have been described in some detail:

- One relates to the role of reticulospinal neurons in the reflex by which painful cutaneous stimuli lead to limb withdrawal.
- The other relates to reticulospinal function in **Renshaw inhibition.**

Reticulospinal tract and Renshaw inhibition

- When medullary reticulospinal fibers are active, then
  - — Renshaw cells are inhibited and
  - — Ia inhibitory interneurons are disinhibited.
  - — The resulting Ia interneuron activity **inhibits alpha motor neurons of the antagonist muscle.**
- If the reticulospinal tract were cut, then
  - — Renshaw cells would be activated by collaterals from agonist alpha motor neurons. As a result,
  - — Ia interneurons would be inhibited, leading to disinhibition of antagonist alpha motor activity.
- The antagonist would fail to relax during agonist contraction.

Descending Tracts

Corticospinal (Pyramidal) Tract

- This tract originates in large pyramidal cells and Betz cells (giant pyramidal cells) of the primary motor cortex.
- Almost all of its nerves terminate on interneurons in the contralateral gray matter (layers V and VI).

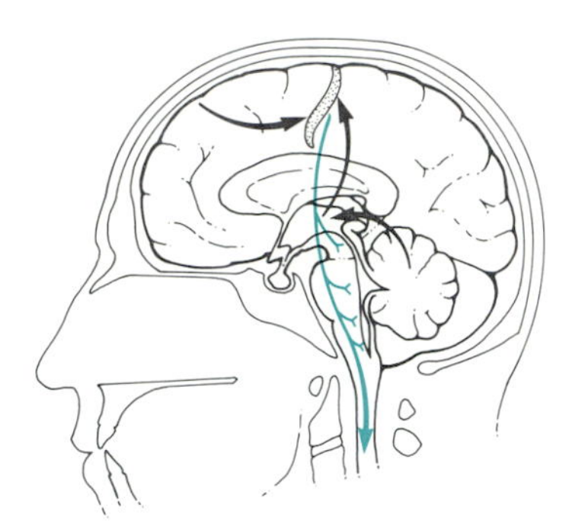

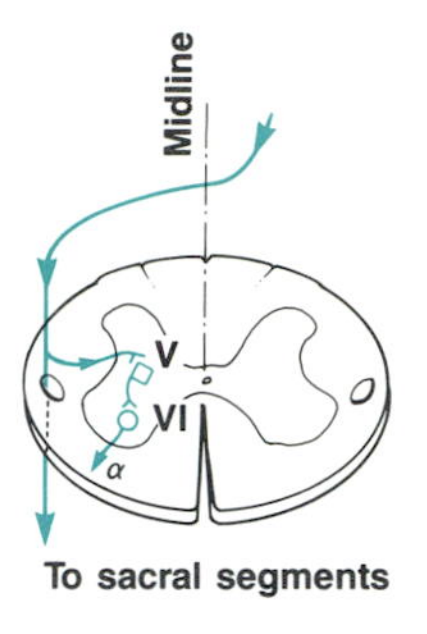

Vestibulospinal Tract

Two tracts originate from different regions of the vestibular nuclei:

- The superior vestibular nucleus projects to the oculomotor nuclei.
- The lateral and medial ventricular nuclei project to the spinal cord.

These vestibulospinal tracts on each side of the cord originate in
—the ipsilateral Lateral Ventricular Nucleus (Deiter's nucleus) and
—the contralateral Medial Ventricular Nucleus. The medial tracts descend only as far as the thoracic level. They influence cervical motor nuclei.

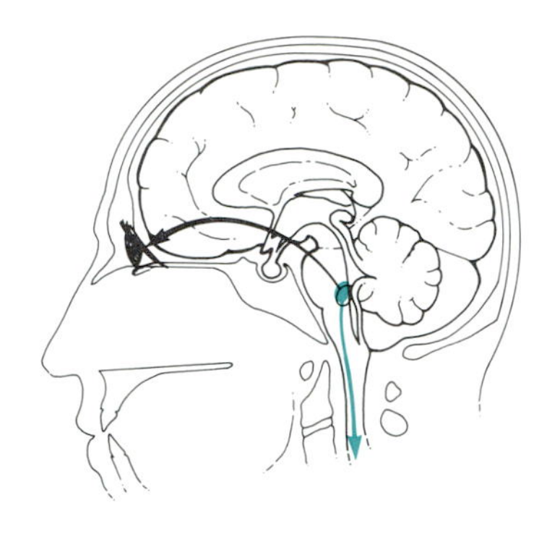

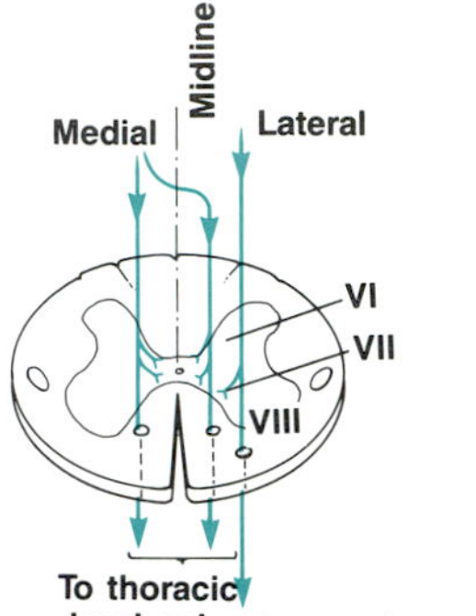

Reticulospinal Tract

The nerves in this tract originate from regions in the medial reticular formation of the pons and the medulla.
—Pontine neurons project only ipsilaterally.
—Medullary neurons project both ipsilaterally and contralaterally.

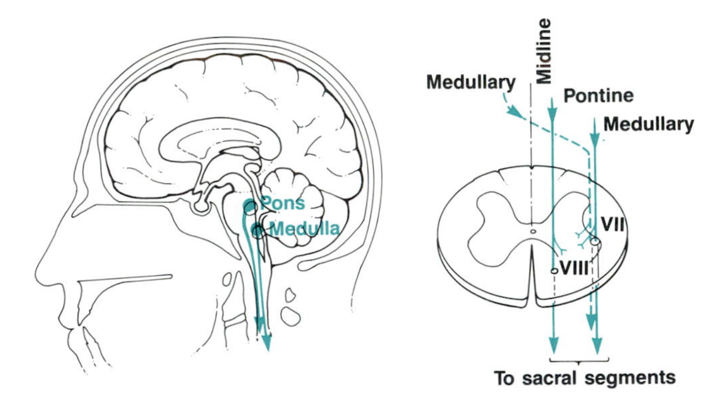

Renshaw Inhibition

Renshaw cells are inhibitory to Ia inhibitory interneurons. They are innervated

- by branches of the medullary reticulospinal tract (inhibitory) and
- by collaterals from alpha motor neurons (excitatory).

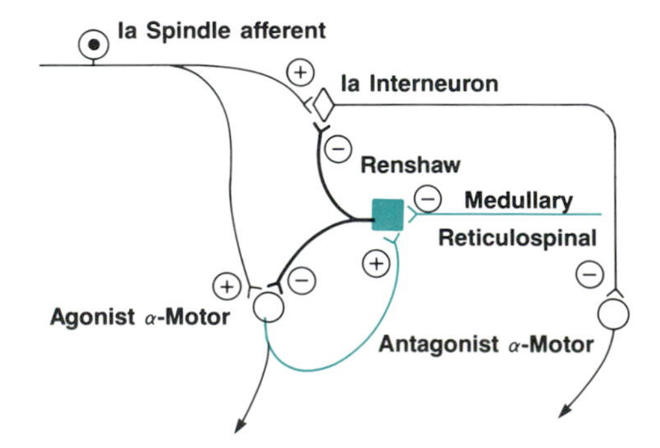

Coordinated Muscle Movement

Movements are accomplished by sequences of alpha motor neuron firing patterns that activate selected motor units to contract or relax at a certain rate and to a certain tension or length. The sequences are called **motor programs.**

Motor programs

These sequences are

- **learned and refined by repetitive use, allowing feedback mechanisms to make corrections whenever outcome does not match desired goal.**

Motor programs for complex movements, involving many joints, are broken down into subprograms, and the progression from one subprogram to the next is governed by sensory feedback.

Motor programs are initiated by **command interneurons** that act like simple on-off switches.

- They **are activated by the central nervous system after it has evaluated a constellation of sensory and motivational inputs.**
- They are located in area 7 of the posterior parietal cortex.
  —Lesions in that area are accompanied by inability to make some complex movements involving limbs contralateral to the lesion, even though other complex movements of those limbs can still be performed.

- Motor programs are assembled and stored in the supplementary motor area.
  - Lesions in the supplementary motor area are associated with an inability to perform movements that require a specific contribution pattern from several muscles in a contralateral limb, even though the individual muscles can be activated on command.

Selection and execution of motor programs

Following assessment of the sensory environment within which a given motor program is intended to function, four central nervous areas are primarily involved in the selection and execution of motor programs:

Command interneurons
Supplementary motor area
Premotor cortex
Motor cortex

The events are these:

1. A command interneuron conveys instructions that allow the supplementary motor area to select the correct motor program. The basal ganglia supply to the supplementary area information that
   — allows selection of programs for appropriate preparatory movements and
   — ensures efficient sequencing and deployment of muscle activity.
2. The premotor cortex
   — receives assembled global instructions from the supplementary motor area and
   — fills in the details and activates specific motor cortex neurons.
3. The motor cortex
   — uses action information from the premotor cortex and
   — uses timing information from the cerebellum
   to activate specific motor units.

Coordinated Muscle Movement

Motor Programs

These are sequences of alpha motor neuron firing patterns that activate selected motor units to contract or relax at a certain rate and to a certain tension or length.

Selection and Execution of Motor Programs

Execution of every motor program is preceded by an **assessment** of the sensory environment. This occurs in the **somatosensory cortex** and the **visual cortex.**

On the basis of the environmental assessment,

- A **command interneuron** instructs the supplementary motor area which motor program is appropriate.
- The **supplementary motor area** also receives extensive input from the basal ganglia (via ventrolateral, ventroanterior, and mediodorsal thalamic nuclei). Its purposes are
  —selection of preparatory movement programs and
  —planning the most efficient deployment of muscle activity.

Thus, the supplementary motor area assembles global instructions for movements that involve participation of many muscles in a specific pattern and sequence. It issues these instructions to the premotor area.

The **premotor cortex** fleshes out the details of smaller components of the global instructions received from the supplementary motor area and then activates specific motor cortex neurons.

Motor cortex neurons activate specific motor units. The timing of the activation-deactivation sequences is determined by cerebellar activity, reaching the motor cortex via ventrolateral thalamic nuclei.

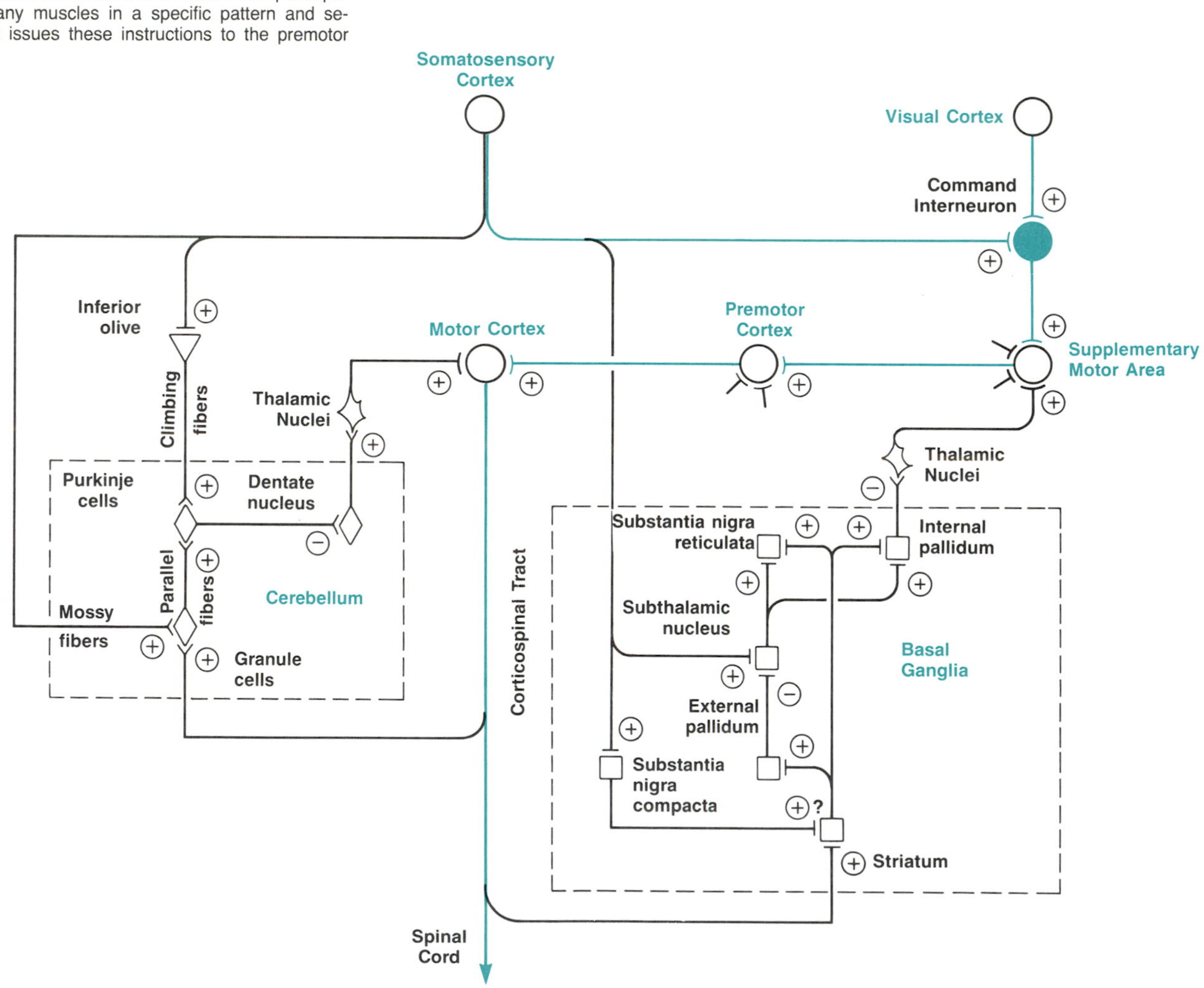

The consequences of motor unit activation are detected by mechanoreceptors in the affected muscles. They are conveyed to the somatosensory cortex and from there

- to the cerebellum
  —via mossy and climbing fibers;
- to the basal ganglia
  —to the subthalamic nucleus and substantia nigra compacta;
- to the command interneuron area (area 7) of the cerebral cortex.

This **reafferentation** gauges the relative success or failure of the motor act and initiates appropriate modifications in subsequent motor unit activity.

MENTAL PROCESSES

Motivation, Learning, and Memory

When a new sensory event is first encountered, it leads to arousal and subsequent motor responses that depend on the nature of the stimulus:

- If the event brings neither pain nor pleasure, then its repeated occurrence leads to a reduction in the complexity and magnitude of the responses **(habituation).**
- **If pleasure or pain is associated with the event, then habituation does not occur. Instead, the event is selected for storage in memory for later retrieval.**

Selection of items for memory

In general, three aspects determine which sensory events will be learned and committed to cognitive memory:

- Emotional factors
- Motivational factors
- Drive-determined factors
  — Hunger
  — Thirst
  — Pleasure
  — Pain avoidance

If motor responses are involved, then

- **frequency of repetition** of sensory inputs and their associated motor responses determines whether a given input-response pattern will become a motor program (i.e., part of **motor memory**).
- **The limbic system is involved in selecting items for memory.**

Storage of items in memory

Short-term memory and long-term memory are probably handled differently.

Short-term memory (lasting up to 2 hours)

- This does not involve changes in protein synthesis.
- It may involve network rearrangements (in view of the observation that permanent structural changes can occur in the nervous system within an hour).
- **The limbic system may be involved**
  — by setting up local reverberating circuits such as the **Papez circuit.**

Long-term memory

- Although the limbic system is not involved in the actual storage of items in long-term memory,
  —it selects them from its short-term memories, and
  —it consolidates these memories by playing them like a continuous tape from hippocampus and amygdala to the relevant sensory cortical areas.

- Long-term memories are probably stored diffusely (rather than in individual neurons) in the sensory, association, and motor areas to which they relate.
- Long-term memory involves two factors:
  — Network rearrangement
  — **Alterations in protein synthesis,** which lead to structural changes such as dendritic growth or receptor development

Motivation, Learning, and Memory

The Limbic System (Hippocampus and Amygdala)

The limbic system is involved in memory item selection. It provides mostly motivational aspects in motor memory, but provides other aspects in the selection of items for cognitive memory. In the human limbic system, hippocampus and amygdala are of particular importance.

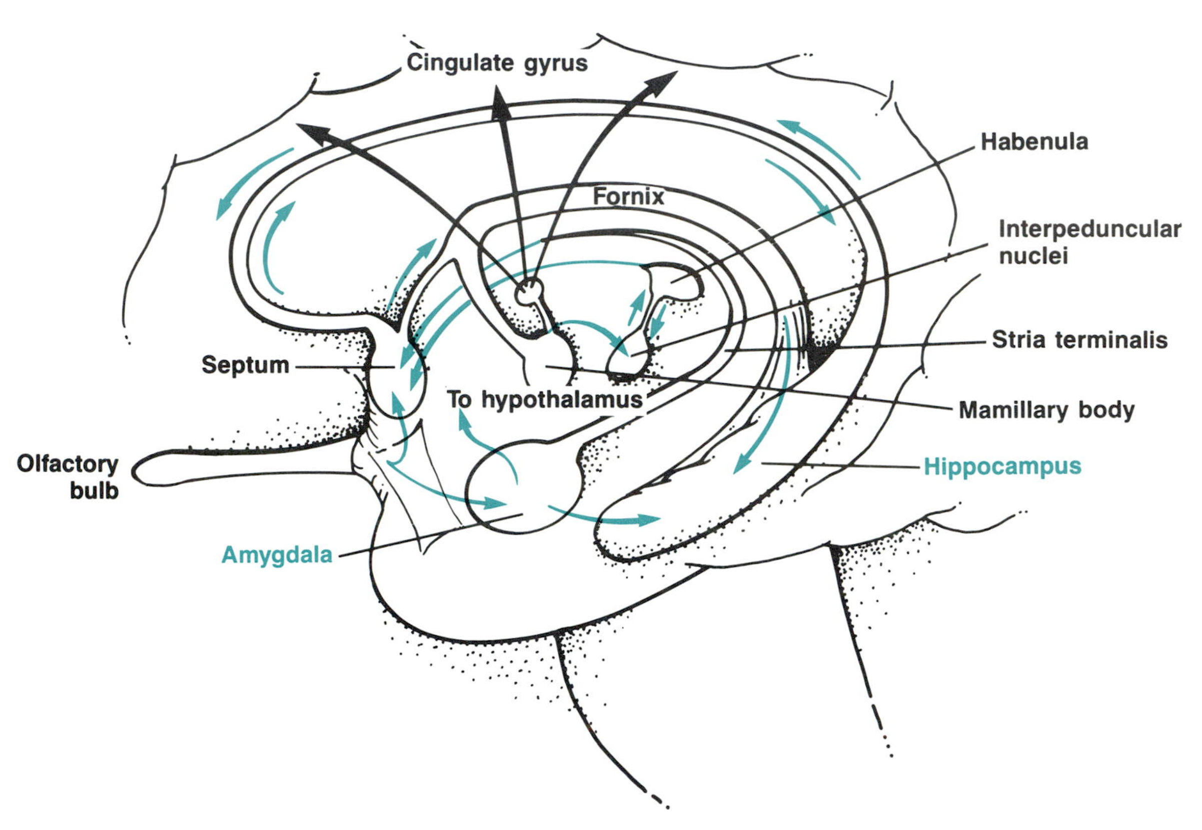

Amygdala

- The amygdala is important for making associations among stimulus modalities (e.g., a certain fragrance often elicits an associated visual image).
- It appears to be responsible for the influence of emotional states on sensory inputs.
  —This produces a spectrum of sensory perceptions from apparently identical stimuli (e.g., the sound of one's own motorcycle is never perceived as noise).

Papez Circuit

This circuit is identified by colored arrows in the diagram above.

- It is local reverberating circuit
- It is involved in short-term memory

Hippocampus

This portion of the limbic system is important for three major purposes:

- Memory of verbal items or items that can be encoded verbally (e.g., telephone numbers)
- Establishing relationships between language and concepts
- Learning spatial relationships among objects (i.e., the landmarks by which a certain place is recognized)

The hippocampus appears to function as a "table of contents" for items in memory.

Language and Speech

Articulation, the forming of speech sounds, is represented bilaterally in the motor areas. However,

◆ **language analysis and speech formulation take place in most individuals in regions of the left hemisphere only.**

Two regions are involved: **Broca's area** and **Wernicke's area.**

Broca's area

This region assembles the motor programs of speech and writing.

◆ Patients with lesions in Broca's area
- understand language perfectly and
- may be able to write perfectly, but
- seldom speak spontaneously.
  — When they do, they utter only monosyllabic sounds.

Wernicke's area

This region is responsible for the analysis and formulation of language content.

◆ Patients with lesions in Wernicke's area
- are unable to name objects;
- are unable to understand the meaning of words;
- articulate speech readily, but usually nonsensically.

Language and Speech

Language analysis and speech formulation take place in

- Wernicke's area and
- Broca's area.

Both are in the left hemisphere.

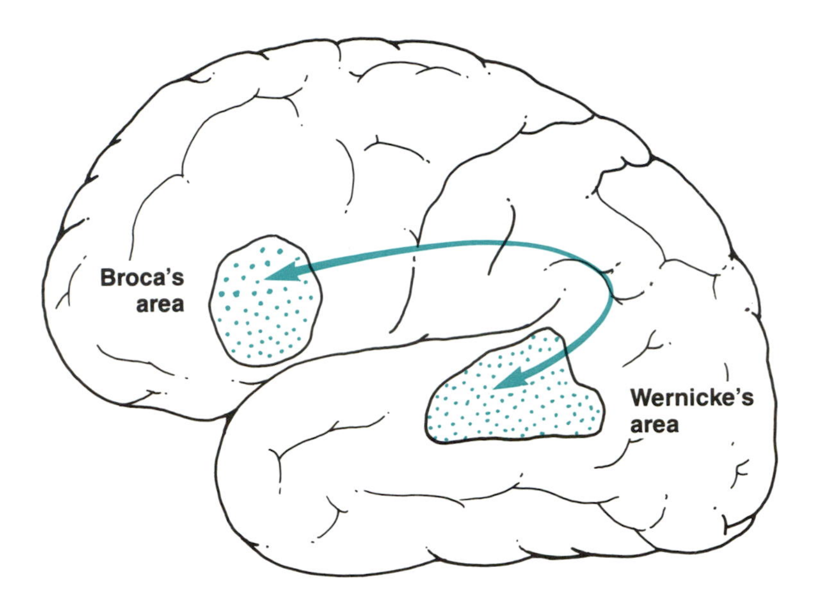

Broca's area is located just in front of the voice control area of the left motor cortex.

Wernicke's area is a part of the auditory and visual association cortex.

Wakefulness and Sleep

Circadian rhythms

Several generators of circadian rhythms exist. They influence

- wakefulness-sleep cycles,
- feeding behavior,
- body temperature, and
- other cyclic physiologic parameters.

- The factors that determine circadian periodicities are mostly unknown, but cultural factors and light-dark cycles are known to be involved.
- It has not been possible to demonstrate that wakefulness-sleep cycles are driven by periodic accumulation, depletion, or regeneration of melatonin or any other specific group of chemicals that circulate in the blood.

Wakefulness

Although wakefulness is a state in which the individual is in active contact with the environment, it is

- not a state that is driven by sensory systems.

Instead, it is

- **driven by activity in the ascending reticular activating system.**
  - This system is believed to be the origin of the ascending stream of activating impulses that are necessary to maintain the level of excitation that characterizes wakefulness.

Sleep

A night of normal sleep consists of

- several cycles of deepening and lightening sleep.
  - Each cycle culminates in a period of rapid-eye-movement (REM) sleep before deep sleep is resumed again.
  - Deep sleep is characterized by dreaming and by descending waves that inhibit spinal motor neurons and cause postural atonia.
  - The periods of light sleep get longer as the night progresses.

Human studies have been inconclusive, but it has been shown in animals that

- serotonergic ascending tracts from the raphe nuclei are responsible for non-REM sleep;
- noradrenergic ascending tracts from the locus ceruleus have been implicated as causing REM sleep.

EEG Patterns of Alertness and Sleep

The electroencephalogram (EEG), recorded from external scalp electrodes, is a good monitor for the state of wakefulness.

- **Alertness and excitement** are accompanied by low-amplitude, high-frequency EEG patterns (beta waves).
- **Relaxation while the eyes are closed** is accompanied by large-amplitude, periodic patterns (alpha waves) that appear to be driven by oscillators in the thalamus.
- **Deep sleep** is characterized by large-amplitude, low-frequency waves (delta waves).

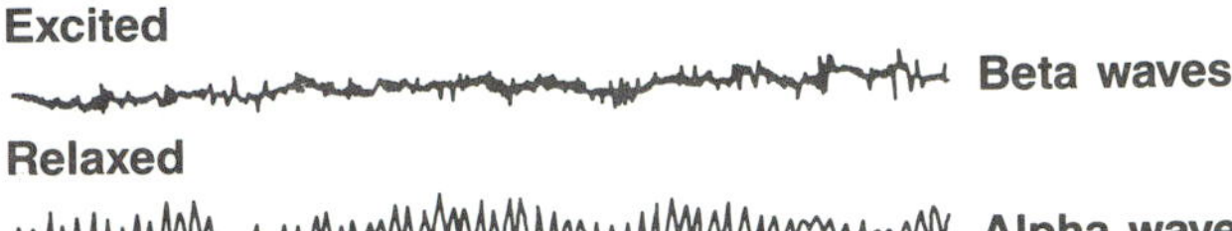

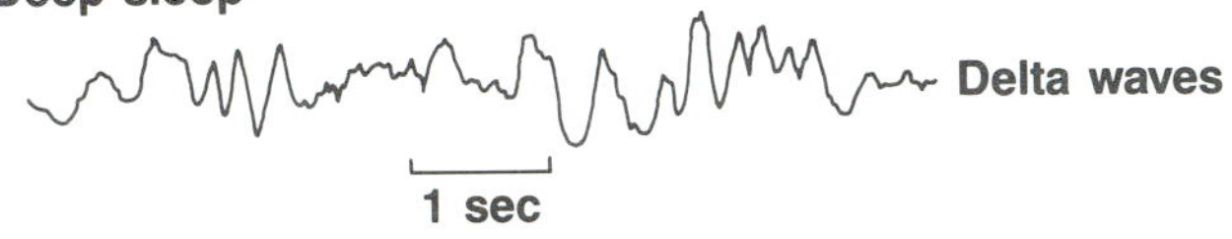

Generators of CNS Patterns of Wakefulness and Sleep

Wakefulness-sleep cycles are normally driven by oscillating activity in the suprachiasmatic nucleus of the hypothalamus. This nucleus

- receives direct afferents from the retina;
- projects to the preoptic nucleus of the hypothalamus;
- has a light-dark circadian rhythm that can be detected in sympathetic fibers that innervate the pineal gland from the superior cervical ganglion. As a result,
  - —there is light-entrained rhythmicity to the secretion of melatonin and its precursor, serotonin, from the pineal (more is secreted in darkness). (Little is known about which organs take cues from this pineal clock. A role for melatonin in sleep facilitation has been inferred from its effect on electroencephalogram patterns, but it has not been possible to demonstrate that wakefulness-sleep cycles are driven by periodic accumulation, depletion, or regeneration of melatonin.)

Wakefulness is driven by activity in the ascending reticular activating system. This consists of the **brainstem reticular formation** and its projections

- to the thalamocortical system, i.e.,
  - —centromedial nuclei,
  - —midline nuclei, and
  - —reticular nuclei of the thalamus;
- to the subthalamic nucleus and
  - —from there to the cortex via ventrolateral and mediodorsal nuclei in the thalamus;
- to the hypothalamus, i.e.,
  - —mammillary body,
  - —periventricular nucleus, and
  - —lateral nucleus;
- to the basal forebrain, i.e.,
  - —septum and adjacent region.

The cycles of deepening and lightening sleep that characterize a night of normal sleep are believed to be controlled by the **suprachiasmatic circadian generator.** The suprachiasmatic nucleus projects to the hypothalamic preoptic nucleus. This, in turn, projects

- to the cortex via the basal forebrain and
- to serotonergic neurons in the Raphe nuclei.
  - —The Raphe nuclei send mostly descending tracts to spinal motor nuclei, but also some ascending tracts.

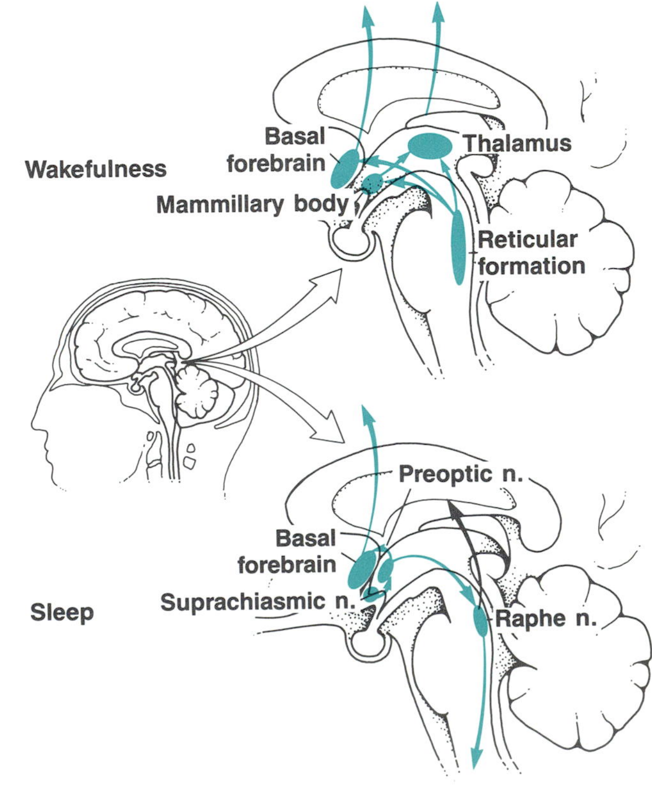

Emotion

Emotion has three major aspects:

- Perception and evaluation of sensory stimuli
- Integration and correlation of sensory stimuli with memory
- Autonomic reactions to sensory stimuli

Memory consolidation circuits appear to be especially important. They are found

- in the limbic system,
- in the amygdala,
- in the orbitofrontal cortex, and
- in the hypothalamus.

Their importance derives from the fact that the experiences that are stored in memory are those that initially aroused an emotion.

- **The principal brain regions that are involved in the regulation and expression of human emotion are**
  - **the orbitofrontal cortex,**
  - **the amygdala,** and
  - **the hypothalamus.**
- A broad outline of central nervous function relating to emotions can be surmised from the general functions of the subregions:
  - The hypothalamus organizes and integrates autonomic reactions to fear, rage, anger, and pleasure, and the amygdala acts as a brake on the hypothalamus.

Personality

The prefrontal cortex (areas of cortex located forward of the premotor area) is used for thought and behavior strategy. Its tangible actions include

- **inhibition of motor areas so that only those motor programs that are socially and behaviorally appropriate are executed.**

Emotion

Frontal lobe cortical regions behind the eye sockets are the locus of feeling and emotion. They are closely interconnected with the amygdala and the hypothalamus.

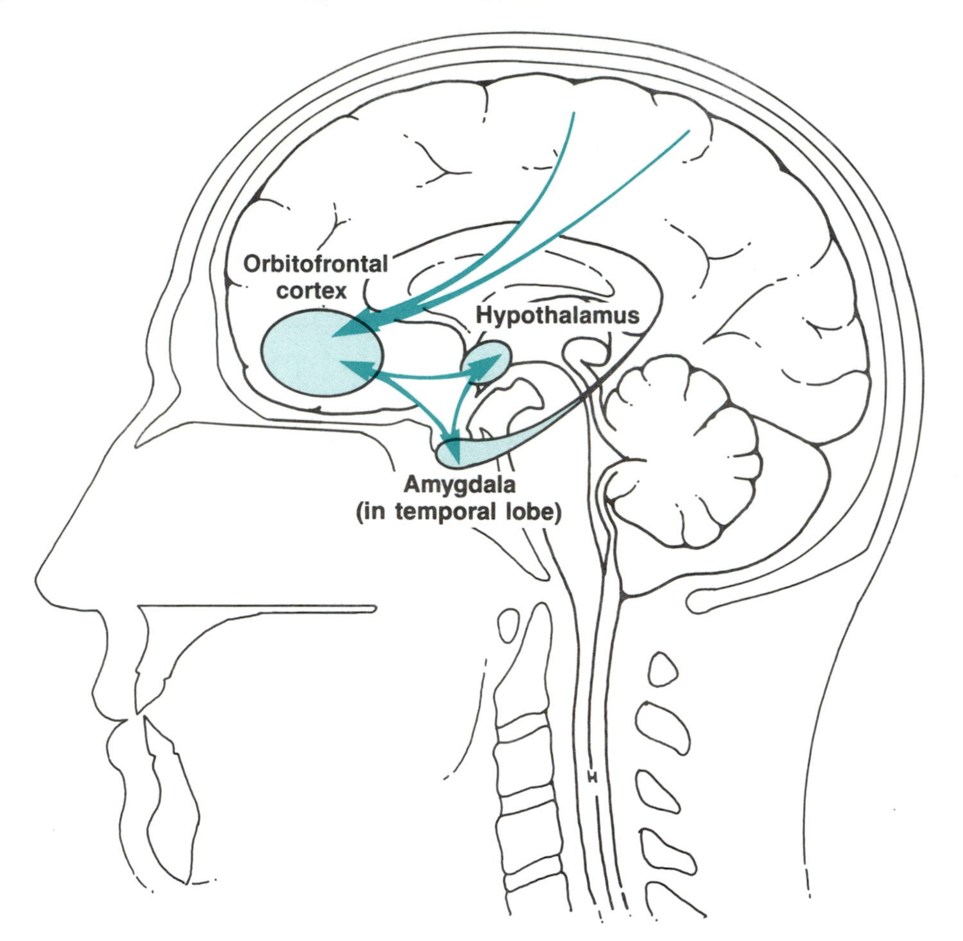

9

ENDOCRINOLOGY

MECHANISMS OF HORMONE ACTION

The Chemical Nature of Hormones

Certain **secretory cells** in the body synthesize and release chemical agents (**hormones**) for the purpose of mediating biologic responses in distant **target cells.** Hormones are derived from one of three sources:

- A single amino acid (e.g., the catecholamines)
- Chains of amino acids (e.g., the peptide hormones of the hypothalamus)
- Cholesterol (i.e., the steroids)

Forms of Hormonal Communication

Three forms of hormonal communication are recognized:

Endocrine communication

Hormones are secreted into the blood to regulate the function of a **distant target cell.** The response of the target cell can be the elaboration of another hormone.

Neuroendocrine communication

Special nerve cells secrete hormones via one of two paths:

- **Directly into the blood** (e.g., norepinephrine secreted from sympathetic nerves)
- **Into brain interstitial space,** from which it is drained by a portal circulation and transported via the blood to target cells (e.g., vasopressin is synthesized in the supraoptic nucleus of the hypothalamus, is released into the portal circulation of the posterior pituitary, and has its primary target cells in the renal collecting tubules)

Paracrine communication

Endocrine cells secrete into the surrounding extracellular space, and **the target cells are neighbors of the secreting cells.** The hormone reaches its target cells mainly by diffusion.

Cellular Mechanisms of Hormone Action

Overview

Hormonal interaction with target cells begins with reversible binding to highly specific **receptors.**

Interactions with membrane receptors

The general sequence is that the hormone-receptor complex activates a coupling mechanism (commonly a G-protein) that influences a catalytic enzyme whose role it is to transform a precursor into a **second messenger.** The second messenger, in turn, acts on an intracellular effector (commonly a **kinase**) to bring about the target cell response.

The known catalytic enzymes are **adenylate cyclase, guanylate cyclase,** and **phosphodiesterase.**

The known second messengers are **cAMP, cGMP, diacyl glycerol** (DG), **inositol diphosphate** (IP_2), and **Ca^{++}.**

Interactions with nuclear receptors

Hormones that can penetrate the cell membrane and form intracellular receptor complexes (steroid and thyroid hormones) **bind to DNA and alter the expression of specific genes.** As a result, the time scale of action of these hormones is in the range of a few days rather than several minutes.

Cellular Mechanisms of Hormone Action

Overview

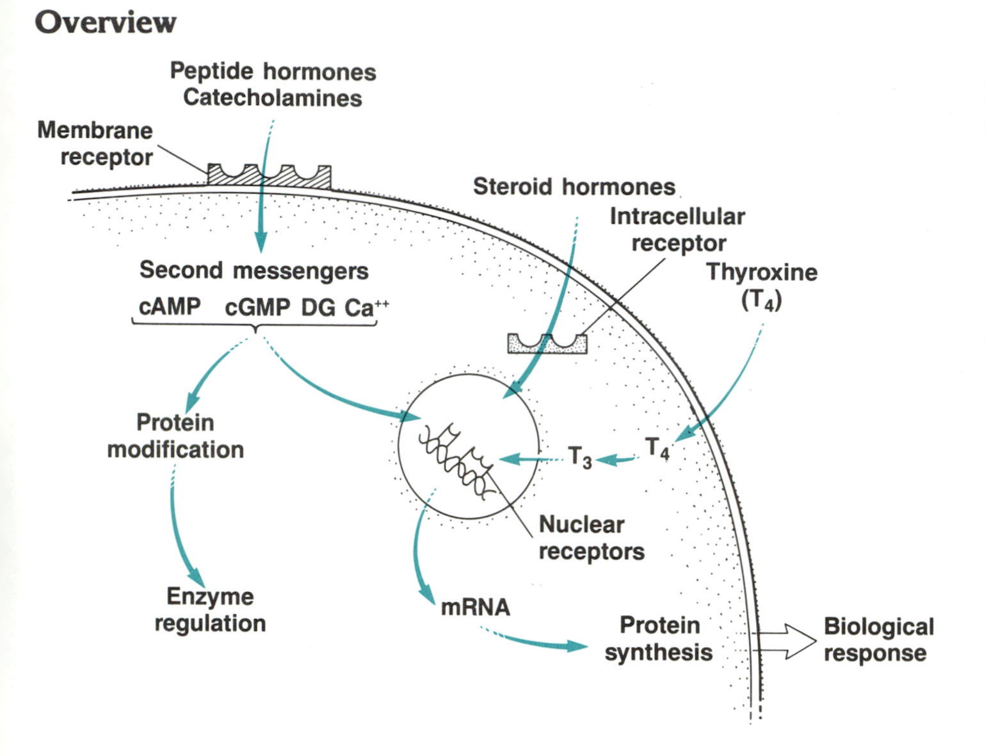

Receptors for peptide hormones and for the catecholamines are located in the external membrane of the target cell.

Receptors for steroid hormones are located in the cytoplasm of the target cell.

Thyroxine, the major thyroid hormone, is first converted to T_3 in the target cell cytoplasm and then interacts with a T_3 receptor in the nucleus.

Interactions with membrane receptors

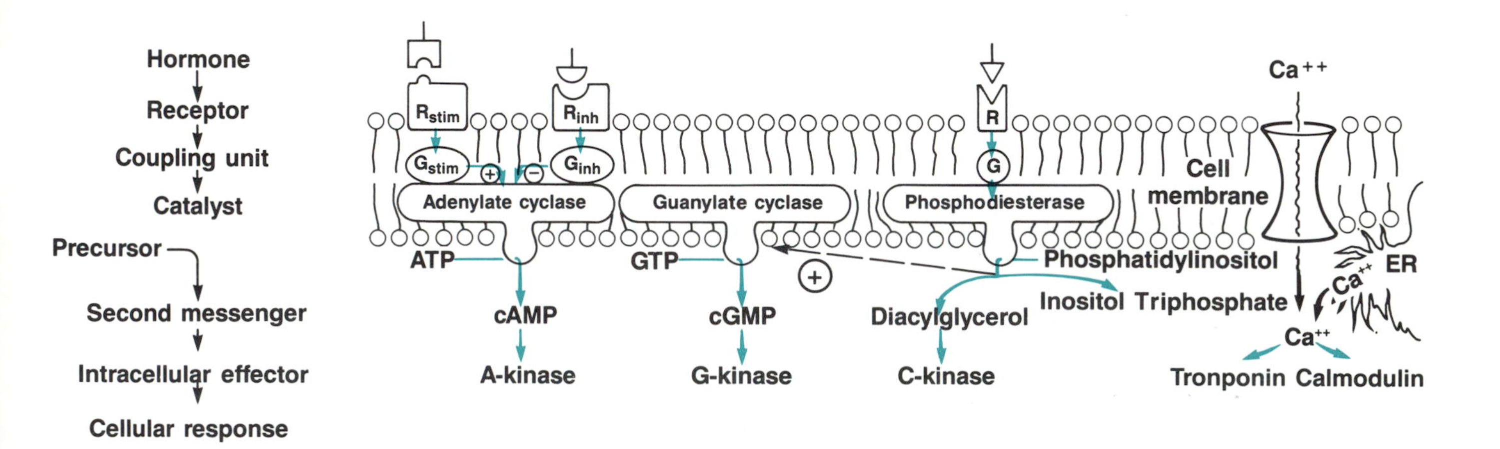

HYPOTHALAMUS AND ANTERIOR PITUITARY

Anterior Pituitary Hormones

Growth hormone

Human growth hormone (GH) is formed in pituitary somatotrophic cells. Its synthesis is stimulated by growth hormone releasing hormone (GHRH) (provided that thyroid hormone is present) and inhibited by somatostatin (SS). Its two major functions are **regulation of metabolism** and **stimulation of body growth.** The balance between these two functions is determined by the nutritional state of the body. When that is adequate and insulin is active, then GH promotes growth. On the other hand, when nutrition is inadequate and insulin levels are low, then the metabolic effects of GH dominate.

Metabolic effects

- Inhibition of glucose uptake and glucose utilization
- Increased lipolysis and consequent increase in the plasma level of free fatty acids

Growth effects

- These are not due to GH directly, but to the **somatomedins,** a group of insulin-like growth factors that are synthesized, under the control of both growth hormone and thyroid hormone, in the liver and in fibroblasts.
- All tissues that are capable of growing are affected in cell size and rate of cell division.
  — Before puberty:
  While androgen levels are low, the major effects are on the longitudinal growth of cartilage and bone.
  — After puberty:
  When androgens have caused ossification and closure of the epiphyseal growth plates of the long bones, the somatomedins continue to stimulate **circumferential growth in all bones,** but promote **longitudinal growth only in membranous bones, which have no epiphyses.**

Thyroid stimulating hormone

Thyroid stimulating hormone (TSH) is the major regulator of thyroid function.

- It is synthesized in and secreted from pituitary thyrotrophic cells at a rate that is increased by thyrotropin releasing hormone (TRH) and decreased by SS.
- The thyrotroph is also controlled by feedback inhibition from the plasma levels of thyroxine (T_4) or tri-iodothyronine (T_3), the active form of thyroid hormone.

Prolactin

The major hypothalamic influence on pituitary lactotrophic cells is inhibition by dopamine. TRH exerts a mildly stimulating influence, but the major stimulus for prolactin synthesis during pregnancy is the level of plasma estrogen. **The principal role of prolactin is to act on the mammary epithelium to increase milk production.**

ACTH and related peptides

Corticotropin releasing hormone (CRH) is synthesized in hypothalamic cells near the supraoptic and paraventricular nuclei. Its target cells are the corticotrophic cells of the anterior pituitary, where it promotes synthesis and secretion of a precursor protein whose cleavage yields adrenocorticotropic hormone (ACTH), gamma lipotropic hormone (γ-LPH), and β-endorphin. **ACTH acts principally on the adrenal cortex to promote the production of steroid hormones.** The roles of γ-LPH and β-endorphin appear to be central nervous rather than peripheral.

FSH and LH

Synthesis and secretion of follicle stimulating hormone (FSH) and luteinizing hormone (LH) are stimulated by gonadotropin releasing hormone (GnRH).

HYPOTHALAMUS AND ANTERIOR PITUITARY

The major role of the hypothalamus is coordination of whole-body function. This involves regulation of autonomic nervous activity as well as regulation of endocrine activity. The hypothalamus accomplishes the latter function via trophic secretions.

Anatomy

A significant feature of the hypothalamus-pituitary unit is its vascular system. The anterior pituitary receives blood from two sources:

- Arterial blood from the superior hypophyseal arteries (branches of the internal carotid arteries)
- Venous blood from a portal system that originates in capillaries of the median eminence

—Blood from these capillaries is collected into a series of parallel veins that drain into the sinusoidal capillaries of the anterior pituitary.

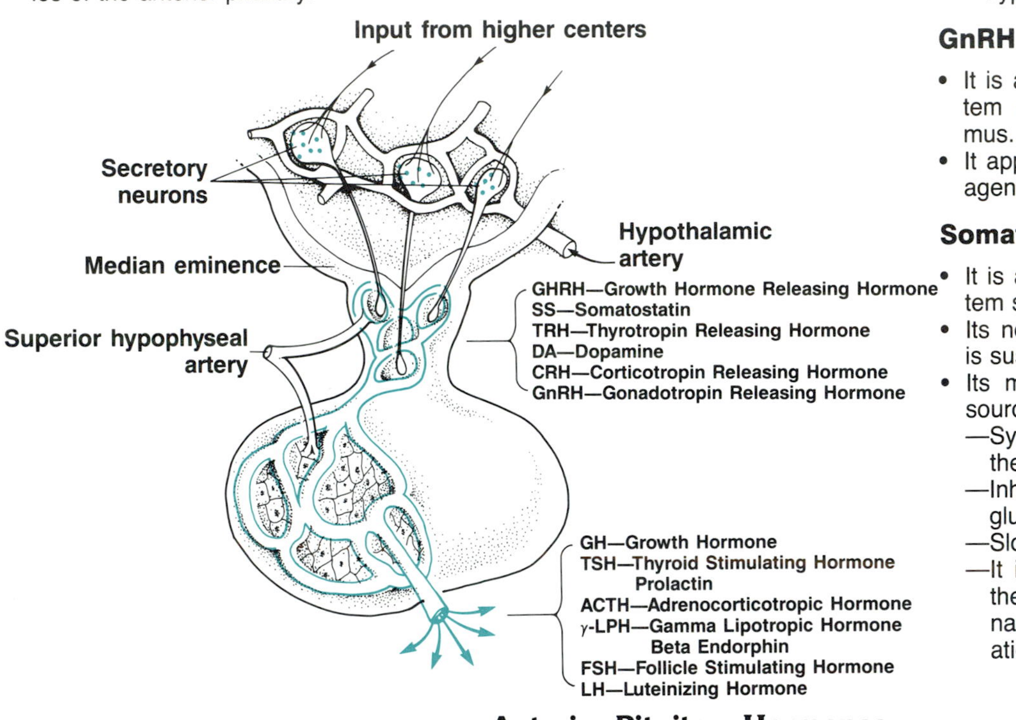

Special secretory neurons in various hypothalamic regions synthesize agents that are transported down their axons and are released from the axon terminals into the local portal system, but not into the general circulation

- The rate of release is determined by action potentials in the secretory neurons. This electrical activity is governed

—partly by higher nervous centers and
—partly by hormonal feedback reaching the neurosecretory cells via the hypothalamic artery.

GnRH

- It is also found in central nervous system areas other than the hypothalamus.
- It appears to act as a neurotransmitter agent in those areas.

Somatostatin

- It is also found at central nervous system sites other than the hypothalamus.
- Its neurotransmitter role at those sites is suspected.
- Its most important extra-hypothalamic source is the GI system:
  —Synthesized in the gut as well as in the D-cells of the pancreatic islets
  —Inhibits the secretion of insulin and glucagon
  —Slows a variety of digestive functions
  —It is believed, therefore, that one of the actions of somatostatin is coordination of GI absorption with pancreatic function.

Anterior Pituitary Hormones

Several hypothalamic factors act as trophic secretions that stimulate or inhibit hormone synthesis in the anterior pituitary.

Hypothalamic Factor	GHRH ⊕	SS ⊖	SS ⊖ TRH ⊕	TRH ⊕ DA ⊖	CRH ⊕	GnRH ⊕	
Anterior Pituitary Hormone	Growth hormone, (Somatotropin)		TSH	Prolactin	ACTH μ-LPH ß-endorphin	FSH LH	
Target Organ	Muscle Fat	Liver	Thyroid	Lacteals	Adrenal cortex	Ovaries	Testes
Target Organ Response	↑ Fatty acid mobilization and metabolism	Somatomedins	Thyroxine (T_4)	Milk	Steroids	Progesterone Estrogen	Testosterone

HYPOTHALAMUS AND POSTERIOR PITUITARY

Posterior Pituitary Hormones

Only two hypothalamic hormones, **vasopressin** and **oxytocin,** are passed through the posterior pituitary.

Synthesis and transport

Vasopressin and oxytocin are small peptides (nine amino acids). They

- are synthesized by cells in the **supraoptic and paraventricular nuclei** of the hypothalamus;
- remain attached to a **neurophysin** during their axonal transport toward the posterior pituitary and during storage in secretory granules within the nerve terminals.
- (Neurophysin is a portion of the large precursor molecules of each.)

Release

Each hormone is released **along with its neurophysin** in response to nerve action potentials.

- It is rapidly split from the larger carrier polypeptide.
- The subsequent physiologic functions of neurophysin are unknown.

Functions

Vasopressin

- The epithelium of the **renal collecting tubule** is the most important target organ.
- Vasopressin **increases the permeability of the tubule to water** and, thereby, regulates body fluid osmolality and body fluid volume.
  - — This function is thought to be mediated by cAMP-induced changes in membrane proteins as well as in cytoskeletal structures that are anchored to the external cell membrane.
- High-affinity receptors for vasopressin are also found in **vascular smooth muscle.**
  - — **Here the cellular response is constriction.** However, at physiologic vasopressin levels this constriction results in systemic blood pressure changes only in circumstances when other pressure-regulatory mechanisms have lost their effectiveness. Hence, it was believed for a long time that cardiovascular effects of vasopressin required very high doses of the hormone.

Oxytocin

- **Oxytocin acts specifically on uterine smooth muscle and on the smooth muscle cells that surround the distal portion of the mammary gland duct system.**
- Its biologic effect is to **cause muscle contraction.**
  - — In the mammary glands, muscle contraction results in transport of milk to the lactiferous sinuses and subsequent milk ejection.
  - — Responsiveness of the uterus to oxytocin depends on many factors, including the presence of estrogen (which stimulates contraction) and progesterone (which inhibits its contraction).

HYPOTHALAMUS AND POSTERIOR PITUITARY

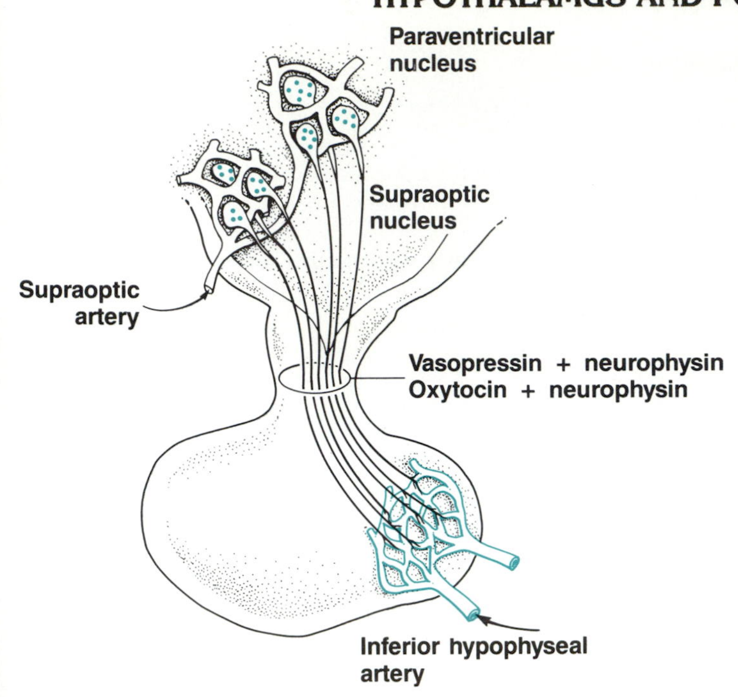

The posterior lobe of the pituitary gland produces no biologic responses to hypothalamic agents. However,

- it acts as a way station for transferring two hypothalamic neurosecretions to the bloodstream for transport to peripheral target organs.

The portal network that is formed in the posterior pituitary by capillaries of the inferior hypophyseal artery provides a point of contact with the terminal boutons of the secretory neurons and is, therefore, a vital anatomical feature.

THE THYROID GLAND

Calcitonin

Synthesis and secretion

C-cells, the source of calcitonin, are found in other tissues, but they are located chiefly in the thyroid gland, where they are also called **parafollicular cells.**

- The **rates of synthesis and secretion** of calcitonin **vary linearly with serum levels of ionized calcium.**

Biologic effects

Calcitonin appears to act mostly as a **short-term regulator of serum Ca^{++} concentration.**

- Its immediate effect is to decrease osteolytic activity in osteocytes. This favors calcium deposition in the rapidly exchanging pool of bone calcium salts ($CaHPO_4$).

Although calcitonin inhibits the formation of osteoclasts (large bone-reabsorbing cells that are normally active over a small portion of bone surfaces), its **long-term effects on serum Ca^{++} are small in adult humans because of compensatory changes in osteoblastic activity and in parathyroid hormone secretion.**

Thyroxine (T_4) and Tri-iodothyronine (T_3)

Synthesis and secretion

T_4 and T_3 are **iodinated derivatives of the amino acid tyrosine.**

During synthesis they are attached to **thyroglobulin** (TGL), a large glycoprotein whose special feature is a structure that forces tyrosine residues into an orientation that allows iodine attachment.

Secretion of T_4 and T_3 into the general circulation is regulated by thyroid stimulating hormone **(TSH)** and is stimulated most potently by **exposure to a cold environment.**

THE THYROID GLAND

Anatomy

The thyroid elaborates two classes of hormones from histologically distinct cell populations:

- Parafollicular cells (also called C-cells) synthesize calcitonin.
- Follicular cells synthesize a mixture of thyroxine (T_4) and tri-iodothyronine (T_3).

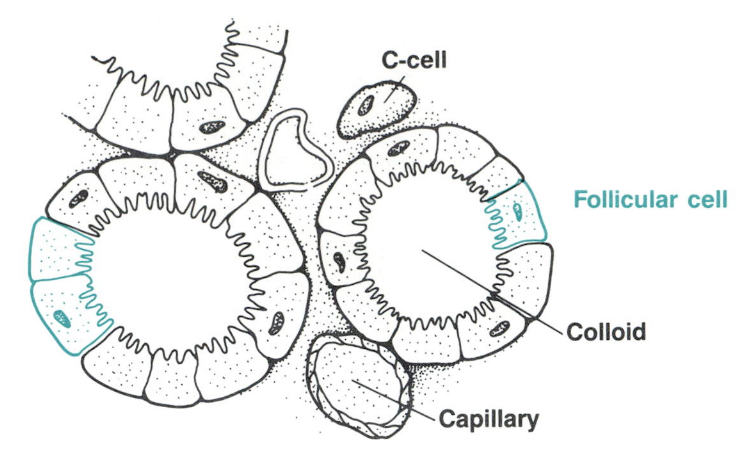

Synthesis and Secretion of T_4 and T_3

- The external membrane of follicular cells absorbs iodide (I^-) by an active transport mechanism.
- I^- is oxidized by intracellular peroxidase and is incorporated into tyrosine (TYR) residues of thyroglobulin (TGL) to yield TGL with embedded clusters of mono-iodotyrosine (MIT) or di-iodotyrosine (DIT).
- A certain proportion of these iodinated, attached clusters will couple to form tri-iodothyronine (T_3; a coupling of MIT with DIT) and tetra-iodothyronine (T_4; a coupling of DIT with DIT).

T_3, T_4, DIT, and MIT are all attached to TGL, and this complex is stored as colloid in the core of the thyroid follicles.

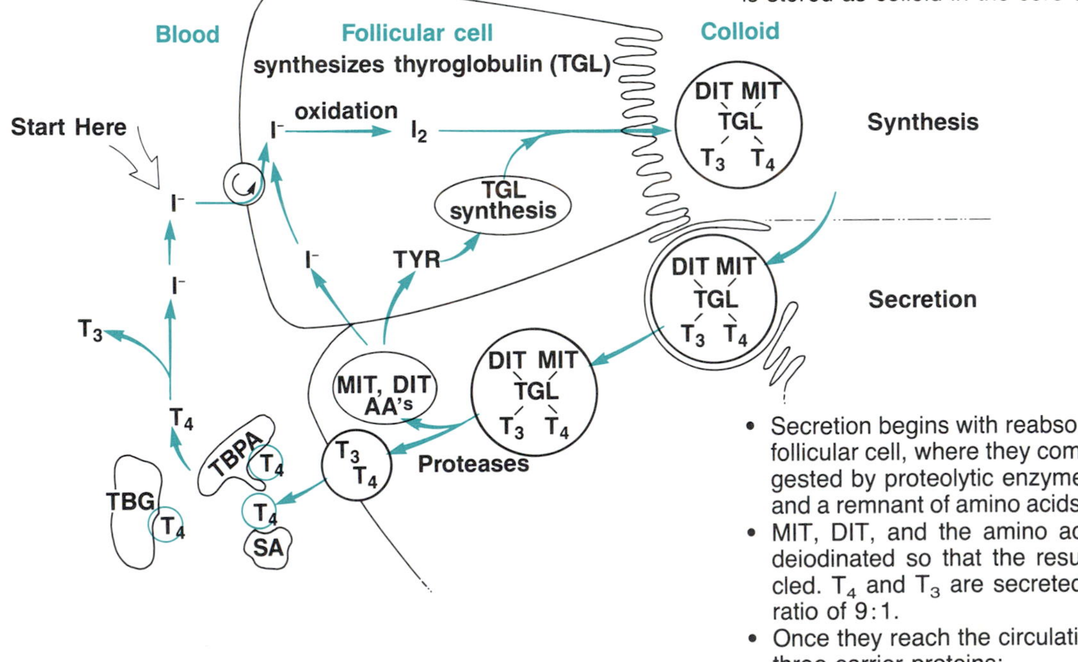

- Secretion begins with reabsorption of colloid droplets into the follicular cell, where they combine with lysosomes and are digested by proteolytic enzymes. This yields MIT, DIT, T_4, T_3, and a remnant of amino acids.
- MIT, DIT, and the amino acids remain in the cell and are deiodinated so that the resulting components can be recycled. T_4 and T_3 are secreted to the extracellular space in a ratio of 9:1.
- Once they reach the circulation, they are reversibly bound to three carrier proteins:
  —Thyroid-binding globulin (TBG; highest binding affinity)
  —Thyroid-binding prealbumin (TBPA)
  —Serum albumin (SA; lowest binding affinity, but present in large concentration)
- Free T_3 is the active form of thyroid hormone. Most of it is produced in the periphery by deiodination of T_4.

Biologic effects of T_3

Many systems and functions are affected.

Growth and development

T_3 is of special importance during fetal life and in the postnatal period because it **promotes body growth as well as normal development of nervous tissue.** These actions result from

- its ability to promote protein synthesis;
- its synergism with GHRH in the secretion of growth hormone and somatomedins;
- its synergistic influence on somatomedins at the epiphyseal growth plate in bone.

Energy metabolism

In nearly all tissues,

- T_3 stimulates **the basal rate of metabolism, oxygen consumption, and heat production.** This effect is strong enough to cause the metabolism of fat and carbohydrates to increase to such an extent that caloric intake cannot be increased sufficiently to maintain body weight.

Cardiovascular and respiratory effects

- Increased metabolic activity and oxygen consumption are stimuli for cardiorespiratory activity, leading to
  — **increased cardiac rate,**
  — **increased cardiac performance,**
  — **cardiac hypertrophy,** and
  — **increased ventilation.**
- T_3 **induces the synthesis of β-adrenergic receptors.** This causes peripheral vasodilatation and further cardiac stimulation.
- T_3 may exert a direct stimulatory effect on cardiac muscle activity by **promoting expression of the more active isoenzyme form of cardiac myosin ATPase.**

Central nervous effects

Central nervous system effects of T_3 are suggested by these characteristic findings in hyperthyroid individuals:

- Diffuse anxiety
- Extreme nervousness
- Frequent movement

The mechanisms by which T_3 affects nervous function are not yet known.

Endocrine effects

T_3 affects a variety of hormonal systems. The most notable consequences of thyroid imbalance are

- **altered menstrual patterns**
  - Lack of thyroid hormone is associated with excessive and frequent menstrual bleeding.
  - Excess thyroid hormone causes reduction or cessation of menstrual bleeding.
- **altered sexual function**
  - Lack of thyroid hormone is associated with diminished libido.
  - Excess thyroid hormone causes impotence in men.

Biologic effects of T_3

All biologic effects of thyroid hormone depend on the binding of free T_3 to nuclear receptors and are ultimately caused by changes in gene expression.

In adult life the main physiologic role of T_3 is the regulation of energy metabolism, presumably within the framework of body temperature control. The mechanisms of these effects are primarily

- enhanced protein synthesis and
- increased mitochondrial size and number.
  —As a result, the rate of formation of ATP and of a number of cellular enzymes is increased, leading to increased metabolism of fats and carbohydrates.

In addition, T_3 has important, but mostly undefined, roles in

- the central nervous system and
- the endocrine system.

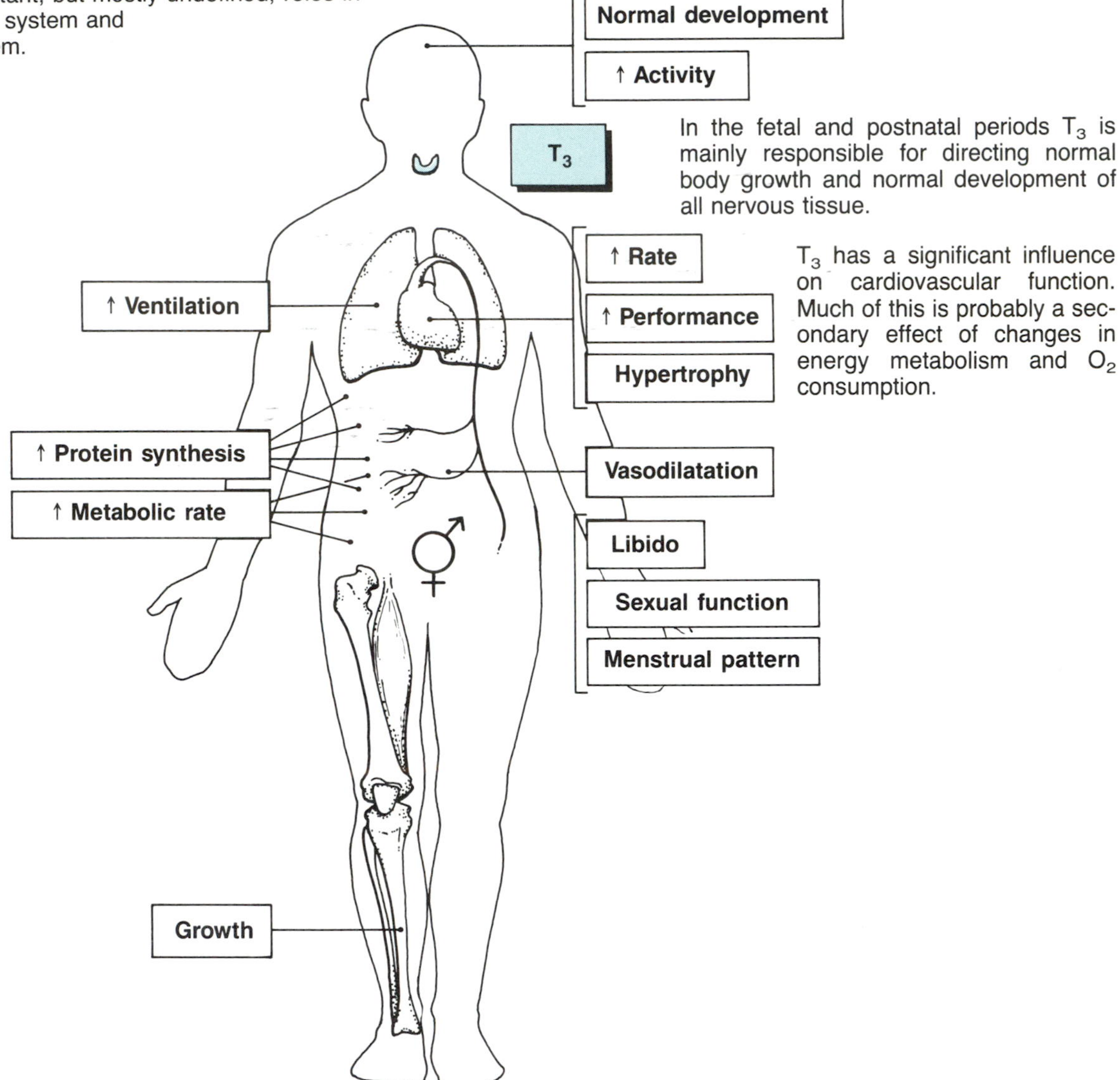

In the fetal and postnatal periods T_3 is mainly responsible for directing normal body growth and normal development of all nervous tissue.

T_3 has a significant influence on cardiovascular function. Much of this is probably a secondary effect of changes in energy metabolism and O_2 consumption.

THE PARATHYROID GLANDS

Parathyroid Hormone

Parathyroid hormone (PTH) is the major endocrine product of the parathyroid glands. It is synthesized in the **"chief" cells** of the parathyroid.

Synthesis and secretion

PTH is **synthesized in response to decreased serum levels of ionized calcium,** and virtually all of the synthesized product is secreted rather than stored.

Biologic activity

PTH plays a vital role in the **homeostasis of body calcium and phosphate.** Its target organs are **bone** and **kidney.**

In bone

The effects of PTH are

- **rapid removal of calcium salts** from bone crystals;
- more slowly progressing **activation of existing osteoclasts** and **formation of new osteoclasts.**
  - — **The effects of increased osteoclast activity are to dissolve bone and to increase serum [Ca^{++}].**

The direct and immediate **effect of PTH on osteoblasts is to depress their activity.** This is opposed, in the long term, by compensatory factors that enhance osteoblast activity in an effort to slow the rate of bone wasting. These mechanisms are, however, not sufficient to compensate completely for the PTH-induced bone lysis, and, therefore, the **net effect of sustained PTH elevation is continued bone wasting.**

In the kidney

The renal effects of PTH are

- to **enhance Ca^{++} reabsorption in the distal tubule;**
- to **decrease PO_4^{3-} reabsorption in the proximal tubule;**
- to **promote the conversion of vitamin D to its biologically most active form,** 1,25-dihydroxycholecalciferol [1,25-$(OH)_2D_3$]. This configuration of vitamin D acts on the intestinal epithelium to promote reabsorption of Ca^{++}. (These aspects are described more fully in Chapter 10, Bone and Connective Tissue.)

THE PARATHYROID GLANDS

Anatomy

The parathyroid glands are four pill-sized structures, embedded in the upper and lower thyroid poles. They are composed mainly of "chief" cells, which synthesize and secrete parathyroid hormone, a polypeptide of 84 amino acids.

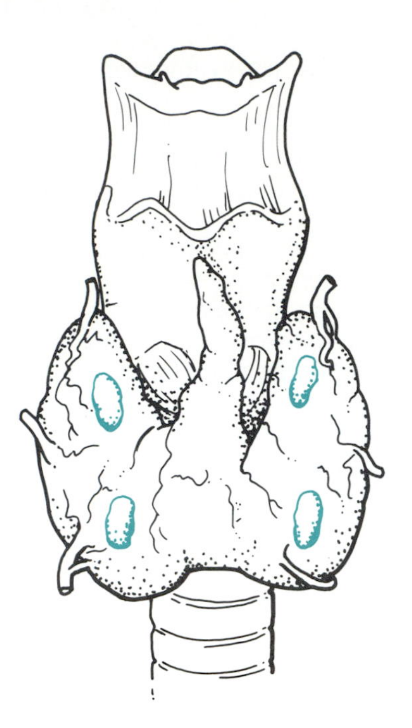

Parathyroid Hormone

Biologic activity

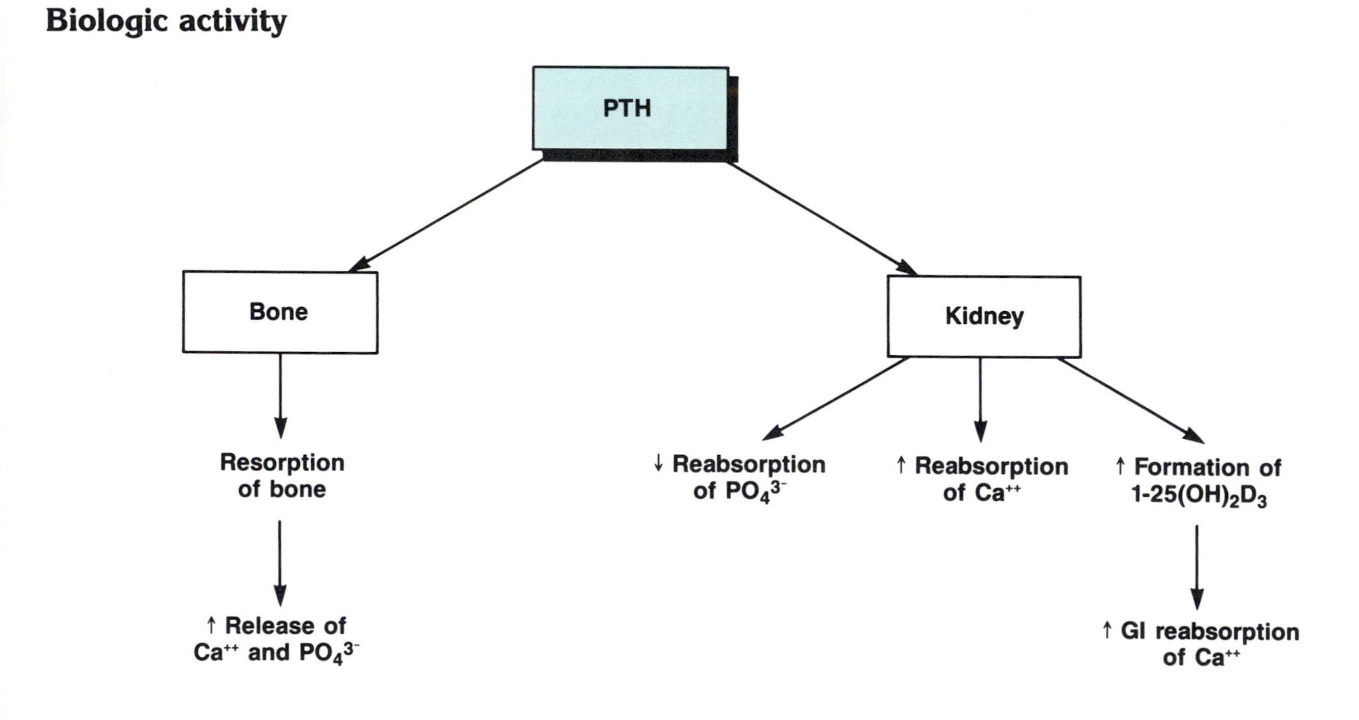

THE ADRENAL CORTEX

Synthesis of Steroid Hormones

Steroids are synthesized from **cholesterol,** which derives from two sources:

- It is present within adrenal cortical cells in the form of lipid droplets, and
- it is brought to these cells in the form of low-density lipoprotein particles (see Chapter 7, Gastrointestinal Physiology).

Within the mitochondria of adrenal medullary cells, **cytochrome P-450 side-chain cleavage enzyme** removes a side chain from cholesterol. This yields **pregnenolone,** the first product in the synthesis chains.

Pregnenolone quickly leaves the mitochondria and is serially modified in the cytoplasm and finally in the mitochondria again, to yield three classes of adrenocortical steroid hormones:

- **Mineralocorticoids**
- **Glucocorticoids**
- **Androgens**

Mineralocorticoids

The principal mineralocorticoids are **deoxycorticosterone** and **aldosterone.** Both exist mostly in the free form rather than being bound to plasma proteins. As a result, they are quickly metabolized (in the liver), and their plasma half-life is only 10 to 20 minutes.

Glucocorticoids

Corticosterone possesses glucocorticoid activity, but its plasma concentration in humans is generally too low for significant biologic effects. Therefore, **cortisol** is the dominant glucocorticoid.

- Most of the plasma cortisol is bound to corticosteroid-binding globulin, making its renal clearance low and its plasma half-life five to ten times that of the mineralocorticoids.

Androgens

In adult humans the adrenals represent a minor source of androgens when their output is compared with that of the ovaries or testes. The principal androgens secreted by the adrenals are

- dehydroepiandrosterone and
- androstenedione.

These have little biologic potency and become active only after peripheral tissues have converted them. The chief product of this peripheral conversion is **testosterone.** Peripheral conversion of testosterone to **estradiol** also occurs. These conversion products are tightly bound in blood to specific globulins.

THE ADRENAL CORTEX

Anatomy of the Adrenal Glands

The adrenals are two pyramidal organs situated on the upper poles of the kidneys.

- They synthesize steroid hormones from cholesterol.
- Approximately 90 percent of an adrenal is cortex, surrounding an inner core, called the adrenal medulla.

In the cortex three zones can be recognized on the basis of both microscopic anatomy and secretory products:

- Zona glomerulosa cells synthesize aldosterone.
- Cells in the zona fasciculata and the zona reticularis synthesize cortisol and precursors for androgens.

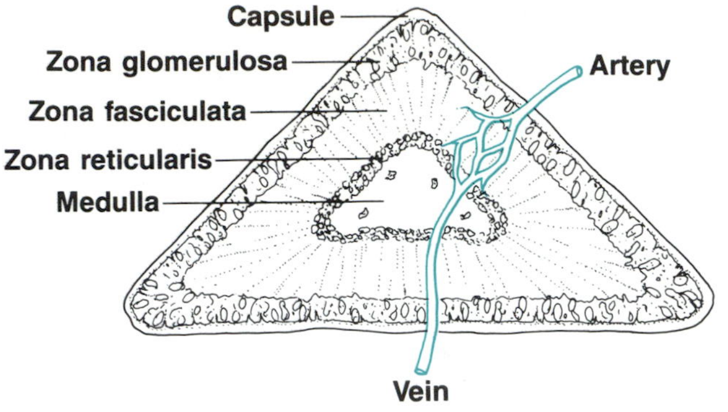

The anatomy of the vascular supply is an important determinant of the nature of the products that will be synthesized at different sites in the glands

- Arteries enter the outer portion of the cortex.
- Successive branchings take the vessels toward the central medulla, where they unite in the collecting veins.

As a result of this arrangement,

- synthesized hormones are transported toward the medulla, and their concentration increases progressively from cortex to medulla.

Along with their increasing concentration, the specific inhibition exerted by these hormones on selected steroidogenic enzymes increases progressively toward the core. The net result is that **the vascular anatomy ensures regional specificity in the synthesis of adrenal hormones.**

Synthesis of Steroid Hormones

Free cholesterol is transported to the inner membrane of the mitochondria, where cytochrome P-450 side-chain cleavage enzyme removes a side chain and thereby produces **isocaproaldehyde** and **pregnenolone.**

Pregnenolone quickly leaves the mitochondria. Within the endoplasmic reticulum it undergoes a sequence of modifications, each step altering the three-dimensional shape of the molecule so that access by the next enzyme is possible.

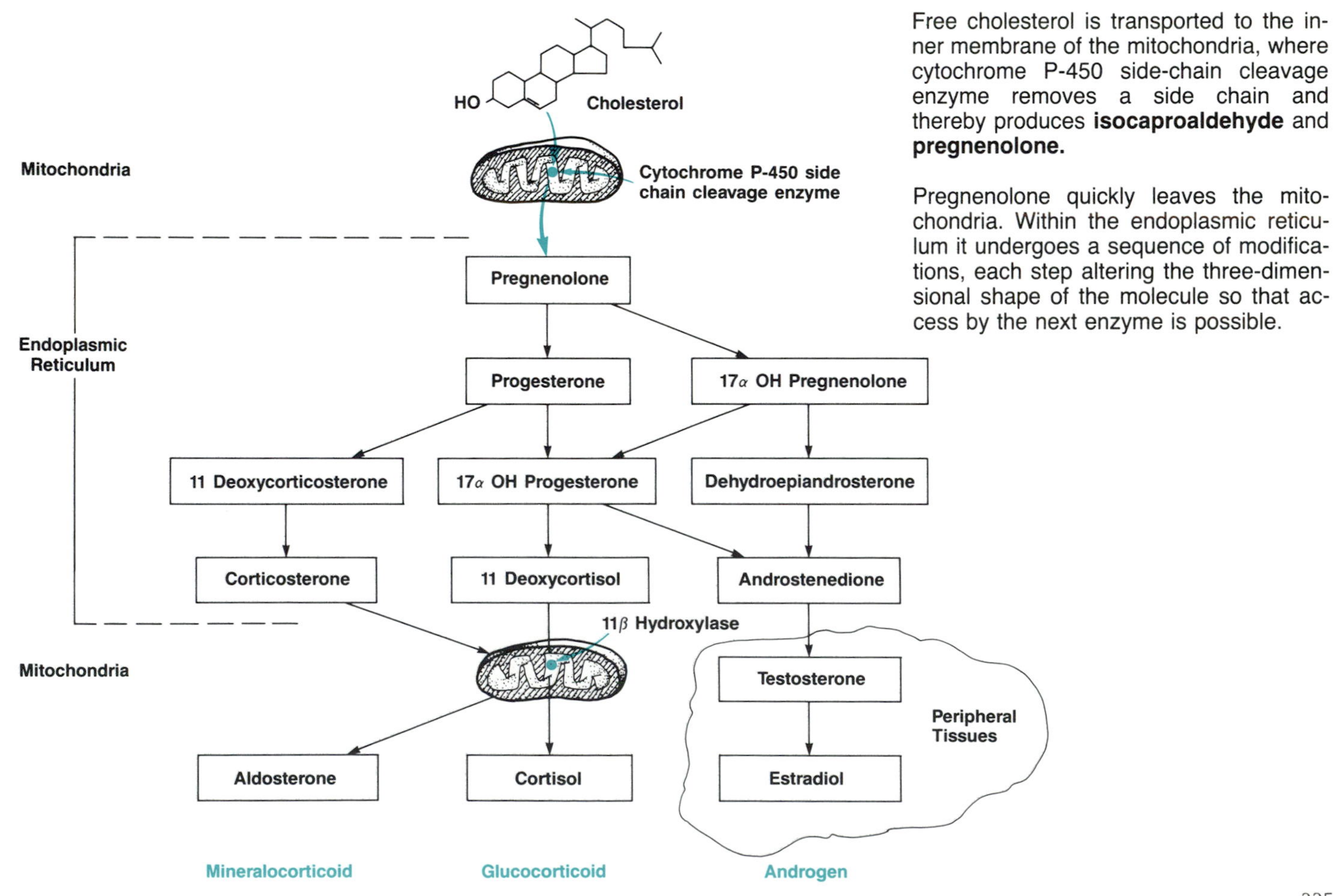

PHYSIOLOGY OF STEROID HORMONES

Regulation of Synthesis

Mineralocorticoids

Aldosterone production occurs mainly in zona glomerulosa cells of the adrenal cortex. It is regulated by three factors:

- Serum $\mathbf{K^+}$ levels
- The **renin-angiotensin** system
- Adrenocorticotropic hormone **(ACTH)**

Glucocorticoids and androgens

The synthesis of these agents is regulated principally by ACTH.

Biologic Effects

Mineralocorticoids

The homeostatic role of aldosterone is to **regulate the body balance of $\mathbf{Na^+}$ and $\mathbf{K^+}$.** This is accomplished by its stimulatory effect on Na^+ and K^+ transport, primarily in the distal nephron but also in sweat glands, salivary glands, and the GI mucosa.

Mineralocorticoids

The influence of angiotensin on aldosterone synthesis is driven by plasma levels of renin, a proteolytic enzyme that is released from juxtaglomerular cells of renal afferent arterioles.

Aldosterone synthesis is governed by three factors:

- Serum K^+ levels
- The renin-angiotensin system
- ACTH

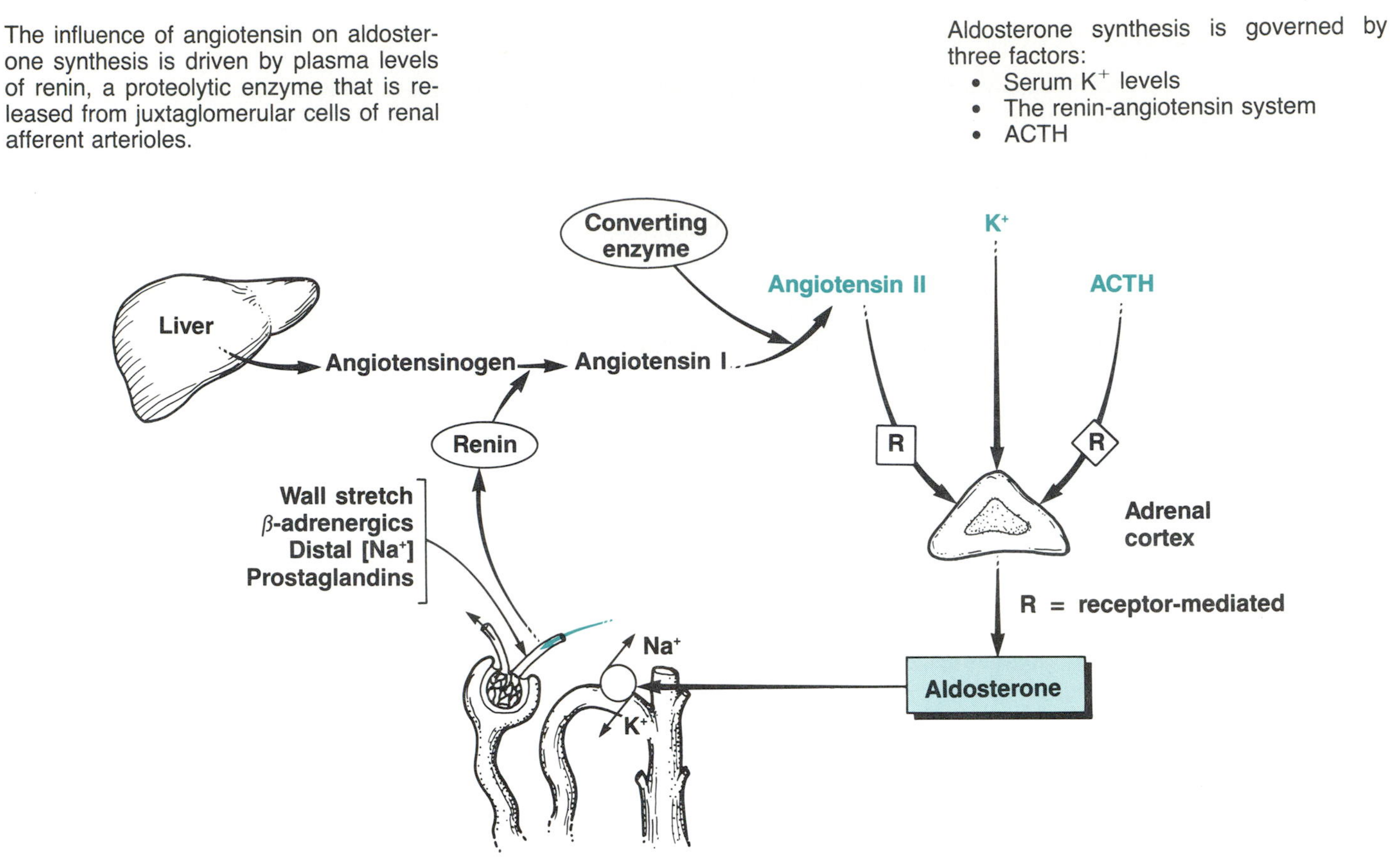

Renin release is increased during cardiovascular hardship by these specific influences on the juxtaglomerular cells (j-g cells) of renal afferent arterioles:

- Decreased stretch of the arteriolar walls
- β-adrenergic stimulation of the j-g cells
- Composition of fluid in the distal nephron
- Prostaglandin levels

- Renin cleaves the decapeptide, angiotensin I, from angiotensinogen (a circulating glycoprotein that is synthesized by the liver).
- Angiotensin I is further processed by converting enzyme. This yields angiotensin II.
  Converting enzyme is widely distributed in vascular endothelium.
  It acts by removing two carboxy terminal amino acids from angiotensin I.

- Angiotensin II, an eight-amino-acid peptide, binds to a specific membrane receptor and possibly uses Ca^{++} as a second messenger to promote the synthesis and secretion of aldosterone from zona glomerulosa cells

Aldosterone acts on active Na^+-K^+ pumps in the distal convoluted tubule to increase Na^+ reabsorption from the tubule and K^+ secretion into the tubule.

- Its action is brought about by receptor-mediated formation of mitochondrial enzymes that promote Na^+-K^+ ATPase activity.

Aldosterone may also stimulate the H^+ antiporter.

Biologic Effects—Cont'd.

Glucocorticoids

Cortisol has a variety of complex effects, some of which are manifest only at pharmacologic levels.

Cortisol at physiologic levels

- regulates the metabolism of fat, carbohydrate, and protein and, thereby, **helps maintain blood glucose levels;**
- **mediates adaptation to stress;**
- has permissive influences for the action and synthesis of some hormones;
- has an important role in normal growth and development.

Cortisol at pharmacologic levels

- suppresses cell-mediated immunity;
- inhibits the mobility of leukocytes and macrophages;
- stabilizes lysosomes;
- depresses the release of vasoactive and proteolytic enzymes such as kinins.

As a result,

◆ **cortisol suppresses all the characteristics of inflammatory responses and delays wound healing.**

Androgens

Testosterone and other androgens have some biologic activity in most tissues at all stages of life.

In fetal life

Androgens determine the development of gender-linked features in **anatomy** and in **patterns of gonadotropin release.**

In adult life

Two effects of androgens are observed, depending on the target organ:

- They stimulate protein synthesis **(anabolic effects).**
- They influence development and growth of male sexual characteristics **(androgenic effects).**

Effects of androgens on gender-specific behavior have been asserted, but the evidence for such effects in humans is conflicting.

PHYSIOLOGY OF STEROID HORMONES—CONT'D.

Biologic Effects—Cont'd.

Glucocorticoids

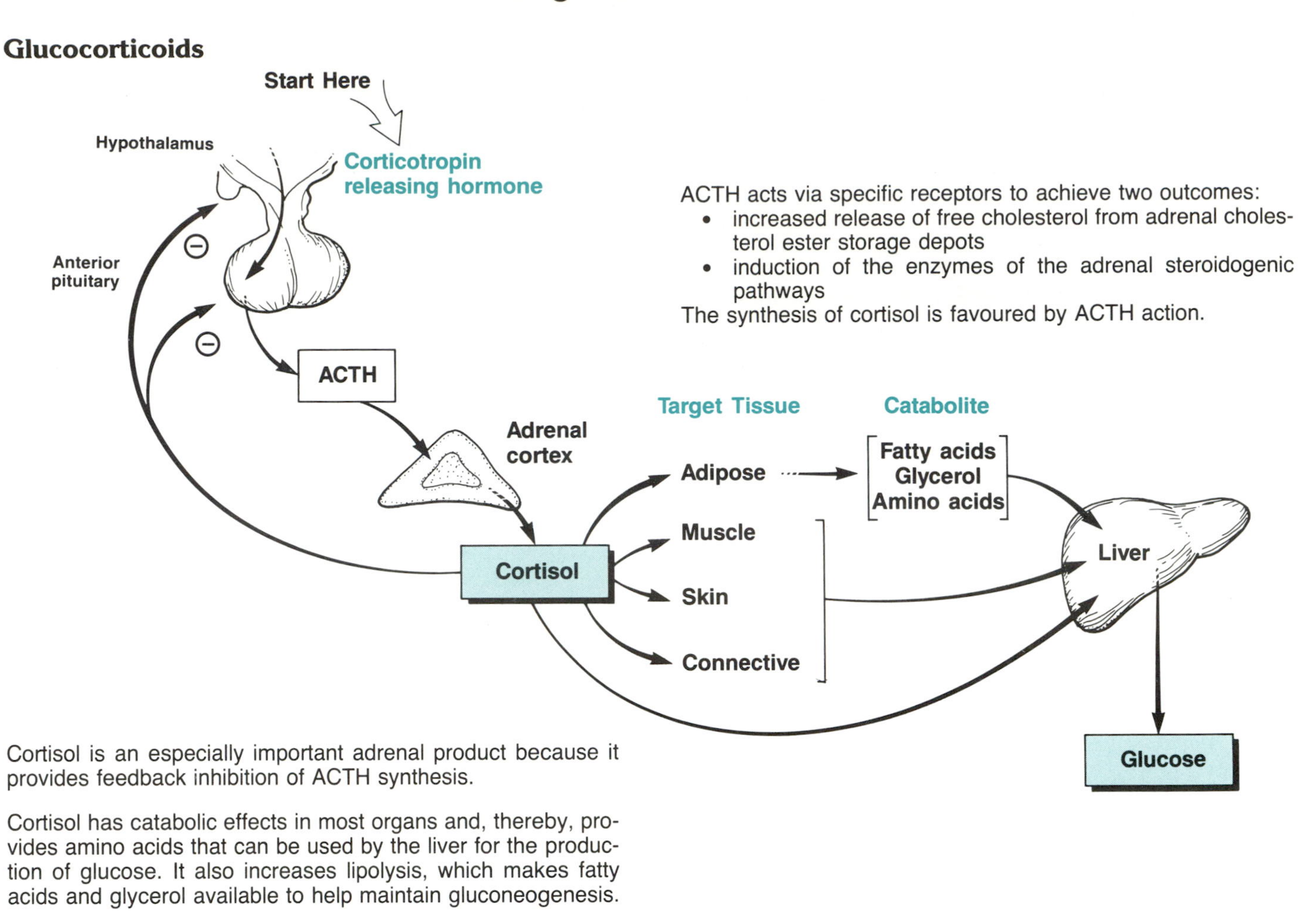

ACTH acts via specific receptors to achieve two outcomes:
- increased release of free cholesterol from adrenal cholesterol ester storage depots
- induction of the enzymes of the adrenal steroidogenic pathways

The synthesis of cortisol is favoured by ACTH action.

Cortisol is an especially important adrenal product because it provides feedback inhibition of ACTH synthesis.

Cortisol has catabolic effects in most organs and, thereby, provides amino acids that can be used by the liver for the production of glucose. It also increases lipolysis, which makes fatty acids and glycerol available to help maintain gluconeogenesis.

Androgens

In Fetal Life	In Adult Life
Androgen concentration in fetal blood during the first 10 weeks determines —whether female or male genitalia (internal as well as external) develop; —whether the hypothalamus will develop a cyclic pattern of gonadotropin release after puberty (female) or a noncyclic pattern (male).	Two androgen effects are observed, depending on the target organ: • Anabolic, i.e., relating to stimulation of protein synthesis • Androgenic, i.e., relating to development and growth of male sexual characteristics, including influences on —the pattern of body growth and muscle development, —the maturation of external genitalia and accessory sexual organs (scrotum, seminal vesicles, prostate), —the size of the larynx and vocal cords, and —the patterns of hair growth and hair loss. • Effects on gender-specific behavior?

THE ADRENAL MEDULLA

The adrenal medulla is an integral part of the sympathetic nervous system. It contains neuronal cells **(chromaffin cells)** that have endocrine function.

◆ They are unique in containing the enzyme that converts norepinephrine to epinephrine (phenylethanolamine-N-methyltransferase).

In healthy adult humans, so much of this enzyme is present that mostly **epinephrine** is released into the circulation when the adrenal medulla is stimulated.

Epinephrine

Synthesis and storage

The important regulators of epinephrine synthesis are

- **sympathetic activity,**
- **ACTH,**
- **norepinephrine,** and
- **cortisol.**

Secretion

Stored epinephrine is secreted into the circulation in response to action potentials in **cholinergic preganglionic nerve fibers** that reach the adrenals via the greater splanchnic nerve.

Once the epinephrine is released, its half-life in the circulation is less than 3 minutes. The key enzymes that are responsible for catecholamine metabolism are **catecholamine O-methyltransferase, monoamine oxidase** (MAO), and **aldehyde oxidase.**

THE ADRENAL MEDULLA

Chromaffin cells are arranged in close relationship with preganglionic cholinergic fibers as well as with venules that drain the adrenal cortex. This anatomic arrangement allows both sympathetic nervous activity and adrenocortical humoral products to influence the synthesis of catecholamines (norepinephrine and epinephrine).

Synthesis and Storage of Epinephrine

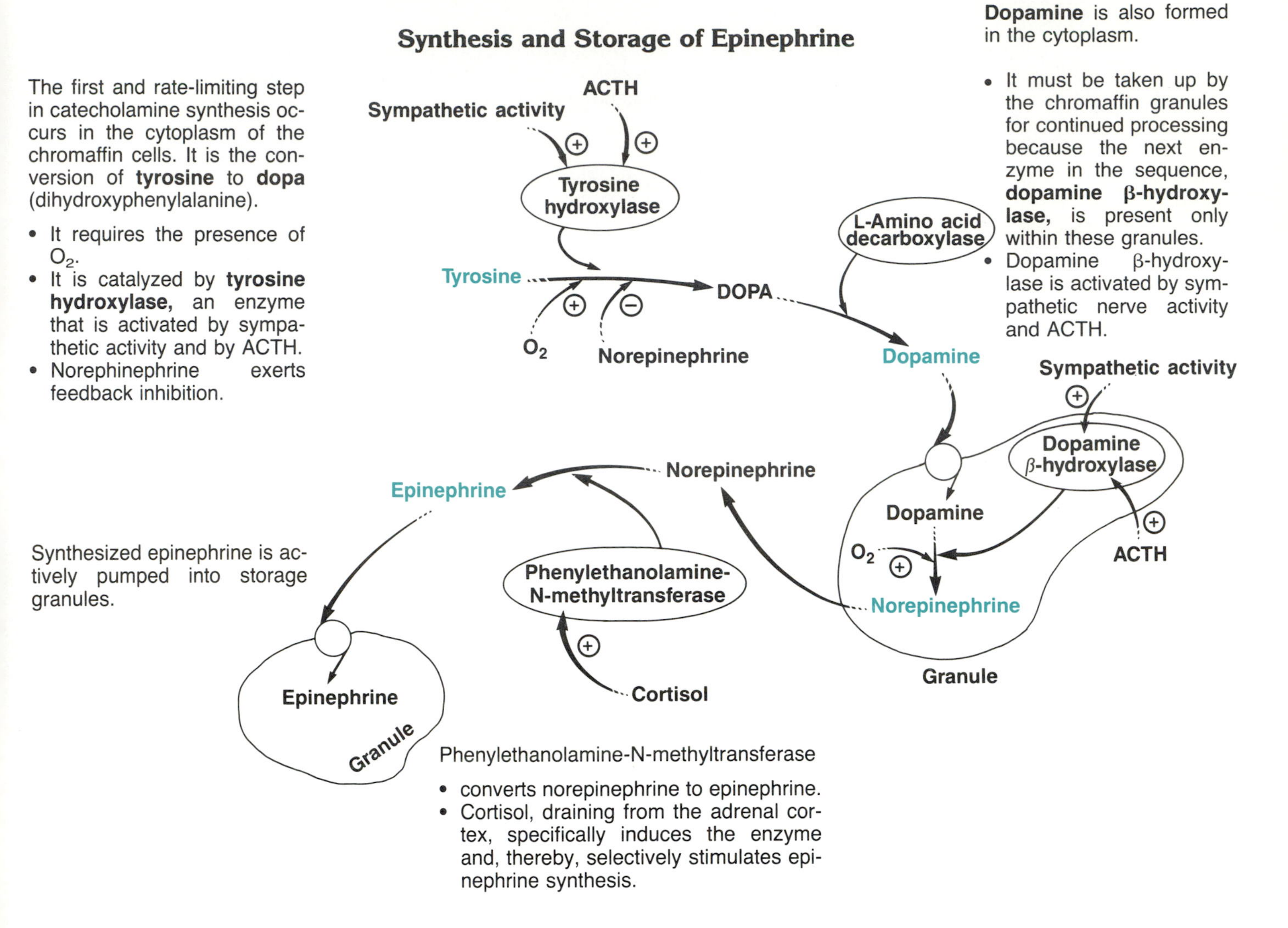

The first and rate-limiting step in catecholamine synthesis occurs in the cytoplasm of the chromaffin cells. It is the conversion of **tyrosine** to **dopa** (dihydroxyphenylalanine).

- It requires the presence of O_2.
- It is catalyzed by **tyrosine hydroxylase,** an enzyme that is activated by sympathetic activity and by ACTH.
- Norephinephrine exerts feedback inhibition.

Dopamine is also formed in the cytoplasm.

- It must be taken up by the chromaffin granules for continued processing because the next enzyme in the sequence, **dopamine β-hydroxylase,** is present only within these granules.
- Dopamine β-hydroxylase is activated by sympathetic nerve activity and ACTH.

Synthesized epinephrine is actively pumped into storage granules.

Phenylethanolamine-N-methyltransferase

- converts norepinephrine to epinephrine.
- Cortisol, draining from the adrenal cortex, specifically induces the enzyme and, thereby, selectively stimulates epinephrine synthesis.

In the presence of oxygen, dopamine β-hydroxylase catalyzes the formation of **norepinephrine** from dopamine. Most of the norepinephrine diffuses out of the granules into the cytoplasm.

Epinephrine—Cont'd.

Metabolic effects

These arise chiefly from interactions with **β-receptors.** Since β-receptors have greater affinity for epinephrine than they do for norepinephrine, epinephrine has the more significant effects on metabolism. Two kinds of metabolic actions are observed:

- **Diabetogenic effects**
  - — Increased production and decreased uptake of glucose and ketoacids occur.
  - — Liver, skeletal muscle, pancreas, and adipose tissue are the major target organs.
- **Thermogenic effects**
  - — Metabolic rate is enhanced.
  - — These effects occur diffusely, in many tissues.

Cardiovascular effects

The general cardiovascular effect is

- to **increase cardiac output**
  — by increasing cardiac rate and cardiac performance.
- An especially important cardiovascular effect of epinephrine is its ability to influence the **selective distribution of cardiac output.**

The fraction of cardiac output that is directed to a tissue is determined by the vascular resistance that is offered by that tissue in comparison to all other tissues. Net tissue vascular resistance, in turn, is determined by the degree of imbalance between local vasoconstrictor factors and local vasodilator factors. Tissues that have a high proportion of vascular β-receptors (e.g., skeletal muscle) show a high potential for vasodilatation in the presence of epinephrine. A physiologically important example is given under "Cardiorespiratory Responses to Exercise" in Chapter 12.

Epinephrine—Cont'd.

Metabolic effects

Diabetogenic

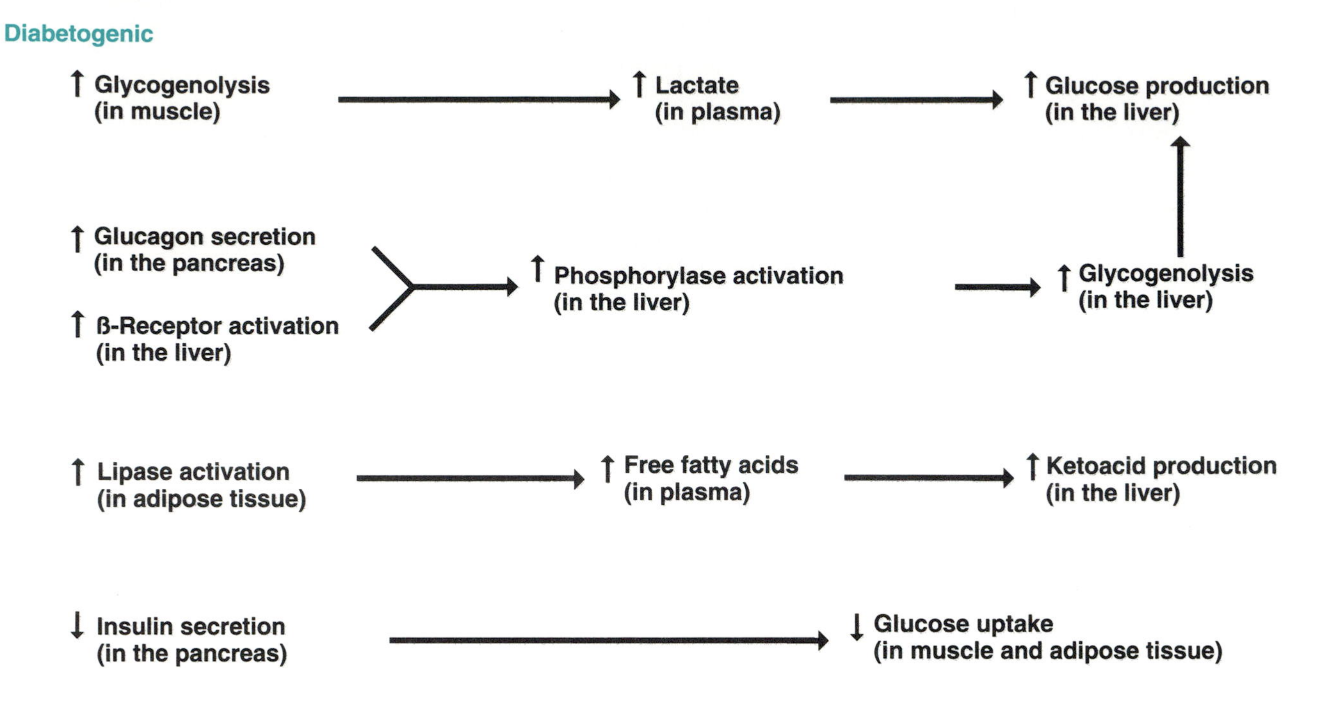

Thermogenic

↑ Basal metabolic rate

THE ENDOCRINE PANCREAS

The **Islets of Langerhans** are the endocrine portion of the pancreas. They are scattered throughout the pancreas, are well supplied with fenestrated capillaries, and are responsible for synthesis, storage, and release of the hormones **insulin, glucagon, somatostatin,** and **pancreatic polypeptide.**

Insulin

Synthesis and secretion

Insulin, a small peptide, is synthesized in **islet B-cells** and is stored in secretory granules within these cells. Its secretion is an active process, requiring cAMP.

The intracellular signal path that leads to secretion involves Ca^{++} and calmodulin, and it is **regulated by metabolic, nervous, endocrine, and ionic factors.**

Metabolic factors

- The most important stimulus for insulin release is an **increase in plasma glucose concentration above its normal level** of 100 mg/dL.
- Ketoacids, arginine, and lysine (resulting from digestion of ingested protein) are stimuli of secondary importance.

Nervous factors

Both sympathetic and parasympathetic nerves supply the islets.

- Insulin release is enhanced by
  — parasympathetic stimulation and
  — β-adrenergic agonists.
- Insulin release is inhibited by α-adrenergic agonists.

Ionic factors

- K^+, Ca^{++}, and Na^+ are needed for normal insulin responses to glucose.
- An excess of Mg^{++} is inhibitory.

Endocrine factors

- Insulin secretion is **directly stimulated** by a number of GI hormones:
  — **Gastric inhibitory polypeptide**
  — Gastrin
  — Secretin
  — Cholecystokinin
  — Glucagon
- Insulin secretion is indirectly stimulated by the effects of other hormones on plasma glucose levels. Among these are
  — cortisol;
  — growth hormone;
  — thyroxine;
  — progesterone, estrogen, and testosterone.
- Insulin secretion is inhibited most importantly by
  — somatostatin and
  — insulin.

THE ENDOCRINE PANCREAS

Islets of Langerhans

A minuscule portion (1 to 2 percent) of the pancreas consists of discrete bodies, the Islets of Langerhans, that are responsible for synthesis, storage, and release of the hormones **insulin, glucagon, somatostatin,** and **pancreatic polypeptide.** Each hormone is secreted from a distinct cell type.

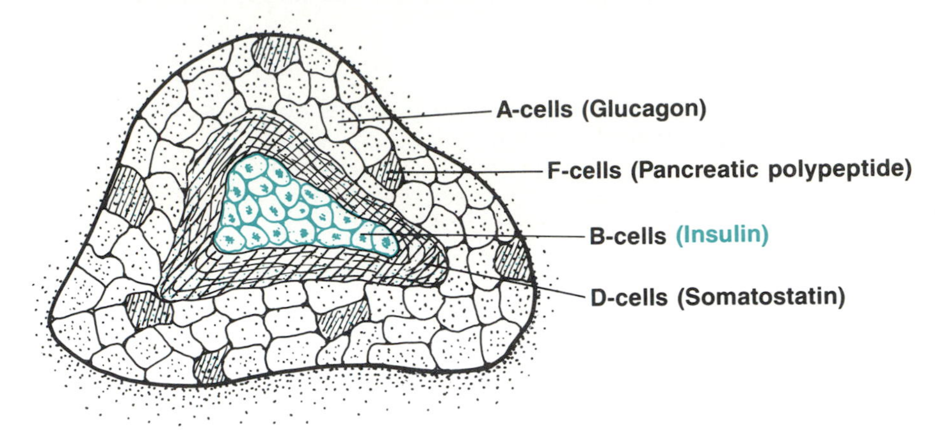

Insulin

Synthesis and secretion

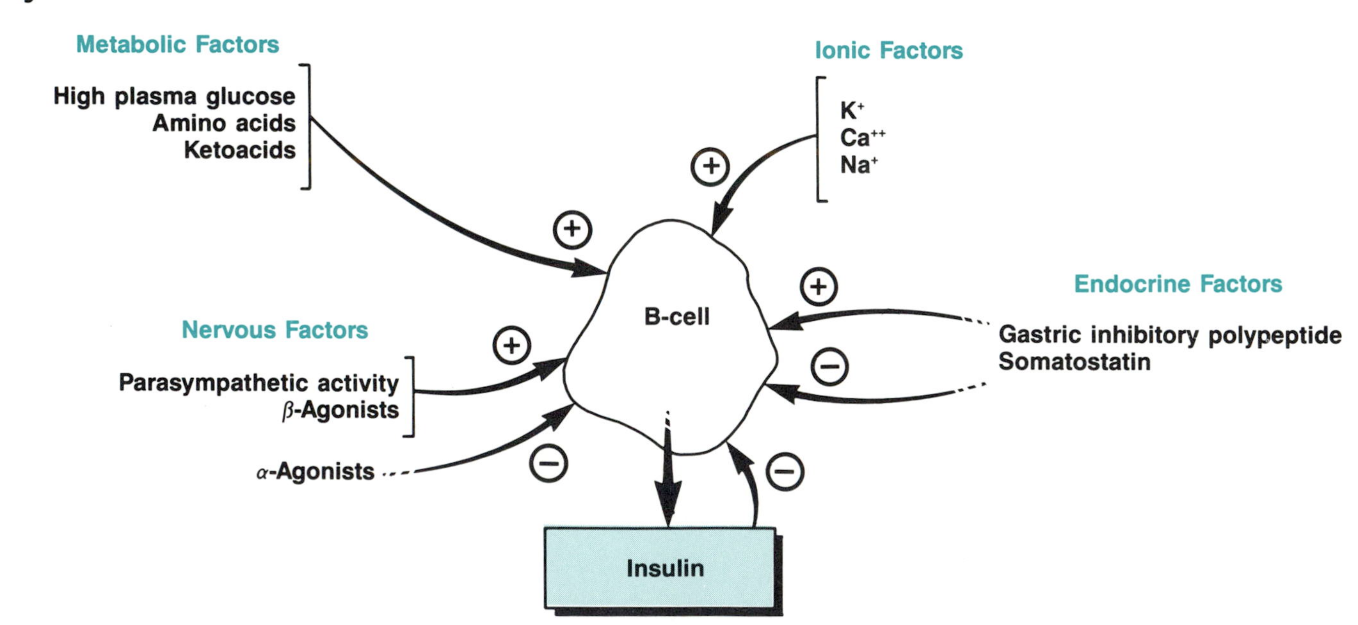

Insulin—Cont'd.

Biologic effects

The insulin-responsive tissues are liver, muscle, and adipose tissue. In each of these target tissues,

- insulin **promotes glucose entry into cells;**
- insulin has profound effects on intracellular enzymes that govern the **rate of metabolism of carbohydrate, fat, and protein.**

Carbohydrate metabolism

Insulin has a dual effect on carbohydrate metabolism:

- it **stimulates glucose utilization and glucose storage,** and
- it **inhibits glucose formation.**

Its **action on the liver depends on plasma glucose levels:**

- When plasma glucose levels are high, then the action of insulin is to convert the liver from a glucose-producing organ into a glucose-storing organ.
- When plasma glucose levels are not high, but insulin levels are (inappropriately) high, then the liver evades the glucose-storing effects of insulin.

Fat metabolism

Insulin promotes fat storage and inhibits mobilization and oxidation of fatty acids. These effects express themselves differently in different tissues:

- **In the blood,**
  — insulin-dependent lipoprotein lipase is activated. This promotes the breakdown of dietary triglycerides to free fatty acids (see Chapter 7, Gastrointestinal Physiology).
- **In adipose tissue cells,**
  — glucose uptake is promoted;
  — intracellular breakdown of triglycerides is depressed and intracellular storage of triglycerides is promoted.
- **In the liver,**
  — conversion of dietary glucose to fatty acids or cholesterol is promoted.

Protein metabolism

Insulin promotes synthesis and sequestration of protein, and it inhibits protein breakdown. This expresses itself

- **in muscle cells**
  — as increased preferential uptake of branched-chain amino acids and
  — as increased RNA synthesis and, therefore, as increased amino acid incorporation into proteins;
- **in liver cells**
  — as decreased uptake of gluconeogenetic amino acids (alanine and glycine).

Ion transport

Insulin stimulates the entry of K^+, PO_4^{3-}, and Mg^{++} into muscle and liver cells. The magnitude of this effect is such that **insulin is considered to be a significant regulator of serum K^+ levels.**

Insulin—Cont'd.

Biologic effects

The biologic effects of insulin are mediated by a membrane receptor. They manifest themselves in two ways:

- As increased entry of glucose, amino acids, K^+, and PO_4^{3-} into the target cell
- As activation of intracellular enzyme systems that promote synthesis of glycogen, protein, and fat

Carbohydrate metabolism

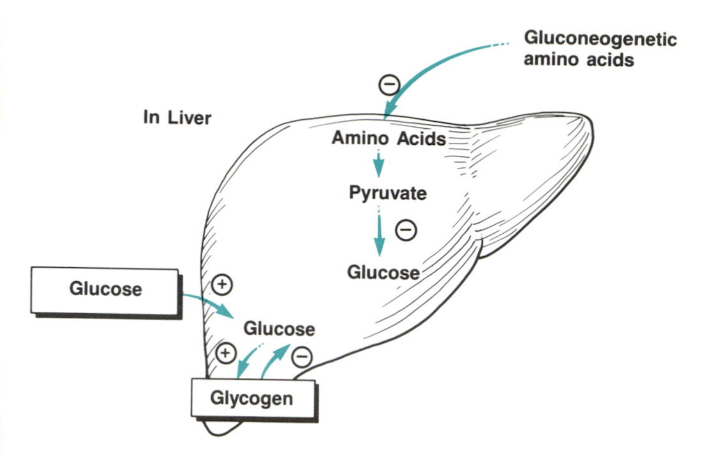

Insulin effects on carbohydrate metabolism in the liver are shown in color. They include

- facilitation of glucose uptake;
- promotion of glycogen formation from glucose;
- inhibition of glycogen breakdown;
- inhibition of the uptake of gluconeogenetic amino acids;
- inhibition of gluconeogenesis.

In **muscle** and in **adipose tissue,** insulin stimulates glucose transport into cells.

- Within muscle cells glucose is converted chiefly to glycogen.
- Within adipocytes glucose is converted chiefly to glycerol.

The liver evades the glucose-storing effects of insulin when plasma glucose levels are not high, but insulin levels are (inappropriately) high. This evasion is mediated by three compensatory factors:

1. A hepatic autoregulatory response to decreased plasma glucose concentration
2. Increased glucose output in response to sympathetic nervous activity that is driven by low glucose levels in the hypothalamus
3. Increased pancreatic secretion of the insulin antagonist, glucagon

Fat metabolism

Effects of insulin on fat metabolism express themselves differently in different tissues. The major effects are shown in color in the following diagrams.

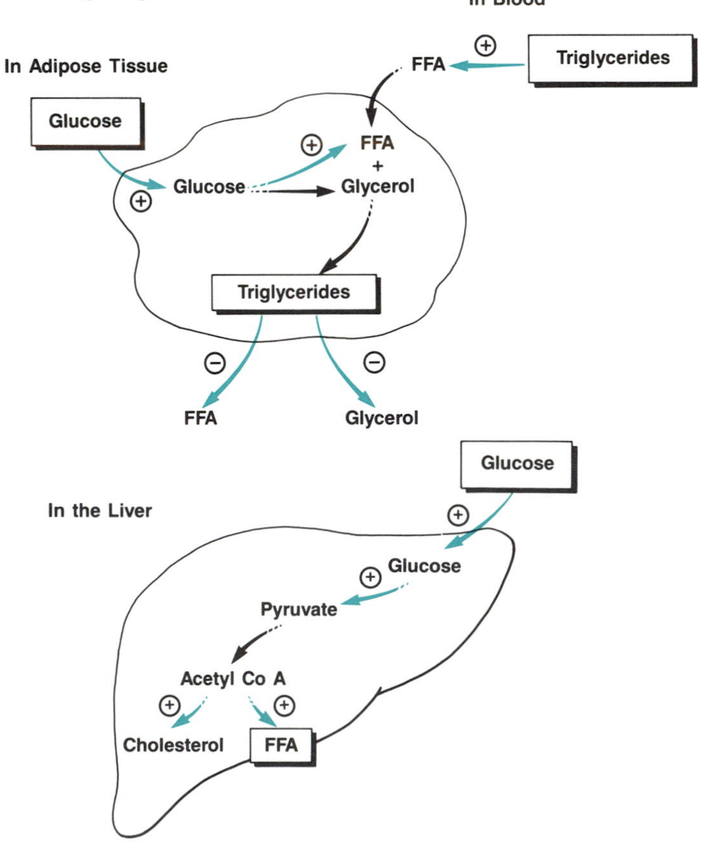

In adipose tissue,

- the intracellular pool of free fatty acids (FFA) is increased
  —partly as a result of increased extracellular FFA concentration and
  —partly by insulin-mediated activation of enzymes involved in the conversion of glucose to FFA;
- the intracellular glycerol pool is increased as a result of enhanced glucose entry;
- increased substrate levels (FFA and glycerol) result in greater triglyceride formation;
- insulin inhibits lipolysis and, therefore, causes formed triglycerides to be stored within adipocytes when glucose is in abundance. They can be made available to supply FFA to the liver during fasting periods.

In the liver,

- breakdown of dietary glucose is stimulated;
- this increases pyruvate levels and, hence, the production of acetyl CoA in hepatic mitochondria;
- acetyl CoA then diffuses to the cytoplasm. There insulin stimulates its incorporation into fatty acids or cholesterol.

Glucagon

Synthesis and secretion

Glucagon is a small peptide. Its major source is the **A-cells of the pancreas.** The gut can also produce some glucagon, but it is not a significant source in humans. The three most important factors influencing synthesis and secretion are the following:

- **A decrease in plasma glucose levels increases glucagon secretion.**
- Stimulation of adrenergic α-receptors (via the sympathetic nervous system or via epinephrine) increases glucagon secretion.
- Somatostatin is the strongest inhibitor of glucagon secretion. (Insulin is a weak inhibitor, perhaps acting via its influence on glucose utilization by the A-cells.)

Biologic effects

These are exerted via a membrane receptor and cAMP as a second messenger.

Glucagon has little effect on peripheral tissues. Its **major target organ is the liver.** There it acts at many of the same biochemical control points as does insulin, but its effects are the opposite:

◆ Its **action is to stimulate and maintain hepatic glucose output.**

Somatostatin

Synthesis and secretion

Somatostatin, the hypothalamic inhibitor of growth hormone release, is also **secreted by pancreatic D-cells,** and it occurs throughout the GI tract. Its secretion is promoted by a variety of dietary and digestive factors.

Biologic effects

Somatostatin concentration is highest within the pancreatic islets. As a result, **its major effect is the local, paracrine inhibition of insulin and glucagon secretion.**

Somatostatin also inhibits digestive function via three mechanisms:

- By inhibiting gut motility
- By inhibiting gastric secretion and pancreatic exocrine function
- By inhibiting absorption at the mucosa

Overall it looks as though one of the physiologic roles of somatostatin is the pacing of foodstuff conversion. This is achieved by coordinated effects on

- gross digestive functions (bulk transport, digestion absorption) and
- cellular disposition of nutrients.

Pancreatic Polypeptide

Secretion is increased by food intake and decreased by hyperglycemia. Its physiologic purpose is not yet known.

Glucagon

Synthesis and secretion

Metabolic influences

The most important metabolic signal is a decrease in plasma glucose levels.

Nervous influences

The sympathetic nervous system acts via α-receptors to increase glucagon secretion. This can lead to significant increases in blood glucose levels after trauma, toxemia, or infection.

Glucagon circulates unbound and has, therefore, a half-life of only a few minutes. It is degraded in the liver and in the kidney.

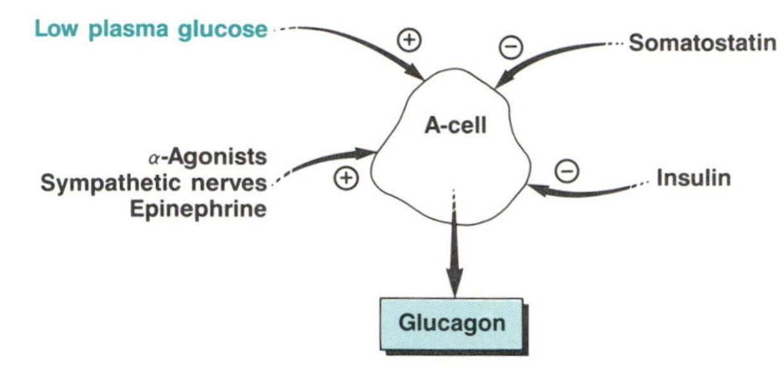

Endocrine influences

- *Epinephrine*
  reaches sufficiently high plasma levels during periods of severe stress to promote glucagon secretion via a receptor-mediated mechanism.
- *Somatostatin*
  is the strongest inhibitor of glucagon secretion.
- *Insulin*
  is a weak inhibitor, perhaps acting via its influence on glucose utilization by the A-cells.
- *Growth Hormone, Cortisol, Estrogen, and Progesterone*
  affect glucagon secretion only via their effects on plasma glucose.

Biologic effects

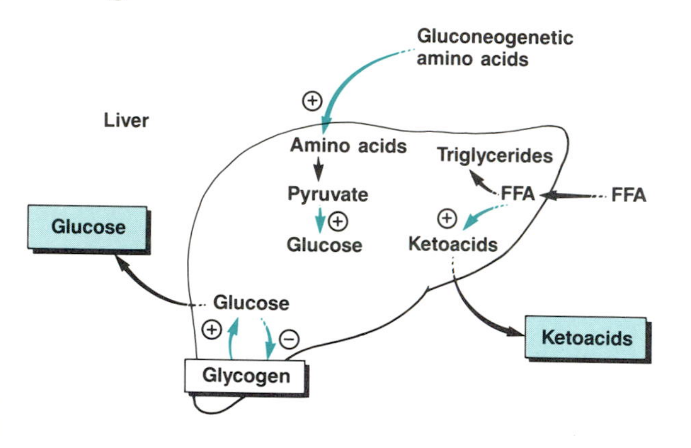

Glucagon acts at many of the same biochemical control points in the liver as does insulin, but its effects are the opposite.

- It stimulates glycogen phosphorylase and, therefore, promotes conversion of glycogen to glucose.
- It inhibits glycogen synthase. This prevents resynthesis of glycogen from glucose.
- It promotes gluconeogenesis by promoting both hepatic uptake of certain amino acids (particularly alanine) and the activities of several enzymes involved in the conversion of pyruvate to glucose.
- It directs incoming free fatty acid (FFA) away from triglyceride synthesis and toward ketoacid formation.
- It inhibits hepatic synthesis of FFA from cytoplasmic acetyl CoA.

Somatostatin

Synthesis and secretion

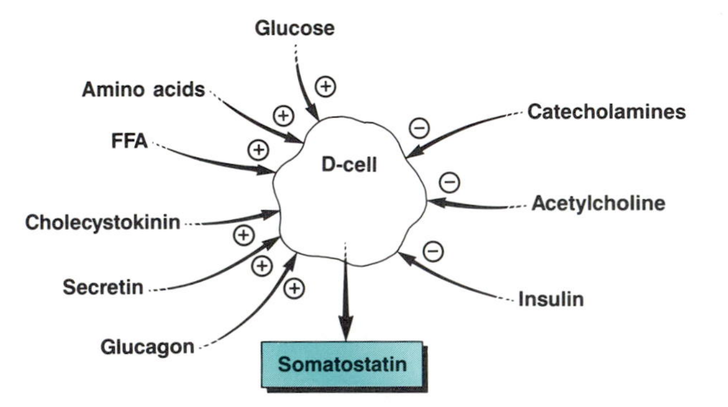

Somatostatin secretion from the adrenal gland is promoted by several dietary and digestive factors.

The autonomic nervous system and insulin inhibit adrenal somatostatin secretion.

Biologic effects

Somatostatin

↓ Motility
↓ Secretion
↓ Absorption

↓ Insulin
↓ Glucagon

The major effect of adrenal somatostatin is the local, paracrine inhibition of insulin and glucagon secretion.

10

BONE AND CONNECTIVE TISSUE

STRUCTURE

OVERVIEW

The basic structural element of all connective tissues is a fine net of **small collagen fibers.** **Proteoglycans** (protein-polysaccharide complexes) or **mineral precipitates** are deposited and held in this net, and they determine the pliability and strength of different tissues.

Cartilage

This tissue is semi-rigid and provides firm, flexible support.

- It is characterized by **deposits of proteoglycans** protein-polysaccharide complexes) in the collagen net.
- It has **no blood supply.**

Bone

This is a specialized, harder form of connective tissue.

- **Calcium and phosphate precipitates** lie within its collagen matrix.
- It has a **blood supply.**

Cartilage Structure

Cartilage is composed of **chondroblasts** and **chondrocytes.**

- Chondroblasts are synthesizing cells, producing matrix and fibers.
- Chondrocytes are mature cartilage cells.

There are three types of cartilage, differing from one another by the fiber content: hyaline, elastic, and fibrous.

Hyaline cartilage

This is the most abundant of the three types. It also represents the initial configuration of newly forming bone. It is found

- in larynx, trachea and bronchi;
- on the articular surfaces of joints.

Elastic cartilage

Differs structurally from hyaline cartilage by a large number of elastic fibers. This feature gives it greater flexibility. It is found

- in the external ear and
- in the epiglottis.

Fibrous cartilage

This form never exists on its own. It forms an intermediate between other forms of connective tissue and contributes tensile strength in addition to flexibility. It is found

- in intervertebral discs and
- in the symphysis pubis.

STRUCTURE

Bone Structure

The mature skeleton contains **trabecular bone** and **cortical bone.**

Trabecular bone (spongy bone)

This is found in a small fraction of the skeleton (ribs, vertebrae, and the ends of long bones).

- It is a network of interwoven partitions (the trabeculae) surrounding cavities that contain red bone marrow.

Cortical bone

The majority of skeletal bone (80 percent) is made up of hard, compact cortical bone. Its central cavity contains yellow marrow.

The fabric of bone is an orderly collagen fiber matrix that has been impregnated with crystals of calcium apatite and with noncrystalline calcium phosphate.

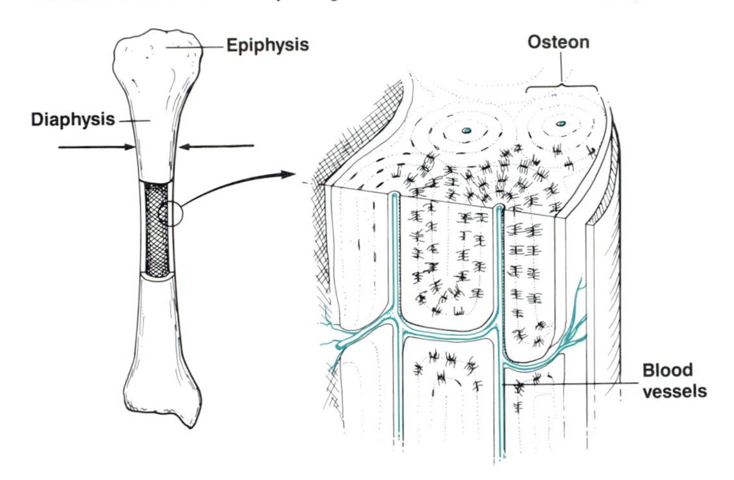

Cortical bone

forms 80 percent of skeletal bone. It is arranged in concentric layers that surround canals (Haversian canals) through which the blood vessels run.

A canal with its concentric lamellae forms an **osteon,** a unit element of bone.

- Within each osteon, small, fluid-filled chambers, called **lacunae,** house one osteocyte each.
- Minute channels, called canaliculi, traverse the bone. The canaliculi allow contact among bone cells and also provide a very large surface area for mineral exchange with extracellular fluid.
- Larger canals (Volkmann's canals) allow blood-vessel interconnections among neighboring Haversian canals.

Bone Cells

The internal and external surfaces of mineralized bone are covered with a layer that contains two types of specialized cells:

- **Osteoblasts** lay down new bone.
- **Osteoclasts** resorb existing bone.

In young, growing bone, the osteoblast-osteoclast layer is about three cells thick. In older bone it is only one cell thick.

A third type of bone cell, the **osteocyte,** is imprisoned in the lacunae.

Bone Cells

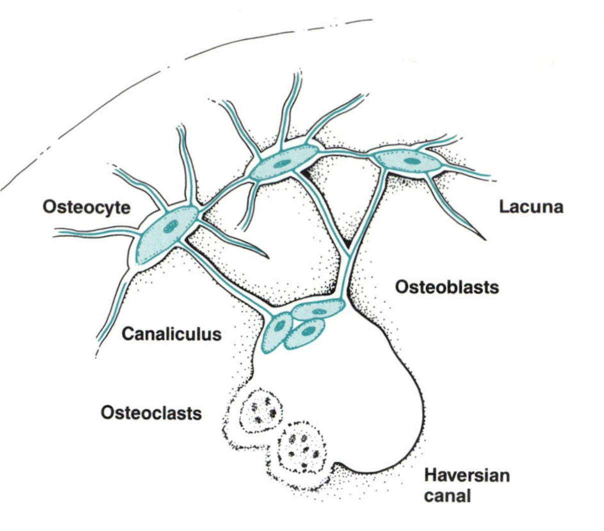

Osteocytes

These are former osteoblasts that have been buried by freshly formed matrix and have lost all synthesizing activity. They are covered by fine branching projections that extend through the canaliculi to communicate with neighboring osteocyte processes and with surface osteoblasts.

- This osteocyte network is closely adjacent to the bone capillary network and communicates directly with the interfibrillar spaces of the bone matrix. It has three probable functions:
  1. Assisting in the exchange of calcium and phosphate between bone and tissue fluid
  2. Maintaining the matrix by extracting plasma constituents and making them available to the matrix
  3. Modifying the matrix to suit physiologic needs

  The precise function of osteocytes is not yet known, but it *is* known that
- when osteocytes die, bone death follows.

Osteoclasts

- These cells derive from the same precursors as circulating monocytes and tissue macrophages.
- They are thought to result from fusion of two or more macrophages. (In consistency with that thought, each of them has between 2 and 100 nuclei.)
- They are highly mobile, and where they attach to a bone surface their secretory activity loosens, fragments, and resorbs bone. (Collagenase, phosphatase, and lysosomal enzymes are involved.)

Osteoblasts

- originate in osteoprogenitor cells, a distinct line of stem cells found in bone marrow;
- synthesize type I collagen and lay down osteoid, a regular array of collagen fibrils and ground substance (mostly chondroitin sulphate and hyaluronic acid).

Formation of new bone results from subsequent deposition of minerals in the osteoid.

SKELETAL HOMEOSTASIS

Bones have two principal functions. They exist to provide **mechanical support for the soft tissues** of the body and to provide the **major body pool of the minerals Ca^{++} and PO_4^{3-}.**

Mechanical Support to Soft Tissues

The mechanical support provided by bone depends on its length, circumference, and composition. Longitudinal growth is an important aspect during childhood. Circumference and composition change throughout life through bone remodeling processes.

Longitudinal bone growth

This occurs only at the ends of long bones by proliferation of epiphyseal cartilage.

- Whereas bone can grow only by apposition (deposition of a new layer upon an existing layer), cartilage grows by both apposition and interstitial enlargement. Both of these result from collagen secretion by **chondroblasts.** Subsequent calcification of the collagen mesh produces new bone.

Chondroblasts

- These are cartilage cells that are analogous to osteoblasts.
- They secrete type II collagen (osteoblasts secrete type I collagen).
- It is believed that **growth hormone is the major stimulus for chondroblast activity.**

When adult height is reached, the epiphyseal plates are sealed from the marrow by a thin plate of bone.

Bone remodeling

Bone undergoes a **continuous process of resorption by osteoclasts and rebuilding by osteoblasts.** It has been suggested that

- remodeling is initiated when a local osteocyte effect alters adjacent bone in such a way that it can be "recognized" by circulating mononuclear phagocytes.
  - Phagocytes accumulate on the bone surface, fuse, and begin resorption.
  - Osteoclast activity continues for about 3 weeks and results in a small tunnel (up to 1 mm in diameter).
  - The tunnel is then invaded by osteoblasts, and their activity over the next several months fills the tunnel with new bone.
- Vitamins, hormones, and mechanical forces influence the rates of activity in osteoblasts and osteoclasts, but the cellular mechanisms of their actions are not yet known.
  - **Growth hormone and mechanical forces are the major long-term regulators; parathyroid hormone (PTH), vitamin D, and calcitonin are the major short-term regulators.**

Bone as a Reservoir for Minerals

The most recently formed bone crystals are a small but rapidly exchanging pool for Ca^{++} and PO_4^{3-}. Mineralized bone, in contrast, is a large but slowly exchanging compartment.

SKELETAL HOMEOSTASIS

Mechanical Support to Soft Tissues

Changes in body size during childhood require both longitudinal and circumferential bone growth.

- Longitudinal bone growth stops when adult height is reached and the epiphyseal plates have been sealed.
- Circumferential adaptations continue at all phases of life.

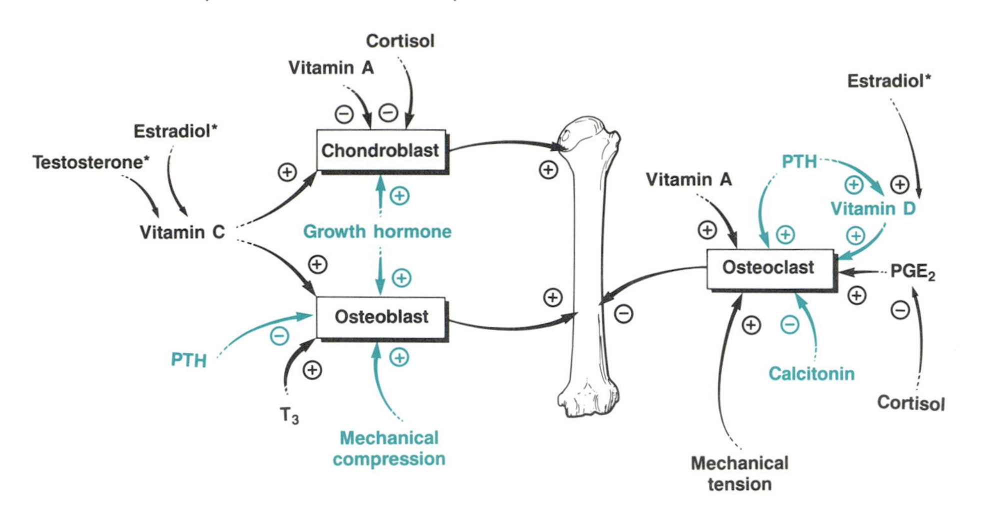

Longitudinal bone growth occurs at the ends of long bones, where epiphyseal cartilage proliferates because of collagen secretion by **chondroblasts.**

***In adults,** gonadotropins exert their effects indirectly, via actions on the metabolism of vitamins D and C. **In children,** they have significant direct stimulatory effects on bone development.

Skeletal homeostasis is achieved by the relative activities of osteoblasts (bone growth) and osteoclasts (bone resorption). Their rates of activity are influenced by vitamins, hormones, and mechanical forces.

PTH = parathyroid hormone
T_3 = thyroid hormone
vitamin D = 1,25-$(OH)_2D_3$ vitamin D
PGE_2 = prostaglandin E_2

Bone as a Reservoir for Minerals

Bone is the major reservoir for body calcium and phosphate.

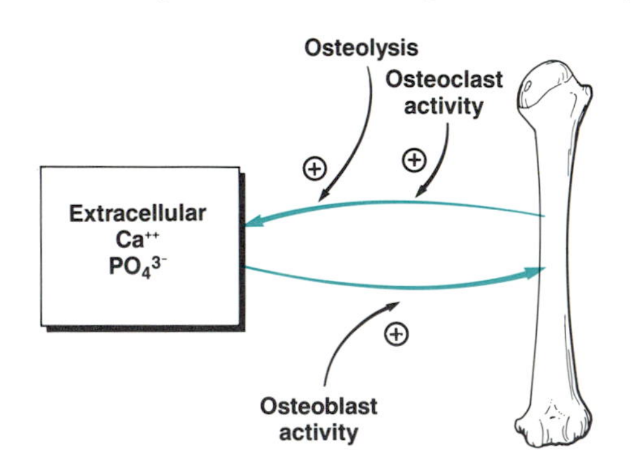

Osteolysis is a process by which osteocyte enzymes extract minerals from the rapidly exchanging mineral pool that surrounds them. Bone resorption by osteoclasts releases minerals from a slowly exchanging pool.

HORMONES IN MINERAL METABOLISM

OVERVIEW

The physiologic mechanisms that regulate mineral metabolism have two concurrent objectives:

- To maintain appropriate plasma concentrations of Ca^{++} and PO_4^{3-}
- To maintain appropriate bone mass

Three hormones are involved:

Vitamin D
Parathyroid hormone
Calcitonin

They act on three tissues:

Intestinal mucosa
Nephron
Bone

The physiology of parathyroid hormone and that of calcitonin are described in Chapter 9, Endocrinology.

Vitamin D

Vitamin D is a vital factor in Ca^{++} and PO_4^{3-} homeostasis. It is **a steroid hormone that is not secreted by an endocrine gland.** It exists in several biochemical forms that vary in their biologic potency.

The liver

This organ concentrates dietary vitamin D (D_2) and vitamin D produced in the skin (D_3). Enzymatic activity produces 25-$(OH)D_3$.

Renal tubular cells

Conversion of 25-$(OH)D_3$ to a more potent form occurs by hydroxylation in renal tubular cells under the control of several factors:

Short-term control

- This is chiefly due to the effects of PTH, Ca^{++}, and PO_4^{3-}. They act by their respective influences on the levels of 1-hydroxylase. This enzyme produces the active hormone, 1,25-$(OH)_2D_3$.

Long-term control

- Estradiol is the major factor. It influences the level of 24-hydroxylase and, thereby, governs the proportion of biologically active vitamin D [1,25-$(OH)_2D_3$] vs. inactive vitamin D [24,25-$(OH)_2D_3$] produced by the kidney.

HORMONES IN MINERAL METABOLISM

OVERVIEW

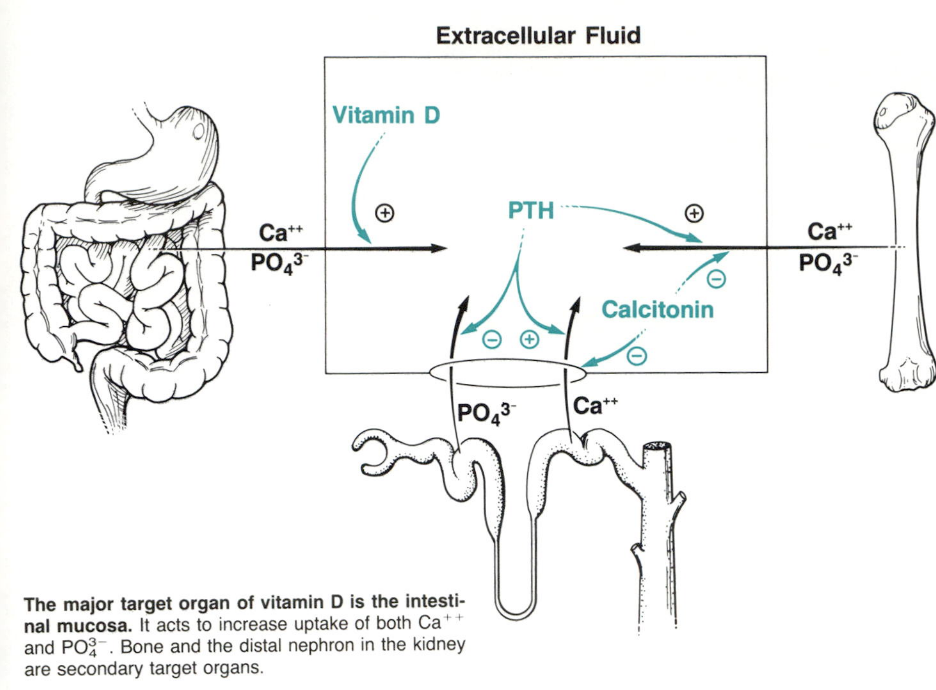

Parathyroid hormone (PTH) acts chiefly on bone and both the proximal and distal convoluted tubules of the kidney. **In the proximal tubule,** PTH depresses PO_4^{3-} reabsorption. **In the distal tubule,** PTH enhances Ca^{++} reabsorption.

Calcitonin depresses bone resorption and nephron reabsorption of both Ca^{++} and PO_4^{3-}. Therefore, it tends to decrease the extracellular concentrations of these minerals. Its effects on plasma concentrations of Ca^{++} and PO_4^{3-} are small under physiologic circumstances.

The major target organ of vitamin D is the intestinal mucosa. It acts to increase uptake of both Ca^{++} and PO_4^{3-}. Bone and the distal nephron in the kidney are secondary target organs.

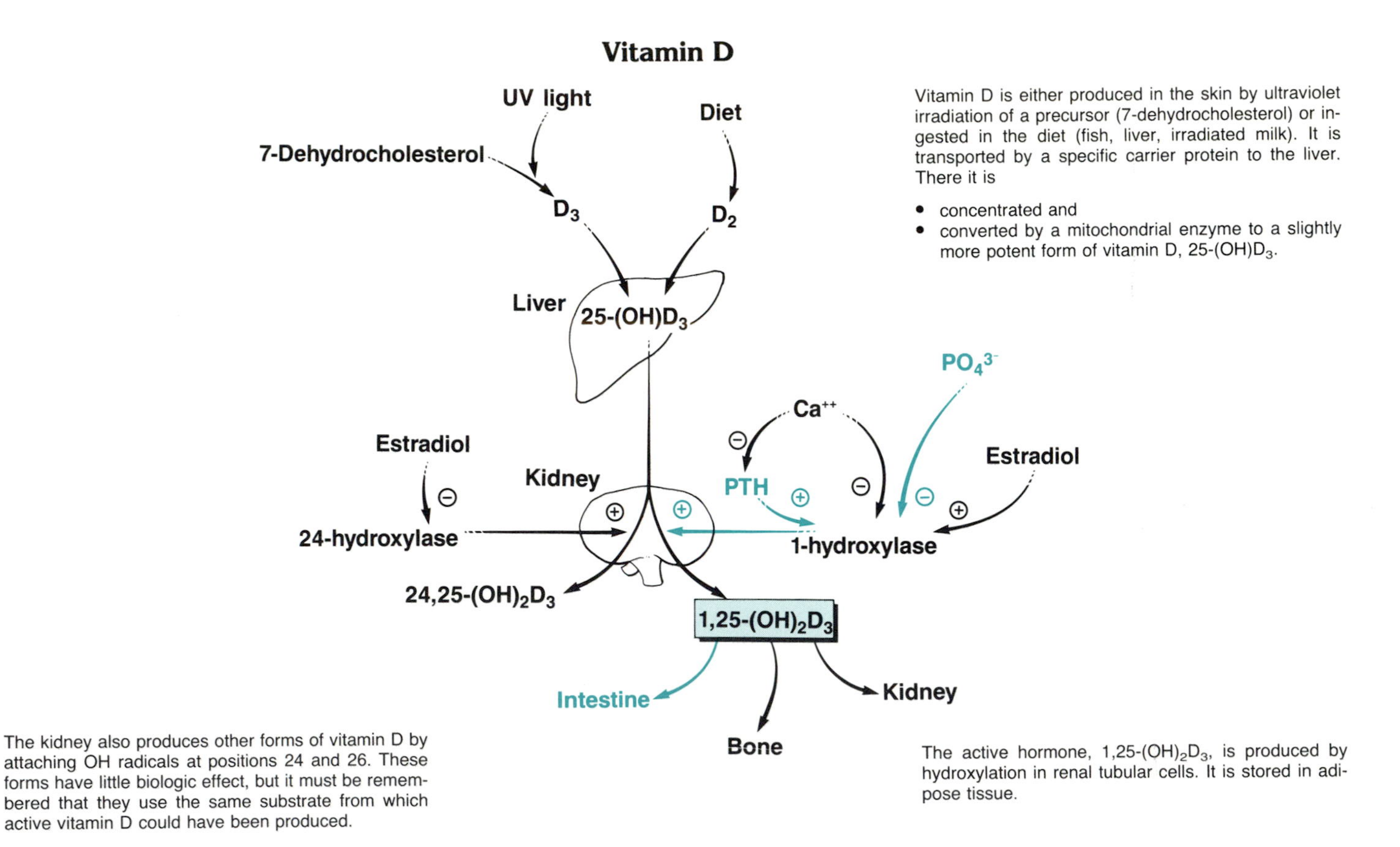

Vitamin D is either produced in the skin by ultraviolet irradiation of a precursor (7-dehydrocholesterol) or ingested in the diet (fish, liver, irradiated milk). It is transported by a specific carrier protein to the liver. There it is

- concentrated and
- converted by a mitochondrial enzyme to a slightly more potent form of vitamin D, 25-(OH)D_3.

The kidney also produces other forms of vitamin D by attaching OH radicals at positions 24 and 26. These forms have little biologic effect, but it must be remembered that they use the same substrate from which active vitamin D could have been produced.

The active hormone, 1,25-$(OH)_2D_3$, is produced by hydroxylation in renal tubular cells. It is stored in adipose tissue.

CALCIUM AND PHOSPHATE

Regulation of Extracellular Ca^{++} Concentration

Parathyroid hormone (PTH), vitamin D, and calcitonin are the major hormones involved in the regulation of ionized calcium in extracellular fluid.

Parathyroid hormone and vitamin D

Regulation of extracellular Ca^{++} concentration is accomplished by an **inverse relationship between plasma Ca^{++} concentration and the secretion of PTH.** As a result of this direct effect of Ca^{++} on PTH,

- a fall in plasma Ca^{++} concentration increases PTH levels;
- PTH, in turn, increases the production of physiologically active vitamin D in the kidney;
- vitamin D stimulates reabsorption of Ca^{++} from bone, intestine, and nephron.

Calcitonin

This hormone acts as an **antagonist to PTH and vitamin D** in calcium homeostasis. However, its effect on Ca^{++} reabsorption is directly related to the basal rate of bone turnover, which is low in adult humans.

Regulation of Extracellular PO_4^{3-} Concentration

The major influence on PO_4^{3-} regulation is exerted by vitamin D. PTH is involved only secondarily.

Vitamin D

Extracellular phosphate concentration inversely affects renal production of biologically active vitamin D. Hence,

- if extracellular [PO_4^{3-}] increases, vitamin D production decreases;
- when vitamin D levels decrease, then phosphate absorption from intestinal mucosa and from bone decreases as well.

Parathyroid hormone

Note that Ca^{++} reabsorption from intestine and bone is also decreased when vitamin D is depressed by high extracellular [PO_4^{3-}]. As a result,

- PTH secretion is stimulated by a decrease in extracellular Ca^{++} concentration;
- PTH acts on renal tubular cells to depress renal reabsorption of PO_4^{3-};
- PTH also promotes renal vitamin D production.

◆ The opposing effects of extracellular [PO_4^{3-}] and [Ca^{++}] tend to normalize intestinal and bone reabsorption.

CALCIUM AND PHOSPHATE

Regulation of Extracellular Ca^{++} Concentration

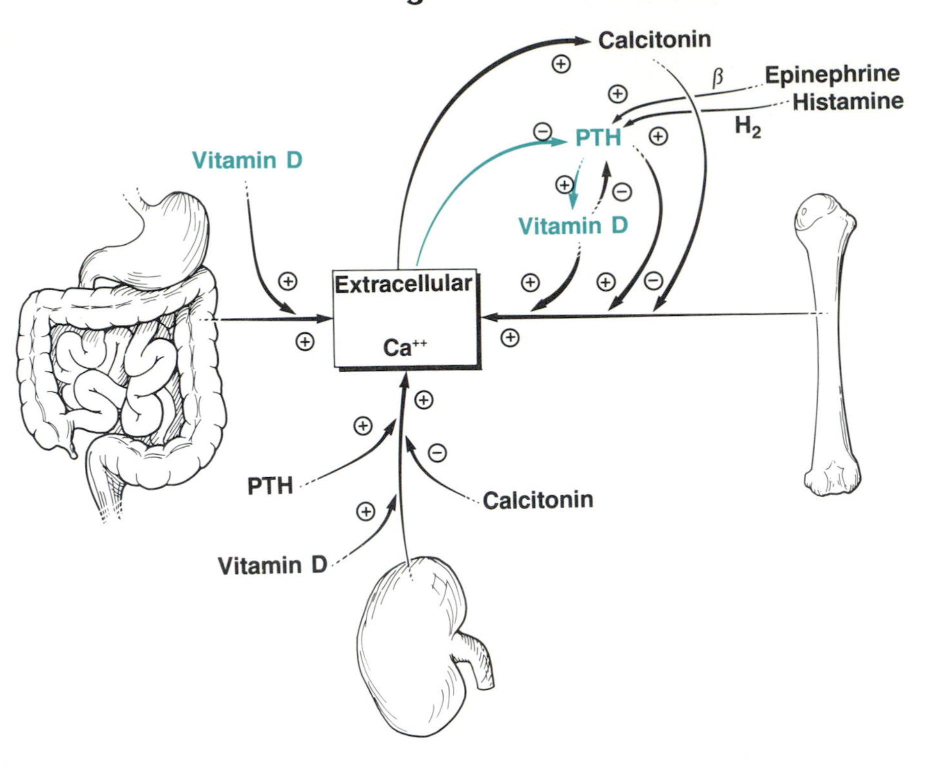

A decrease in extracellular $[Ca^{++}]$ increases the secretion of PTH. PTH, in turn, stimulates production of physiologically active vitamin D in the kidney. **Vitamin D is the most significant promoter of Ca^{++} reabsorption from the GI tract, bone, and nephron.**

Regulation of Extracellular PO_4^{3-} Concentration

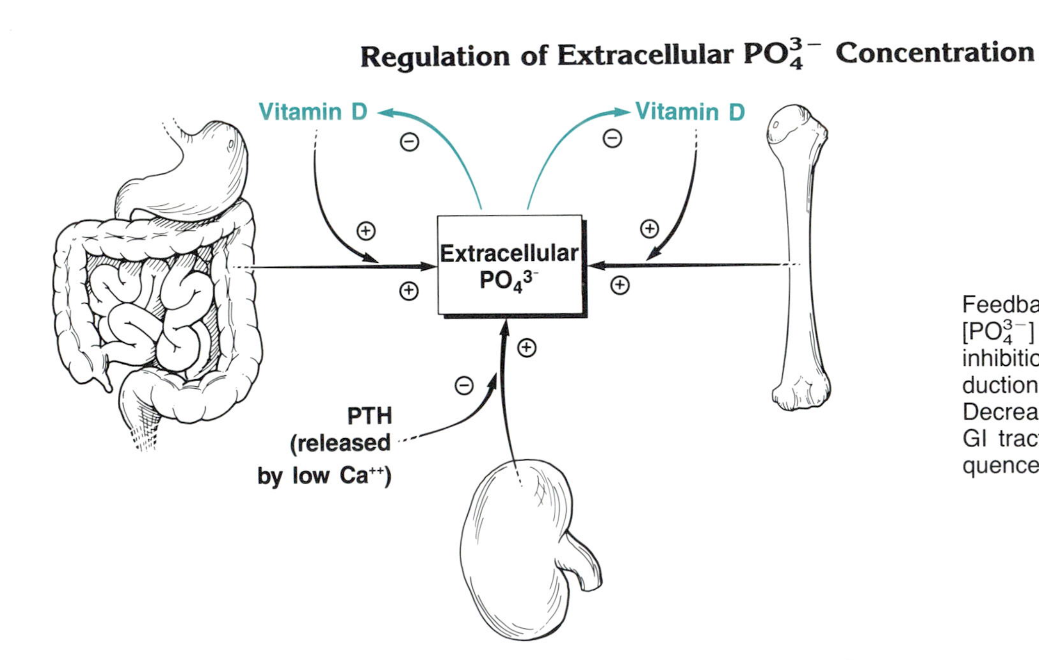

Feedback regulation of extracellular $[PO_4^{3-}]$ involves, in the first instance, the inhibition of renal tubular vitamin D production when extracellular $[PO_4^{3-}]$ rises. Decreased reabsorption of PO_4^{3-} from the GI tract and from bone will be a consequence of diminished vitamin D levels.

11

FETAL AND PERINATAL PHYSIOLOGY

OVERVIEW

Placenta

Fetus and mother are reversibly joined by the placenta. It is a fetal organ.

Formation

The placenta is formed by apposition of the chorion (a highly vascularized membrane that surrounds the embryo) to the epithelium of the uterine mucosa. It begins to form by about the 20th day of gestation.

- Chorionic villi invade the uterine mucosa.
- The villi penetrate and erode the endometrial capillaries and come to lie within the resulting network of interconnecting vascular pools, which has replaced the discrete capillary channels.
- Microvilli develop on each chorionic villus.
- The microvilli progressively fill with fetal capillaries.

Functions

The three main functions of the placenta are **circulatory, endocrine,** and **membrane.**

Circulatory function

- Fetal blood is channeled through pools of maternal blood.

Endocrine function

At least four hormones are synthesized by the human placenta:

- Chorionic gonadotropin
- Chorionic somatomammotropin
- Progesterone
- Estrogen

Membrane function

The placenta acts like a complex biologic membrane.

- It protects the fetus from being rejected by an invasion of maternal T-cells.
- It selectively transports immunoglobulin G to the fetus for protection in the immediate postnatal period.

Fetal Age and Growth

Both body weight and crown-rump length increase linearly up to the normal term (40 weeks). Then the rate of weight gain decreases.

OVERVIEW

Placenta

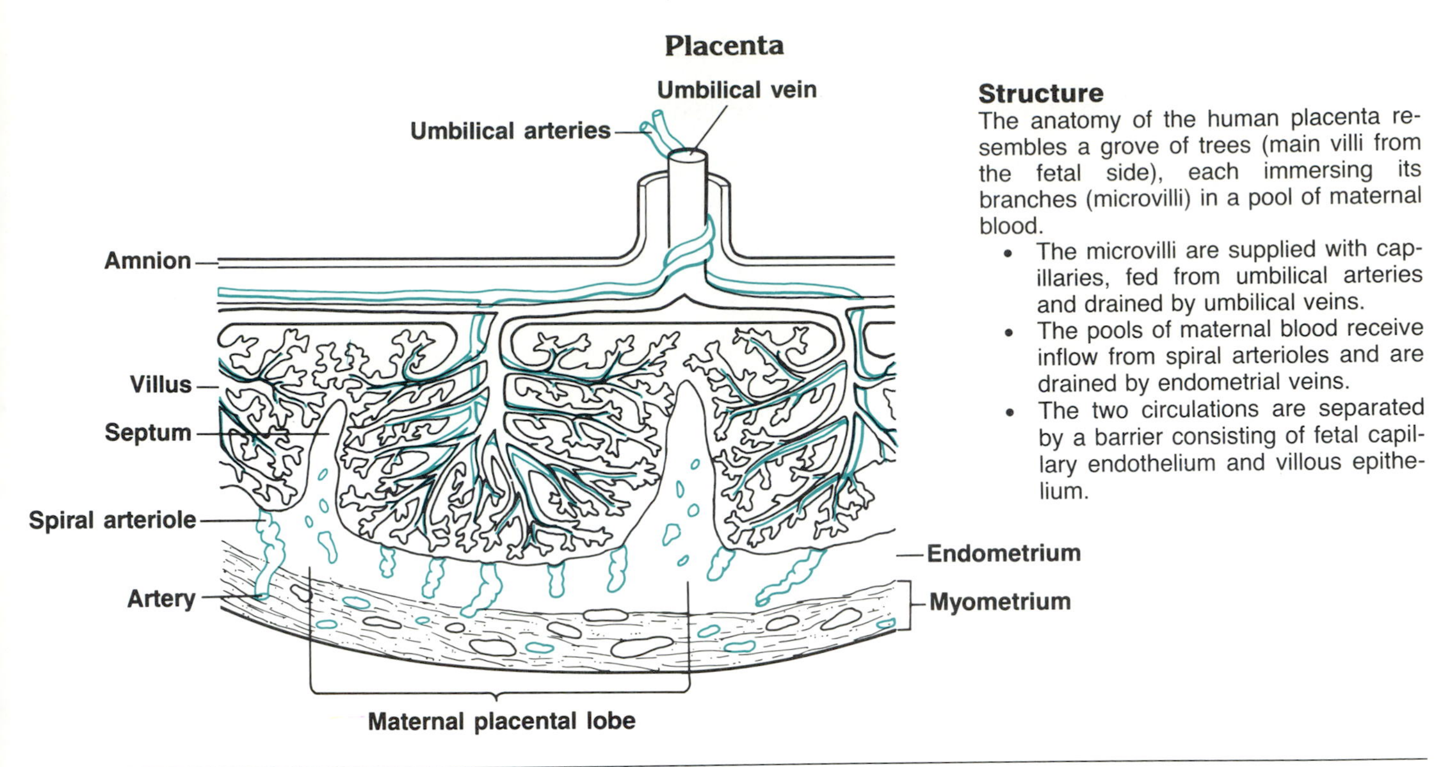

Structure

The anatomy of the human placenta resembles a grove of trees (main villi from the fetal side), each immersing its branches (microvilli) in a pool of maternal blood.

- The microvilli are supplied with capillaries, fed from umbilical arteries and drained by umbilical veins.
- The pools of maternal blood receive inflow from spiral arterioles and are drained by endometrial veins.
- The two circulations are separated by a barrier consisting of fetal capillary endothelium and villous epithelium.

Modified from Walsh SZ, Meyer WW, Lind J. The human fetal and neonatal circulation. Springfield IL: Charles C Thomas, 1974.

Fetal Age and Growth

Fetal age is sometimes given as the time elapsed since the onset of the last menstrual period (menstrual age) and sometimes as the time elapsed since the last presumed ovulation (fertilization age or gestation age). Fertilization age is approximately 2 weeks less than menstrual age.

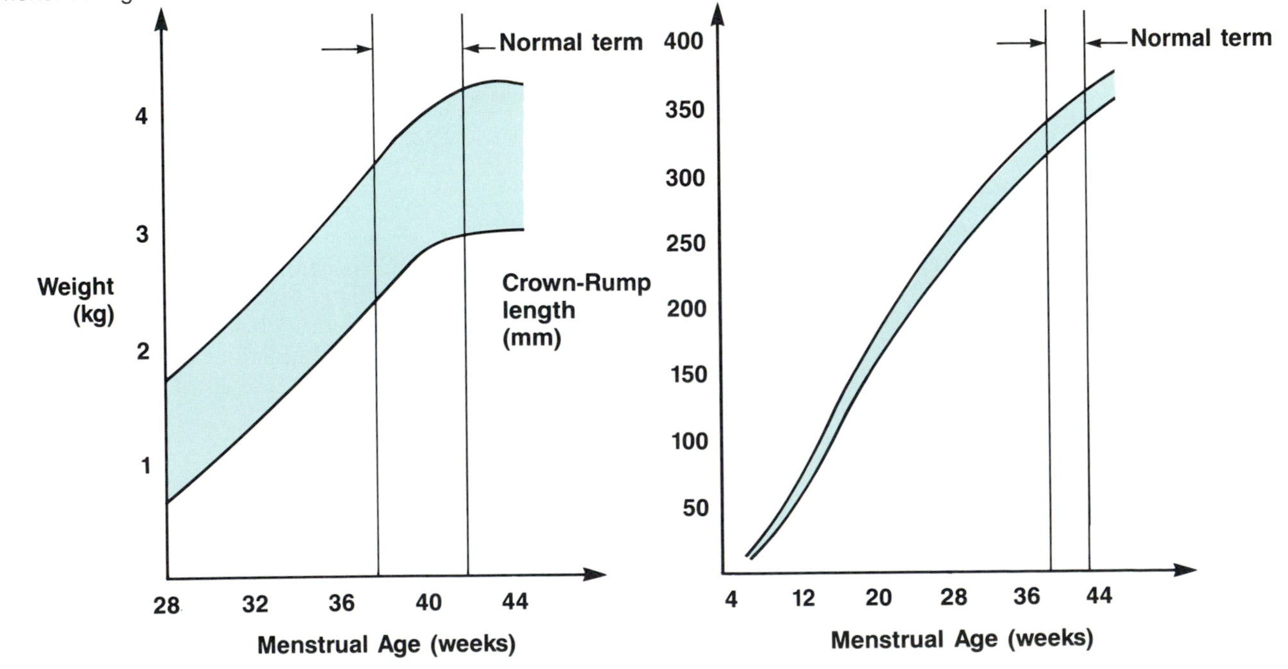

IMMUNITY

T-Cells

The placenta protects the fetus from immunologic stress. Nevertheless,

- **lymphocytes appear in the fetal thymus and blood by the ninth week.**
  - One or two weeks later, some of these cells are identifiable as T-cells.
- By the 15th week, about 65 percent of the lymphocytes in the fetal thymus are T-cells, and T-cells are beginning to appear in other tissues. Their immunologic potential at that point is sufficient to do damage to red cells of another species.

B-Cells

B-cells **appear about a week later than T-cells** (in the liver).

- B-cells synthesizing immunoglobulin M (IgM) appear first,
- then IgG-synthesizing cells, and
- finally IgA-synthesizing cells (to a limited extent and only near birth).

Plasma cells are not normally found in the fetus, and **immunoglobulins produced by the fetus are sparse unless there is exposure to pathogens.** Such exposure can elicit formation of plasma cells and fetal IgM antibodies by the 28th week.

Immunoglobulins

Transfer of immunoglobulins across the placental barrier depends on structural and functional aspects of the placenta as well as on a receptor-mediated mechanism.

- Rate and selectivity of transfer are influenced less by the thickness of the barrier than by the ability of fetal capillary endothelial cells, or of trophoblasts covering the chorionic villi, to transport internalized protein molecules without denaturing them.

Isoimmunization

Antibodies to fetal antigens have been detected in maternal blood, but their concentration is generally low.

- In about 5 percent of mothers who have had multiple pregnancies by the same father, maternal antibodies to the infant's paternally derived IgG occur. Most of these are of a size that can cross the placenta and enter the fetus. There they may suppress fetal IgG levels sufficiently to cause fetal hypogammaglobulinemia, but even after the fourth such pregnancy the levels in maternal blood are low. A noted exception is Rh isoimmunization.

Rh isoimmunization

- Fetal erythrocytes always enter the blood of the mother.
- If the fetal red cells carry Rh-positive antigen and the mother is Rh-negative, then the mother produces antibodies first of the IgM class and eventually of the IgG class.
- IgG antibodies can be transported to the fetus and agglutinate fetal red cells.

- The danger of this hemolytic disease of the newborn is small in a first Rh-mismatched pregnancy because the degree of antibody formation is small. The danger increases greatly with subsequent pregnancies.

Immunoglobulins

In humans, only IgG is transported across the placenta to any extent. Therefore,

- the unstimulated serum profile in the fetus is dominated by IgG from maternal blood.

The extent of IgG transport is such that

- the fetal blood level of IgG at birth is near the adult level even though fetally produced IgG contributes only about 5 percent of that level at birth.

It is IgG of maternal origin that provides immunologic protection to the newborn during the first month after birth, while fetal IgG synthesis is beginning to increase.

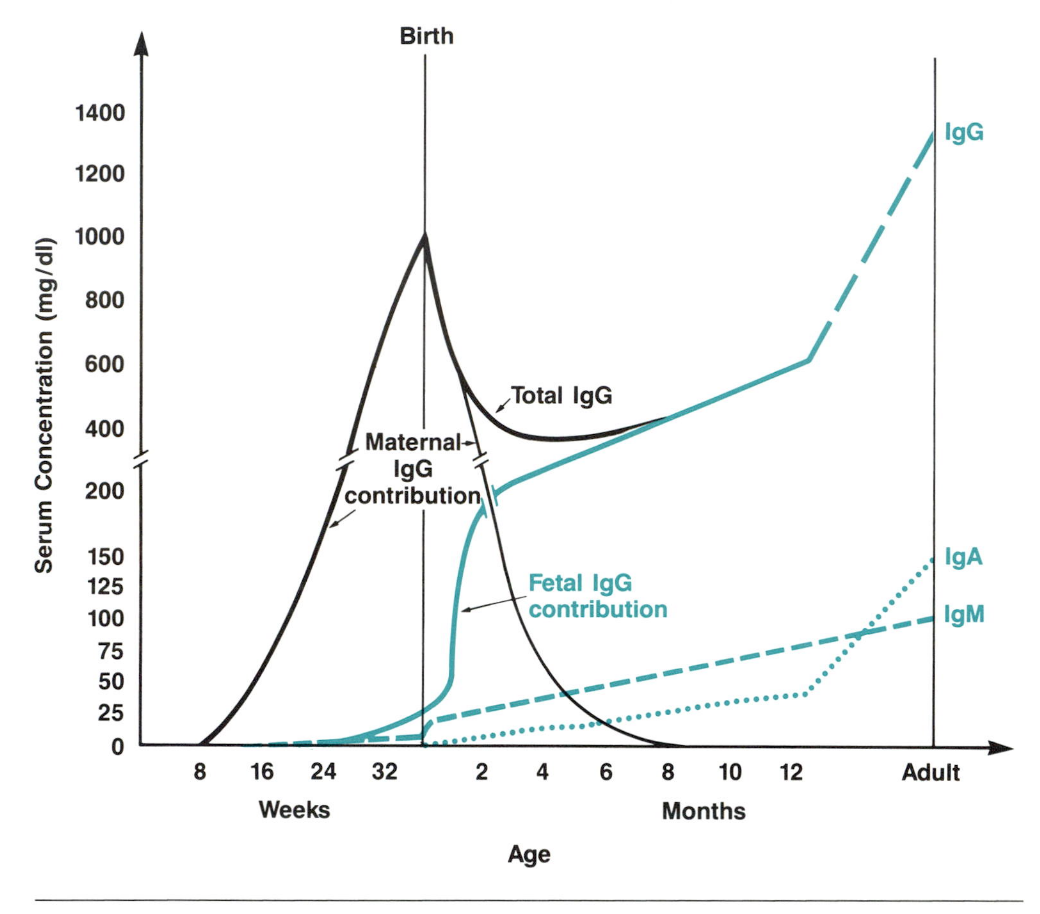

Modified from Stave U. Perinatal physiology, 2nd ed. New York: Plenum Medical Book, 1978.

RESPIRATION

Gas Transport and Exchange

Blood provided to the fetus from the placenta has a P_{O_2} much lower than that of adult arterial blood. Two factors allow the fetus to exist under hypoxic conditions: **fetal hemoglobin** and a relatively high **cardiac index.**

Fetal hemoglobin

Fetal blood has a higher hemoglobin concentration than maternal blood, and fetal hemoglobin has higher O_2 carrying capacity than does adult hemoglobin.

- Fetal red cells have lower concentrations of 2,3-diphosphoglycerate (DPG), the compound that competes with oxygen for heme binding sites. This shifts the fetal O_2 dissociation curve to the left relative to maternal blood and maximizes the amount of O_2 carried by fetal erythrocytes because they return fully saturated from the placenta. The shift to the left is only partly offset by the greater acidity of fetal blood (due to higher P_{CO_2}).

Cardiac index

Fetal tissues are perfused at high rates because fetal cardiac output (per kilogram of body weight and relative to O_2 demands) is two to four times the adult level.

Gas Transport and Exchange

Fetal gas exchange occurs in the placenta because the fetal lung is filled with fluid and cannot perform respiratory functions.

Oxygenated blood is provided by the uterine artery
CO_2-rich blood is removed by the uterine vein.

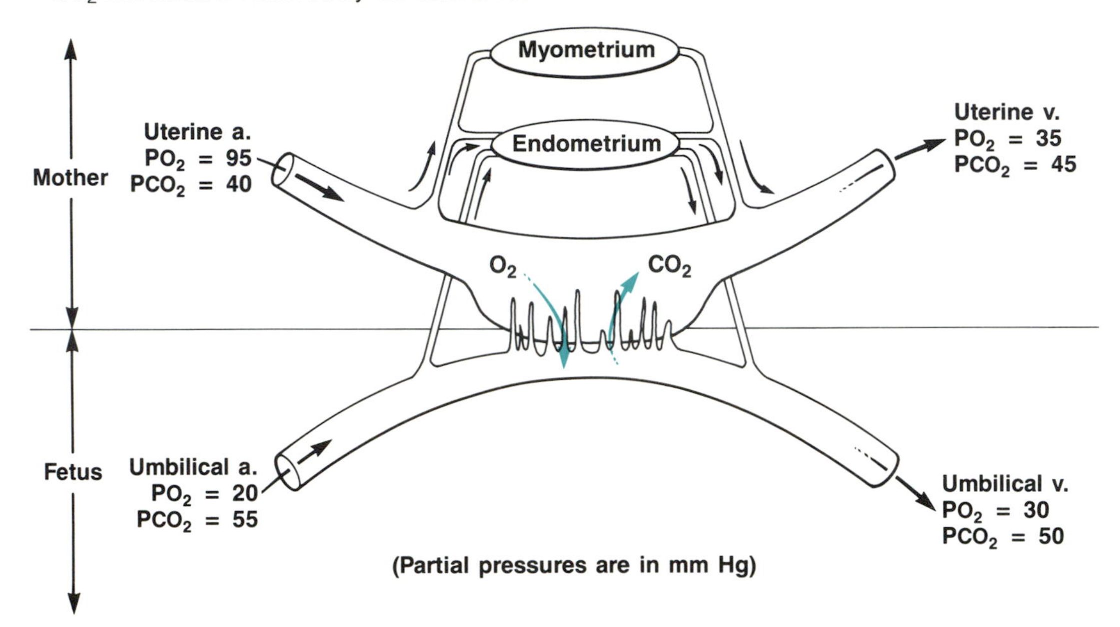

Fetal blood with low Po_2 is brought to the placenta by the umbilical artery.
Oxygenated blood reaches the fetus via the umbilical veins.

The Po_2 of umbilical venous blood is much lower than that of adult arterial blood because of the following:

1. The high rate of O_2 consumption that is associated with high metabolic activity in the placenta causes uterine venous Po_2 to be low.
2. Since O_2 diffuses freely across the whole length of the placental barrier, the nature of the anatomic coupling between maternal and fetal vessels dictates that the Po_2 of umbilical venous blood can, at best, be equal to the Po_2 of uterine venous blood. However, because the rates of blood flow through the two placental circulations (maternal and fetal) are high and because O_2 permeability is not infinite, umbilical venous Po_2 must be lower than uterine venous Po_2.

RESPIRATION—CONT'D.

The First Breath

The fully prepared respiratory mechanisms are activated during birth by several developments. These include

- profound hypoxia and acidemia (resulting from interruption of umbilical blood flow);
- generalized activation of the autonomic nervous system (resulting from cold stimulation, release from submersion, tactile stimuli, and pain).

Maintenance of Respiration

Continuation of respiration involves maintenance of **lung expansion** as well as maintenance of the respiratory **rhythm.**

Maintenance of lung expansion

Once the lungs have filled with air, they are prevented from collapsing by a balance between total alveolar surface tension and pleural surface tension. **Surfactant** plays an important role in this balance.

Surfactant

- Surfactant is a mixture of phospholipids, neutral lipids, and proteins.
- It is secreted from alveolar cells from about the eighth week of fetal life and increases steeply between the 20th and 30th weeks.
- Inadequate surfactant (in a premature baby) or surfactant of the wrong composition may cause lung collapse and lead to "respiratory distress syndrome," a progressive respiratory failure.
- Blood levels of corticosteroids, perhaps acting in synergy with estrogen, exert a major positive influence on surfactant formation.

Maintenance of rhythm

Once initiated, breathing is maintained and regulated by medullary and **peripheral chemoreceptors.**

- Normalization of blood gases and of acid-base balance to adult levels occurs within 5 to 10 days of birth.
- Episodes of apnea, often accompanied by bradycardia, are frequently seen in the newborn, particularly during environmental stresses such as large fluctuations in ambient temperature. This is thought to be a result of early imprecision in homeostatic mechanisms.

The First Breath

Before the first breath occurs

Although respiration is initiated suddenly at birth, all mechanisms associated with breathing have developed to a state of readiness by then.

- Neuromuscular networks that coordinate breathing and swallowing have been developed.
- Pharyngeal receptors and chemoreceptors are ready to respond to their specific stimuli.
- Pulmonary stretch receptors are functional.
- Medullary rhythmicity is established.
- Occasional "respiratory" chest movements, including deep sighs, have taken place in utero.

CARDIOVASCULAR SYSTEM

The cardiovascular system is the first organ system to become fully functional because it is required for the transport of gases, nutrients, and waste products once the embryo is larger than a few millimeters.

The heart begins to pump late in week 3, and at eight weeks its development and that of the major blood vessel trunks are complete apart from an opening in the atrial septum (foramen ovale).

Hemodynamics

- About **30 percent of vena caval flow enters the left atrium directly via the foramen ovale** and, thus, bypasses the lungs.
- The remainder enters the right ventricle and is pumped into the pulmonary trunk.
  - **Only about 40 percent of right ventricular output reaches the lungs** because of their high vascular resistance (hypoxic vasoconstriction as well as high tissue pressure in the collapsed state).
  - The remaining 60 percent is diverted to the aorta by the **ductus arteriosus.** (This distribution of flow indicates that right ventricular systolic pressure exceeds left ventricular systolic pressure in the fetus.)
- **Most of the aortic flow in the fetus comes from the right heart.**

Hemodynamics

The anatomy of the fetal circulation is determined by the requirement for high cardiac output (because of low P_{O_2} in fetal blood) in the face of high pulmonary and hepatic vascular resistance (because these organs are not active in the fetus).

The requirement for high cardiac output is met by low-resistance shunts

- around the liver (i.e., the ductus venosus) and
- around the lung (i.e., the foramen ovale and the ductus arteriosus).

The fetal circulation brings oxygenated blood from the placenta by the umbilical vein to the fetal inferior vena cava partly via the ductus venosus (a large-caliber, direct continuation of the umbilical vein within the right half of the liver) and partly via the portal venous network that vascularizes the liver and enters the hepatic veins.

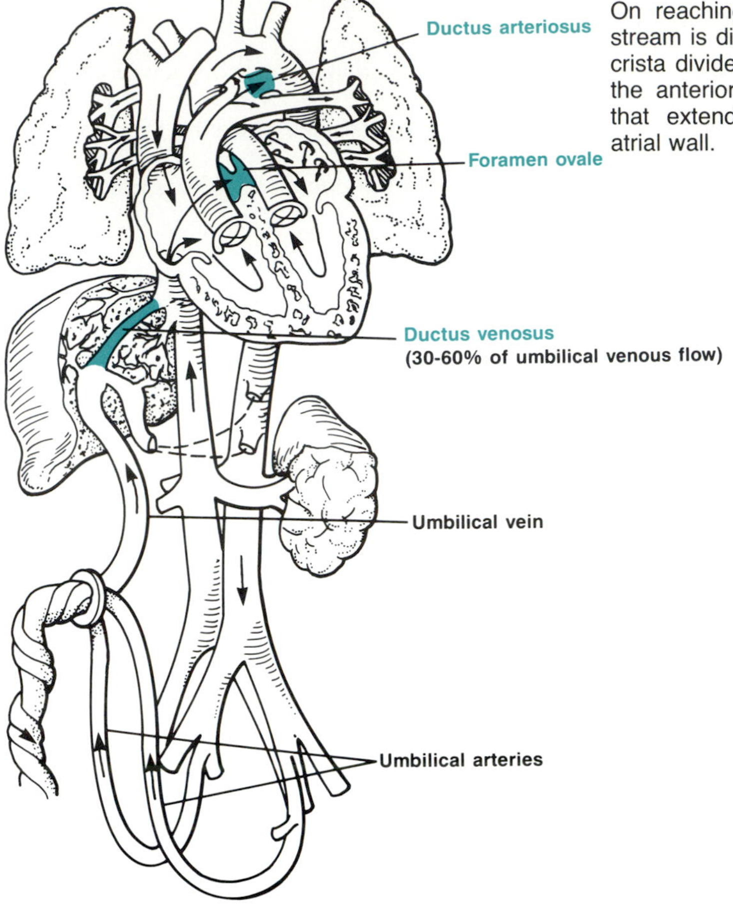

On reaching the right atrium the blood stream is divided into two streams by the crista dividens, an anatomic projection of the anterior edge of the foramen ovale that extends posteriorly almost to the atrial wall.

Modified from Marieb EN. Human anatomy and physiology. Redwood City, CA: Benjamin Cummings, 1989.

Hemodynamics—Cont'd.

Changes at birth

Two events that occur at birth cause significant hemodynamic changes:

- **Cord tying obliterates** the parallel, **low-resistance** placental flow path that was responsible for high fetal caval pressure.
- Exposure of vascular smooth muscle in the ductus arteriosus to **high-Po_2 blood causes intense constriction of the ductus.**

The ductus venosus also closes. Closure of this vessel, which is devoid of vascular smooth muscle, results from retraction and narrowing secondary to tying of the umbilical vein. The causes are not clear.

Cardiovascular Control

Rudimentary mechanisms for short-term circulatory control are in place from an early stage of gestation. They govern

- effective distribution of cardiac output as well as
- adaptive changes in cardiac output.

Distribution of cardiac output

The respective sensitivities of individual vascular beds to local Po_2 determine distribution.

- **Even in the fetus the cerebral and coronary circulations show marked vasodilator responses to hypoxia.**

Autonomic nervous control

Cardiac parasympathetic innervation is complete at birth, but cardiac sympathetic innervation (and possibly sympathetic innervation of other organs) is not.

- Heart rate and rhythm show large swings in the newborn:
  - Rate fluctuations are generally associated with external stimuli such as sounds, hiccups, sneezes, sucking, and so on, which the adult system has "learned" to disregard.
  - Rhythm fluctuations appear to arise from atrial premature beats in conjunction with a higher conduction velocity through the a-v node of the newborn. (The adult is protected from the development of ventricular arrhythmias following premature atrial beats by the delay through the conduction tissue. Infants can conduct atrial rhythms of up to 400 per minute, and in them atrial premature beats can result in repetitive ventricular beats, perhaps even fibrillation.)
- Some degree of autonomic nervous control over stroke volume is established during the last trimester. The degree of control is not sophisticated.
- Changes in skin blood flow do occur after local application of thermal stimuli, but the responses are sluggish.
- **Chemoreceptor and baroreceptor reflexes are present at birth.**

Changes at birth

Although anatomic closure of all fetal vascular shunts may take up to 8 weeks, birth is associated with immediate hemodynamic changes that are initiated by two primary events:

(i) Closure of foramen ovale

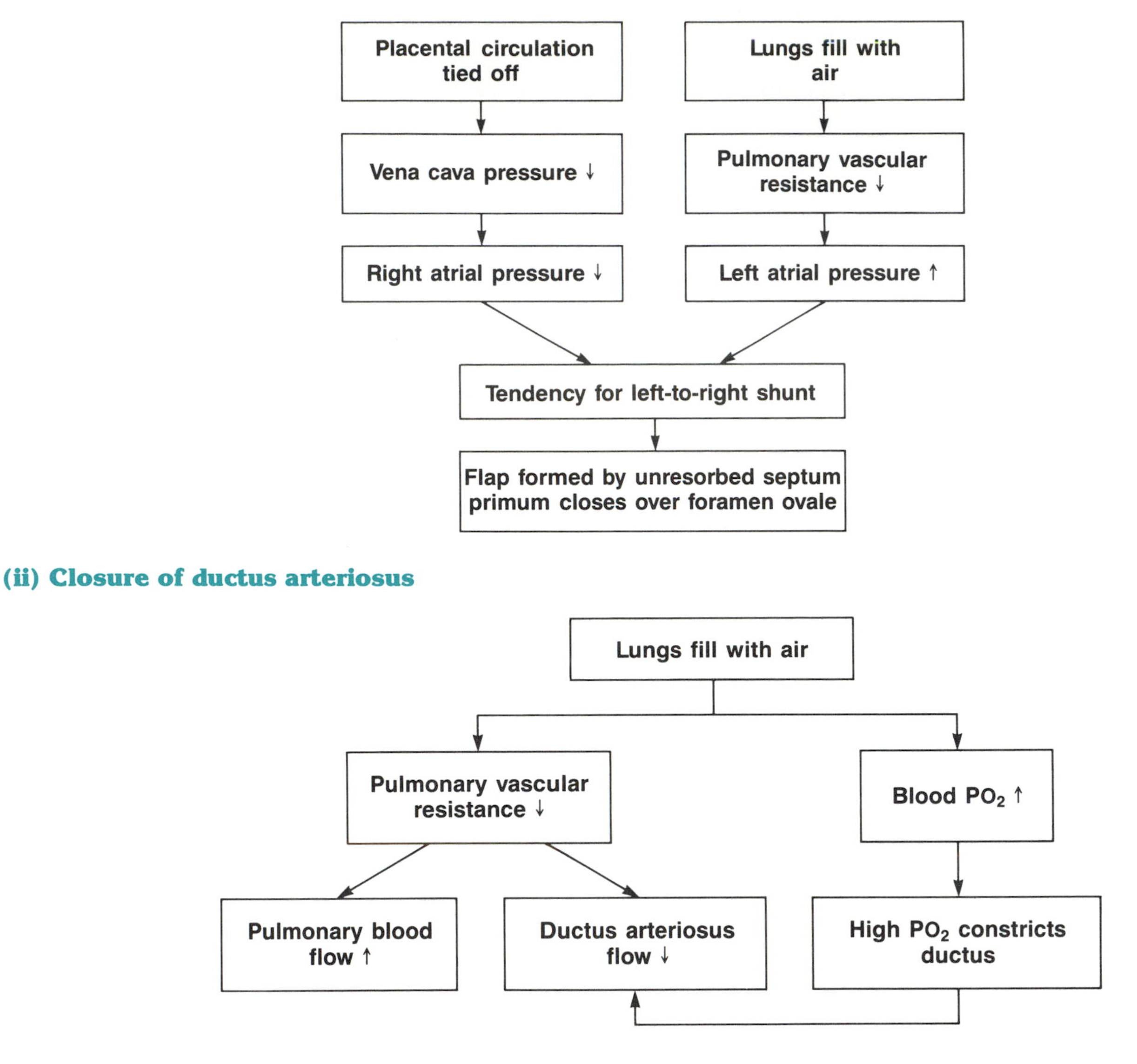

Neonatal vs. fetal system

No placental circulation
Higher pulmonary blood flow
The foramen ovale is closed. Therefore,

- right and left heart pump in series, not in parallel.

Peripheral vascular shunts are closing. Therefore,

- total peripheral vascular resistance is rising. Therefore,
  —systemic arterial blood pressure is rising.

Neonatal vs. adult system

Structure
The right ventricular wall is thicker than the left.

Hemodynamics

- Pulmonary vascular resistance is higher.
- Pulmonary arterial pressure is higher.
- Cardiac output is higher.

Control
Autonomic nervous control is not sophisticated because cardiac sympathetic innervation (and possibly sympathetic innervation of other organs) is not complete.

FLUID AND ELECTROLYTE BALANCE

Fluid Compartments

The fetus receives fluids and electrolytes

- partly via the placental circulation and
- partly by swallowing amniotic fluid.

It can "lose" water and electrolytes (via the skin, the GI tract, or the kidney) only to the amniotic fluid. Therefore, it has no fluid and electrolyte intake or output that is not in some way equilibrated with maternal body fluids. It is clear that

◆ **fetal fluid and electrolyte metabolism is maintained by the mother.**

Kidney Function

Nephrons develop late in fetal life:

- By the fifth month of gestation the kidney contains only 30 percent of its final number of nephrons.
- At birth, nephrogenesis is complete, but the more recently formed nephrons are not yet mature. As a result,

◆ **fetal kidney function is limited, and the kidney of even the newborn is functionally immature.**

The limitations arise

- partly from physical factors (low hydrostatic pressures and short nephrons) and
- partly from immaturity of transport mechanisms (limited responsiveness to aldosterone and to ADH).

ADH insensitivity is responsible for the limited ability of the newborn to produce concentrated urine.

Although kidney function in the newborn is not sufficient to cope with adult fluid and electrolyte stresses during the first few weeks after birth, it is **perfectly suited to the usual fluid and electrolyte loads of the newborn.** The kidneys reach maturity within weeks and then grow until adolescence along with the rest of the body.

FLUID AND ELECTROLYTE BALANCE

Fluid Compartments

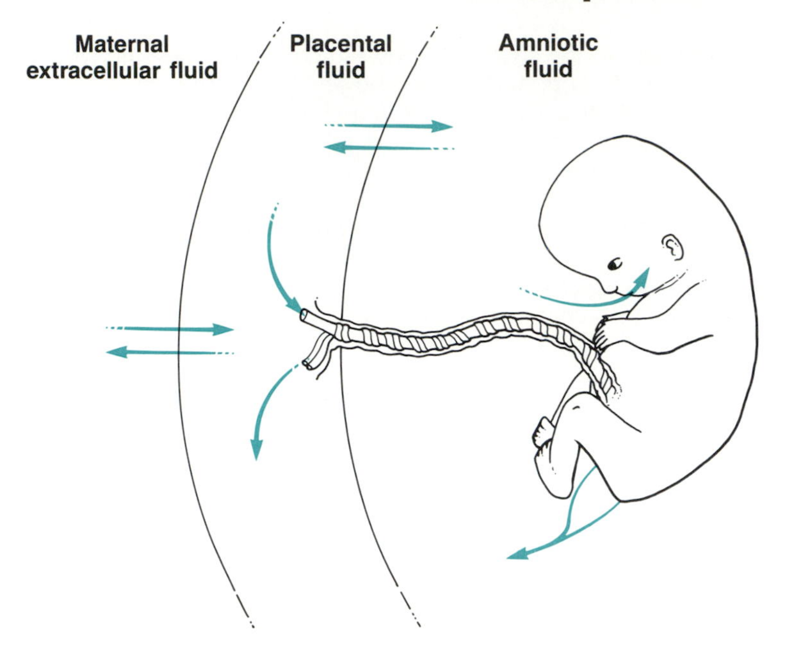

Intrauterine fluids amount to 3½ L at term. They are divided into three compartments:

- Placental fluid
- Amniotic fluid
- Fetal body fluids.

Each is iso-osmotic with maternal extracellular fluid from which it was derived.

Kidney Function

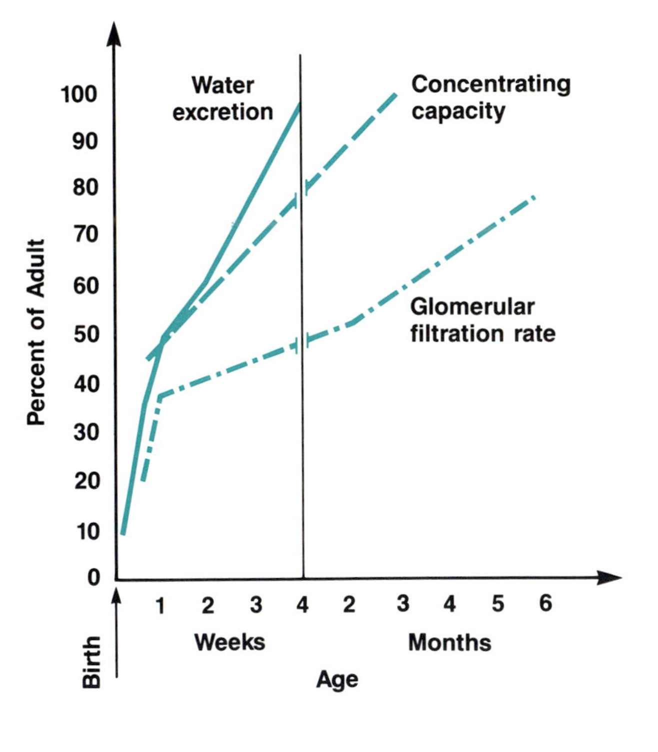

The kidney of the newborn is functionally immature. It has

- a low rate of glomerular filtration and
- a limited capacity to excrete loads of water, sodium, and hydrogen ions.

GASTROINTESTINAL FUNCTION

Motility

Spontaneous rhythmic activity begins in the small intestine by the sixth or seventh week of fetal life and spreads gradually to more distal portions of the bowel.

By about the fourth month the fetus begins to swallow, transport, and digest amniotic fluid. However, the vigor and extent of rhythmic activity are not at adult levels until the enteric plexuses complete their development some time after birth. Therefore,

- **food takes longer to pass through the GI tract of infants than through that of adults.**

Digestion and Absorption

Carbohydrates

The ability to transport monosaccharides and break down disaccharides is established early and reaches its full, adult level by 1 year after birth.

- **Active glucose transport proceeds even under anaerobic conditions in the fetus** (but not in the adult).
  (This important feature of prenatal physiology ensures the maintenance of cell nutrition even under the hypoxic conditions of intrauterine life.)
  - After birth, some enzyme levels continue to increase (e.g., maltase), and others may decrease (e.g., lactase, unless the child belongs to a population in which milk has been a significant dietary component for many centuries).
- **The ability of the infant to digest complex starches is limited.** The reason is that enzymes such as amylase develop slowly.

Proteins

The mechanisms required for protein digestion and absorption are

- proteolytic activity in stomach and pancreas, and
- active transport of amino acids in the small intestine.

These mechanisms are **weak at birth and require about 3 years to develop** to adult levels.

- They develop more rapidly if the infant's diet has high protein levels.

Fats

Digestion of fat requires lipases and bile salts. A newborn has enough of these to absorb about 80 percent of the fats that are normally presented in the diet.

- **Short-chain fatty acids are absorbed effectively during the first month after birth.**
- The ability to deal with long-chain acids develops during the next 3 years.

GASTROINTESTINAL FUNCTION

Motility

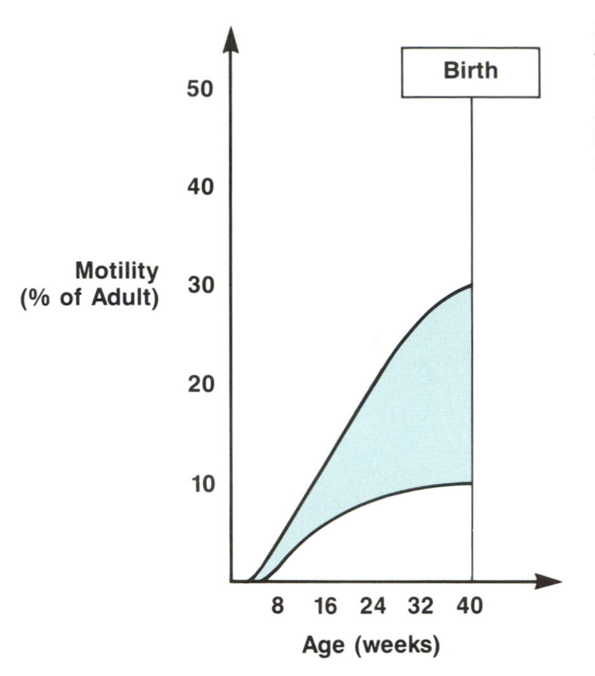

Spontaneous rhythmic activity begins in the small intestine by the sixth or seventh week of fetal life, and motility increases as more distal portions of the bowel become rhythmically active. At birth the vigor and extent of rhythmic activity are only 10 to 30 percent of adult levels. They increase as the enteric plexuses complete their development.

Digestion and Absorption

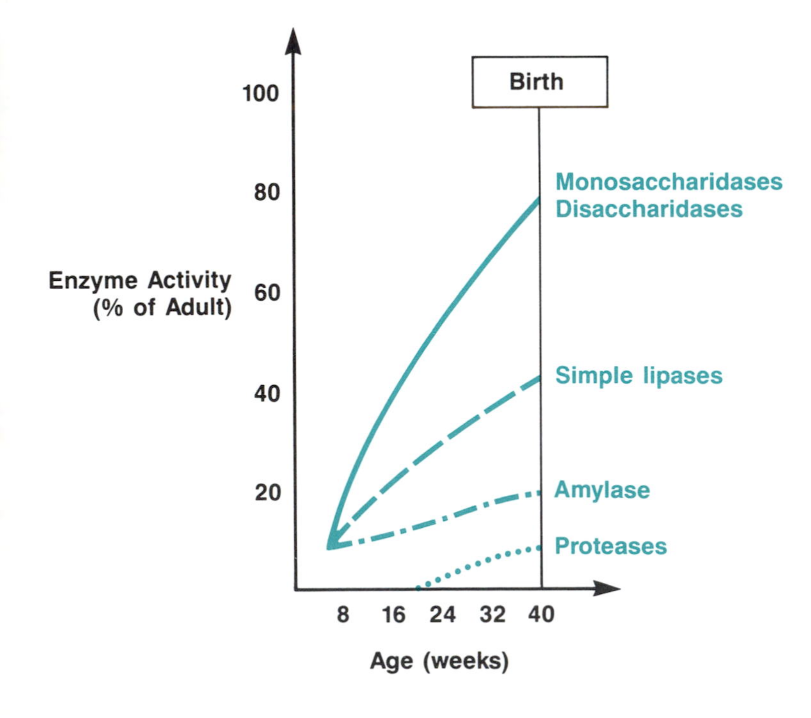

Enzymes for disaccharide breakdown and monosaccharide transport are developed early and reach their full, adult levels by 1 year after birth.

During the first month after birth, enough simple lipases are present to digest about 80 percent of the fats that are normally presented in the infant's diet.

Starch digestion develops slowly because enzymes such as amylase, required for the digestion of complex starches, are present in only small amounts even at 1 year after birth.

The ability to handle dietary protein begins to be developed by the fifth month of fetal age. Development is slow, and proteolytic activity in stomach and pancreas is still weak at birth.

METABOLISM

Metabolism in the Fetus

Fetal metabolism is driven by the substrates that are contained in maternal blood and reach the fetus via the umbilical circulation. They are **dominated by carbohydrates.**

Carbohydrates

Glucose and lactate are the major substrates for the fetus.

- The human fetal liver is able to synthesize and accumulate glycogen from an early age. As a result,
- ◆ excess glucose is converted and stored in the liver as **glycogen.**
- **Galactose** does not assume a significant role until after birth, when there is an increase in the intake of dietary lactose, the principal carbohydrate in mother's milk. (Lactose is a disaccharide composed of galactose and glucose.)

Proteins

The fetus receives large amounts of **amino acids** and uses them for **protein synthesis.**

- The enzymatic pathways for conversion of amino acids to glucose are weak in the fetal liver. Therefore,
- excess amino acids are catabolized, resulting in high urea production.

Fats

Lipids are **not an important fuel for the fetus.**

Metabolic Changes at Birth

Birth causes two important metabolic changes:

- ◆ It forces the fetus to **regulate its own blood glucose concentration.**
- ◆ It brings about a **change in diet** from one that is mostly carbohydrate to one that is rich in **lipid** (from mother's milk).

Regulation of neonatal blood glucose concentration

When the maternal supply of glucose stops abruptly at birth, **blood glucose levels fall within the first 24 hours.** Although hypoglycemia is only a weak stimulus in the pancreas of the newborn,

- pancreatic glucagon release is increased soon after birth.

In addition,

- glucose-6-phosphatase levels increase during the birth process.

The net result of these postnatal changes is

- depressed glycogen synthesis,
- accelerated glycogen breakdown, and
- accelerated glucose formation.
- ◆ **Blood glucose levels are returned to normal in the first few days after birth.**

Lipids in the neonatal diet

Although infants have limited ability to digest long-chain fatty acids, they are **able to digest and absorb about 80 percent of the fats that are present in mother's milk.**

METABOLISM

Overview

Fetus

The major substrates for metabolism in the fetus are glucose and lactate, both supplied by maternal blood.

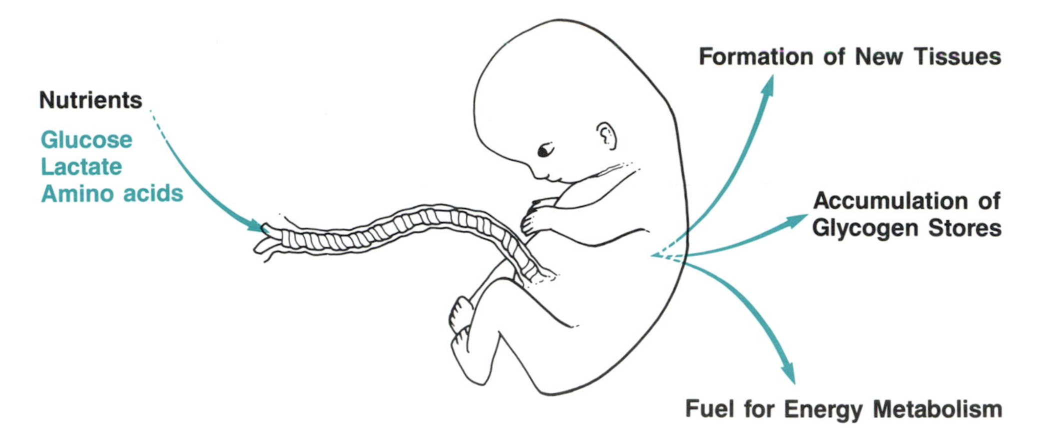

Newborn

At birth the umbilical supply of glucose stops abruptly and dietary intake is dominated by the lipids in mother's milk.

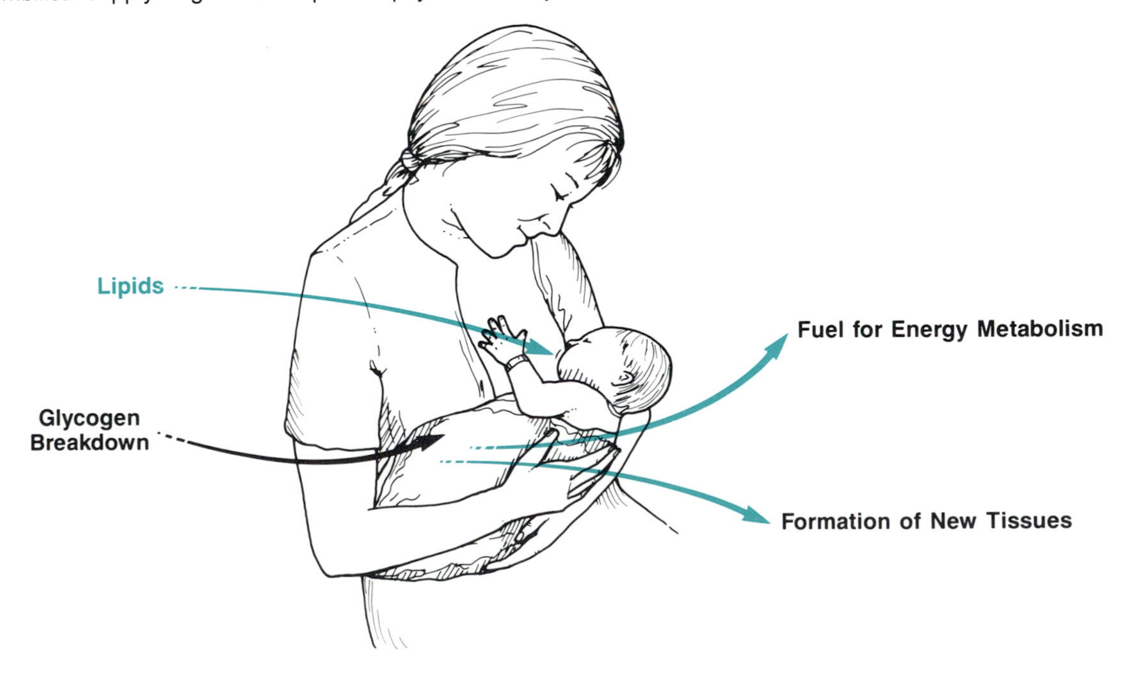

NEUROPHYSIOLOGY

Neuromuscular System

In the fetus

Responses to touch occur only after sensory fibers have grown near enough to the surface to reach the cutaneous epithelium.

By weeks 6 to 10

- **The area around the mouth is the only cutaneous region that is sensitive to tactile stimulation.**

By weeks 11 to 13

- Raw nerve tips contact the surface in a larger area around the mouth as well as on the eyelids, and some of them end on large epithelial cells that will eventually form end disks.
- The fetus begins to develop reflex responses to stimulation of the eyelids, the genito-anal region, the palms, and the soles.
- The ability to squint and scowl appears.
- **Feeding reflexes begin to develop.**

By weeks 12 to 14

- Respiratory chest movements begin.

By week 22

- Spontaneous diaphragm contractions are seen.

In the newborn

Motor behavior of the newborn is characterized by complex motor reactions.

Special Senses

Vestibular, taste, and hearing reflexes

These act at the level of the brainstem and should mature earliest (at about 8 to 13 weeks) because central nervous differentiation begins at cervical levels and spreads toward both the head and the feet. Experimental confirmation of this has not been obtained. However,

- The existence of taste reflexes has been deduced from increased swallowing in response to saccharin in amniotic fluid and grimacing in response to quinine in amniotic fluid.
- The presence of auditory reflexes in fetuses aged 30 weeks or more is suggested by increased movement and heart rate in response to external sounds.

Vision

While the fetus is within the uterus it receives no visual stimuli. In the newborn, visual acuity is poor, but it develops rapidly. The eyes move in response to vestibular stimuli or moving visual stimuli, and there is evidence of visual perception and association. Accommodation is not active until the second month after birth.

Autonomic Nervous System

Peripheral innervation is incomplete and many reflex responses are immature. However, infants have a highly sensitive diving reflex and some of them show an increase in the frequency of apnea and bradycardia during periods of REM sleep. These observations have led to the hypothesis that

- ◆ abnormal or exaggerated autonomic discharges during the neonatal period, arising either spontaneously or in response to the application of cold stimuli to the face, may lead to cardiorespiratory arrest and may be the cause of Sudden Infant Death Syndrome.

NEUROPHYSIOLOGY

Neuromuscular System

In the fetus

The earliest neuromuscular response occurs at about 6 weeks of fetal age. It is a protective avoidance response to a tactile stimulus of the mouth area.

Floor of fourth ventricle

Pons

Trigeminal n.

Spinal tract of accessory n.

Mandibular n.

Maxillary n.

C_1 to C_3

Stimulus:
Perioral touch

Response:
The response includes:
Contraction in cervical muscles

Contraction in cervical portion of trapezius and results in bending of the head to the side opposite to that which is being stimulated

By about 8 weeks of fetal age, the protective avoidance response includes bending the trunk and pelvis to the side opposite the stimulus side. Soon after 8 weeks it includes

- opening of the mouth,
- forceful extension of the arms,
- spreading of the fingers and toes, and
- rotation of the trunk.

By 11 to 13 weeks, some tactile stimuli of the mouth region elicit reflex responses that include bending toward the stimulus. This is considered to be the beginning in the development of **feeding reflexes.**

In the newborn

Motor behavior is characterized by complex reactions such as

The placing-standing response

- This occurs when an infant is lifted while its insteps are touching the edge of a table.
- The normal response is lifting of the feet followed by placing the feet on the table.
- Subsequent laying down of the infant leads to extension of the legs.

The righting response

- A healthy neonate will turn its trunk in the direction in which the head is turned.

The Moro response

- This is a momentary extension of the arms in response to brief dorsal flexion of the head.

ENDOCRINOLOGY

Anterior Pituitary

All the hormones of the adult anterior pituitary—

growth hormone,
adrenocorticotropic hormone (ACTH),
thyroid stimulating hormone (TSH),
luteinizing hormone (LH),
follicle stimulating hormone (FSH), and
prolactin (PL)—

can be released in vitro by the fifth week of gestation and are present in the fetal circulation by the 12th week.

Their trophic actions are believed to be significant from about week 25 on because absence of the pituitary at that point inhibits development and function of target organs.

ENDOCRINOLOGY

Anterior Pituitary

Secretion rates of anterior pituitary hormones are high early in gestation, while the anterior pituitary functions autonomously.

Hypothalamic nuclei develop from 16 weeks on and gradually, by 18 weeks, their inhibitory influences begin to cause serum levels of anterior pituitary hormones to decrease:

- LH and FSH by 18 weeks
- ACTH and GH by 20 weeks
- TSH by 27 weeks

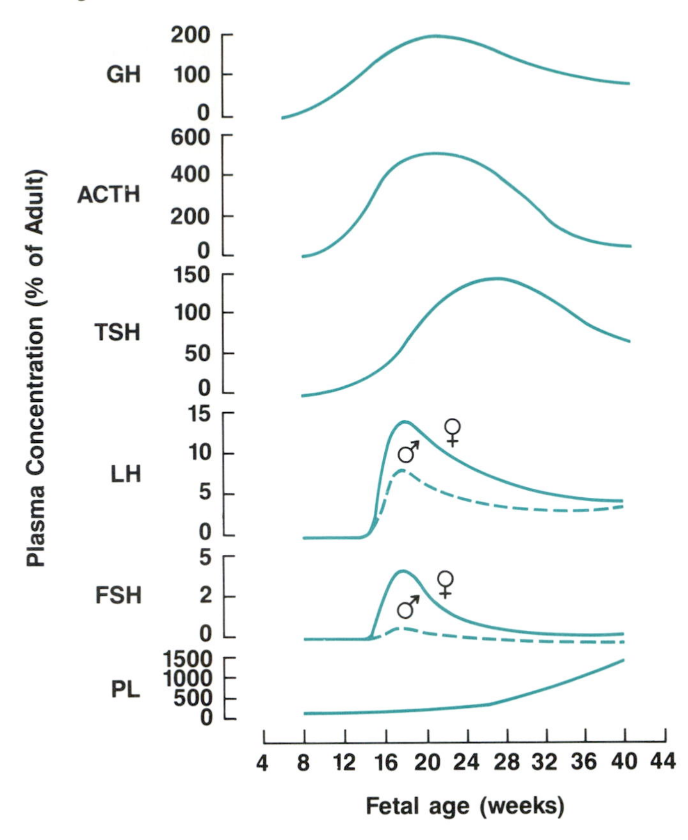

Adrenals

Cortex and steroid synthesis

Two aspects of fetal steroid synthesis differ from the adult pattern: the **adrenal enzyme profile** and **control by ACTH.**

The enzyme profile

Conversion of pregnenolone to progesterone is a two-step process requiring the enzymes 3-β-OH dehydrogenase and $\Delta^5\Delta^4$isomerase. The fetal adrenal cortex lacks one or both of these, and so

- **adrenal cholesterol metabolism in the young fetus is shifted toward the androgens (mainly dehydroepiandrosterone).**

Mineralocorticoids (aldosterone) and glucocorticoids (cortisol) are produced because

- the placenta synthesizes large amounts of progesterone from circulating cholesterol and
- the fetal adrenal is able to metabolize progesterone.

Control by ACTH

ACTH is feedback-inhibited by cortisol, the major steroid produced in the adult adrenal medulla. In the fetus,

- **inhibition of ACTH is weak because cortisol levels are low in fetal plasma.**
  - The reason for this is not lack of synthesis, but rapid dehydrogenation of cortisol to cortisone in the placenta. Cortisone is not capable of inhibiting the pituitary.

Medulla

This is principally a sympathetic ganglion. Its development is not yet complete at birth.

Development

- Blast cells appear by week 5 and begin to concentrate near the blastema of the adrenal cortex.
- **By the ninth week, chromaffin granules,** which are associated with norepinephrine synthesis, **are evident.**
- During weeks 12 and 13, these chromaffin cells penetrate into the adrenal cortex, forming scattered islets of medullary tissue.
- **When cortisol appears, it induces** in chromaffin cells the adrenal medullary enzyme phenylethanolamine-N-methyltransferase, required for the **conversion of norepinephrine to epinephrine.**
- Fusion of the scattered islets of medullary cells into a single centrally located mass occurs after birth.

Adrenal Cortex

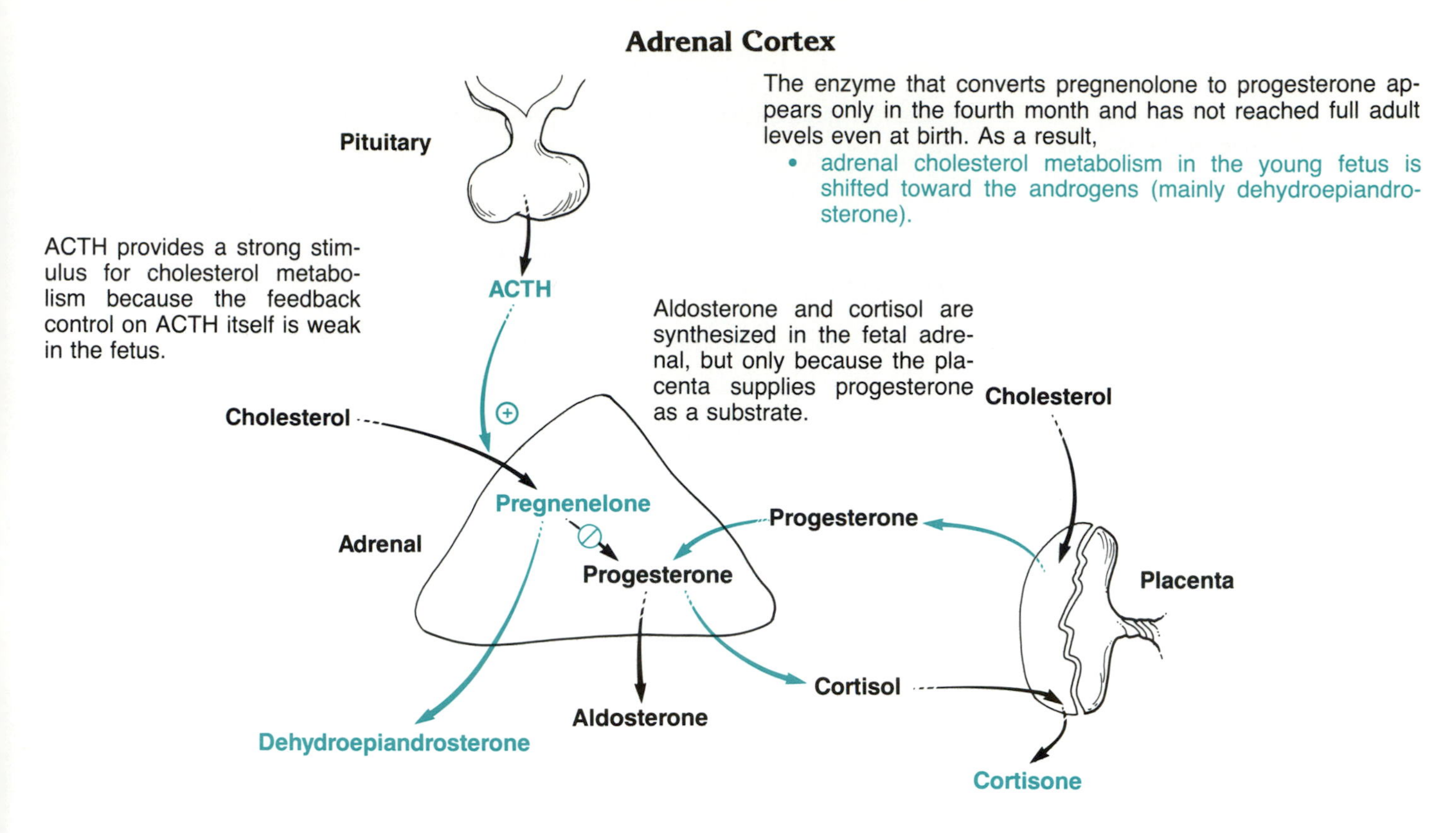

ACTH provides a strong stimulus for cholesterol metabolism because the feedback control on ACTH itself is weak in the fetus.

The enzyme that converts pregnenolone to progesterone appears only in the fourth month and has not reached full adult levels even at birth. As a result,

- adrenal cholesterol metabolism in the young fetus is shifted toward the androgens (mainly dehydroepiandrosterone).

Aldosterone and cortisol are synthesized in the fetal adrenal, but only because the placenta supplies progesterone as a substrate.

Pancreas

The regulation of the plasma glucose level is not the primary function of the fetal or neonatal pancreas.

Fetal pancreas

The pancreas appears by the fourth week of gestation, and A-cells containing abundant glucagon appear by the seventh week.

- Insulin-containing B-cells develop later and are functional from about the 15th week.

Function

- **Acute changes in plasma glucose concentration are not able to affect secretion of insulin or glucagon in the fetus or in the neonate.**

However, fetal A-cells and B-cells do respond appropriately to *chronic* changes in plasma glucose levels, and they do respond with secretory bursts to acute challenges by any factor that is able to raise intracellular cAMP (e.g., ions or amino acids). Therefore,

- **The cause for the glucose insensitivity appears to be an inability of these cells to generate sufficient intracellular cAMP in response to glucose** (enzyme defect? receptor deficit?).

As a result, the physiologic role of fetal insulin is not the regulation of plasma glucose concentration. That function is performed by maternal mechanisms.

- **Fetal insulin is required for the formation of substrate stores in the form of glycogen and fat.**

Neonatal pancreas

Glucose continues to be a poor stimulus for insulin secretion and a poor stimulus for glucagon inhibition in the newborn. Nevertheless, the neonatal period is characterized by

- constant plasma levels of glucose, insulin, and glucagon during the first 24 hours.

Thereafter,

- plasma glucose levels begin to decrease;
- plasma insulin levels remain at a constant (low) level;
- plasma glucagon concentration increases (not because of the hypoglycemia, but because of other, nonspecific factors associated with the cutting of the cord).

Thyroid

The placenta is impermeable to thyroid stimulating hormone (TSH) and practically impermeable to T_4 and T_3.

Fetal period

The hypothalamic-pituitary axis begins to mature at about week 20. After that there is a progressive increase in the plasma concentration of TSH as well as in those of bound and free plasma T_4. (The source for the iodine is maternal blood, and it diffuses freely across the placenta.)

Neonatal period

Within the first hour after birth,

- **Plasma T_3 levels begin to rise in the newborn, and they peak 24 hours after birth.** (The stimulus for this rise is, presumably, the change in ambient temperature that occurs at birth.)
- By 6 to 30 days of age, T_3 levels are at the upper end of normal adult levels.

Pancreas

Role of fetal insulin

A fetal source of insulin is required because insulin cannot cross the placenta. It is particularly needed in the latter portion of the pregnancy, when fetal growth is largely due to fat synthesis. At this stage the transfer of free fatty acids across the placenta is inadequate for fetal needs.

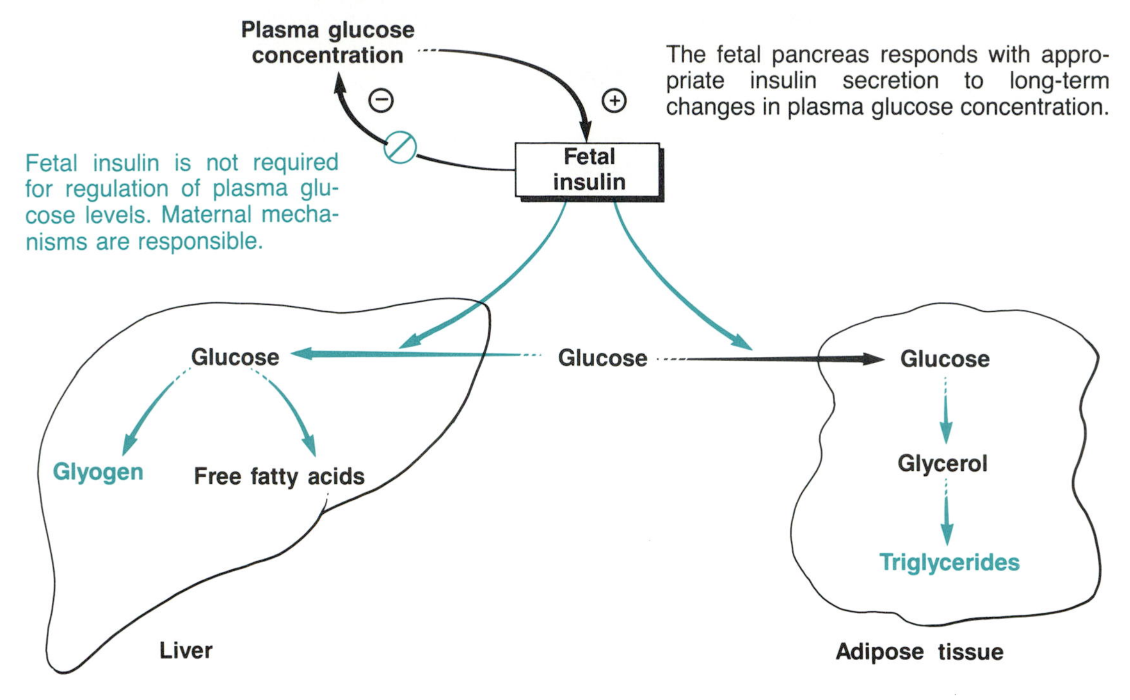
Plasma glucose concentration
⊖
⊕
The fetal pancreas responds with appropriate insulin secretion to long-term changes in plasma glucose concentration.
Fetal insulin is not required for regulation of plasma glucose levels. Maternal mechanisms are responsible.
Fetal insulin
Glucose
Glucose
Glucose
Glyogen
Free fatty acids
Glycerol
Triglycerides
Liver
Adipose tissue

Fetal insulin facilitates hepatic glycogen formation.

Insulin promotes lipogenesis from glucose.

TEMPERATURE REGULATION

The principles of neonatal temperature regulation are identical to those found in the adult:

- **Heat dissipation must balance heat production.**
  - The overall regulatory system is functionally mature at birth in that it exhibits a stable body temperature within a given ambient range. However, this **tolerance range is narrower than that found in adults,** who have experienced repeated or sustained exposure to hot or cold conditions.

Heat Production

The sudden exposure to a cold environment at birth requires from the newborn a sharp increase in heat production so that body temperature can be maintained.

- **Nonshivering thermogenesis,** represented chiefly by metabolism of brown fat, plays a major part in the neonate.
  - Brown fat appears at about the 25th week of gestation. Its total quantity decreases after birth at a rate that is inversely proportional to the environmental temperature.

Heat Loss

Although the respiratory tract plays some role in neonatal temperature regulation via **panting,** the skin is the most significant organ for heat loss.

Heat is brought to the skin by blood flow or by direct conduction and is lost by **convection, radiation, external conduction, and sweating.**

Panting

In contrast to adults, neonates show a sharp increase in respiratory rate just before the onset of sweating.

Convection, radiation, and external conduction

The thinner layer of body fat gives neonates a higher skin surface temperature and, therefore, greater loss of body heat by conduction, convection, and radiation.

- **Poor thermal insulation is the most serious handicap of newborns.**

Sweating

Sweat secretion is the most effective heat-dissipating mechanism in the neonate, just as it is in the adult. Newborns are more richly endowed with sweat glands than are adults, but their maximum secretory capacity is only about 10 percent of that found in adults. Therefore,

- **the maximum tolerated heat stress is relatively low in infants.**

Fever

Fever is thought to result when a leukocyte pyrogen resets the hypothalamic temperature set point and alters heat conservation mechanisms.

- **Many neonatal infections are not accompanied by fever.**
  - This is puzzling in light of the observation that all temperature-regulating mechanisms are operational.
  - It is probably the result of an immunologic "immaturity" by which the neonate fails to produce the leukocyte pyrogen that causes fever in adults.

TEMPERATURE REGULATION

Heat Production

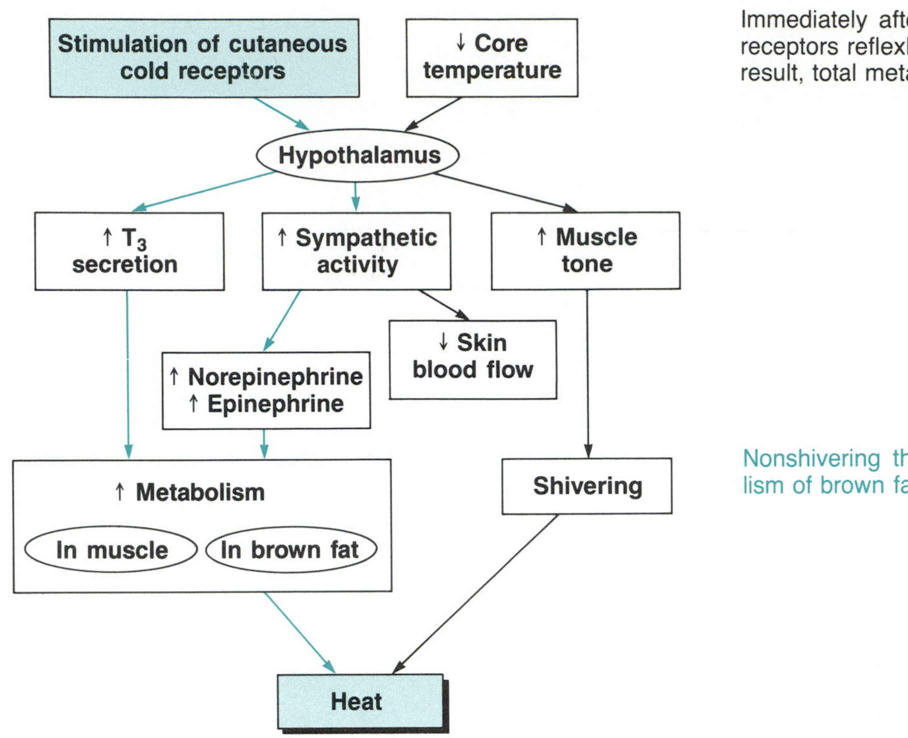

Immediately after birth, signals from supersensitive skin cold receptors reflexly stimulate secretion of thyroid hormone. As a result, total metabolic rate increases up to fivefold.

Nonshivering thermogenesis, represented chiefly by metabolism of brown fat, is the principal heat source in the neonate.

Heat Loss

- Heat can be dissipated only after it has been transferred to an interface with cooler surroundings.
- Skin is the most significant such interface, and once heat reaches it, four processes are involved in the elimination of heat:
  - Convection
  - Conduction
  - Radiation
  - Sweating

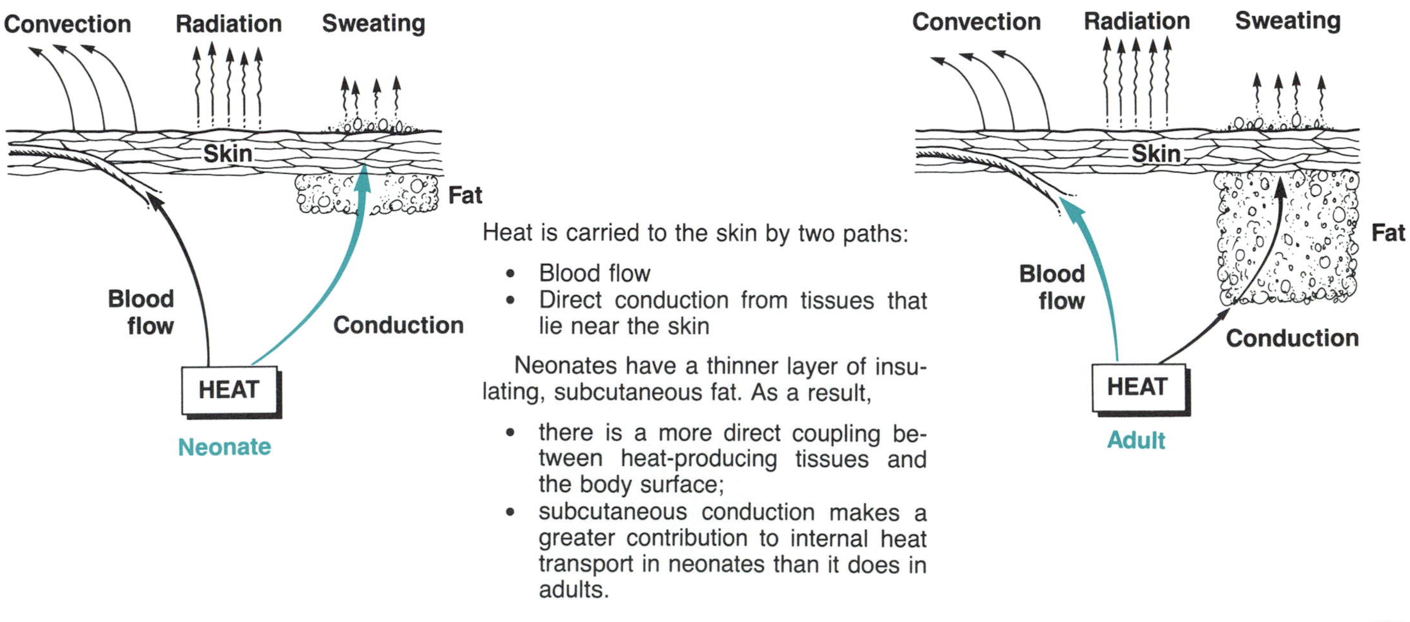

Heat is carried to the skin by two paths:

- Blood flow
- Direct conduction from tissues that lie near the skin

Neonates have a thinner layer of insulating, subcutaneous fat. As a result,

- there is a more direct coupling between heat-producing tissues and the body surface;
- subcutaneous conduction makes a greater contribution to internal heat transport in neonates than it does in adults.

12

INTEGRATIVE PHYSIOLOGY

CONTROL OF FUEL SUPPLY; STARVATION

The Fed State

Energy Sources

Dietary intake and subsequent intestinal absorption predominate, but some energy substrates are provided by the liver.

Glucose

- derives from intestinal absorption (after **dietary intake**) as well as from the liver by **glycogenolysis** and by **gluconeogenesis.**

Free fatty acids (FFA)

- derive from **dietary intake.**

Ketones

- can be used by many tissues such as skeletal muscle and cardiac muscle (and brain under some circumstances);
- are formed from FFA by the liver.

Body weight

Excess caloric intake is accumulated in body energy stores. These amount to approximately 2,600 kcal per kilogram of body weight and are composed of fat (76 percent), protein (23 percent), and glycogen (1 percent).

Physical activity increases the need for energy substrate. Therefore, it is normally accompanied by increased caloric intake to ensure stable body weight.

The mechanisms that maintain body weight through the regulation of appetite depend critically on hypothalamic areas that respond to a variety of stimuli (e.g., plasma glucose level, glucose consumption rate, rate of heat production, brain concentration of gastrointestinal hormones, and so on).

Energy Sources During Starvation

During the first few days of fasting

Glycogenolysis provides the major and immediate source of glucose. However, the total glycogen stored in the body is only about 450 g (1,800 kcal), and so it can serve as a fuel reservoir for only a short time.

During the first week or two of fasting

Nervous tissue, erythrocytes, and leukocytes continue to require glucose. It derives from hepatic gluconeogenesis, using amino acids as a substrate.

Adipose tissue provides ketones and FFA for the other tissues.

When starvation is prolonged

Glucose requirements decline gradually because metabolic adaptations in the brain allow it to use **ketones** as a fuel. As ketone utilization becomes more effective, muscle proteolysis diminishes and **fat reserves** are used to a greater extent until they are near depletion (after about 6 weeks of starvation). At that time, **protein stores** are the only remaining energy substrate and the rate of proteolysis increases again.

Death occurs when metabolic requirements have depleted body protein stores to a level where protein-dependent cellular functions can no longer be maintained (after about 8 weeks of starvation).

CONTROL OF FUEL SUPPLY; STARVATION

Daily Energy Balance

Daily Intake	Daily Usage
Average of 2,500 kcal (10×10^6 J) in the form of carbohydrate, protein, and fat.	Glucose: 670 kcal Fat: 850 kcal —Basal metabolism: 1,200 kcal —Voluntary muscle activity: 1,100 kcal —Food absorption (specific dynamic action of food): 200 kcal

Tissues differ with respect to their preferred energy substrate. Thus, in the fed state,
- —most of the glucose is used by the central nervous system;
- —most of the fat is used by resting skeletal muscle.

The Liver as a Glucose Producer

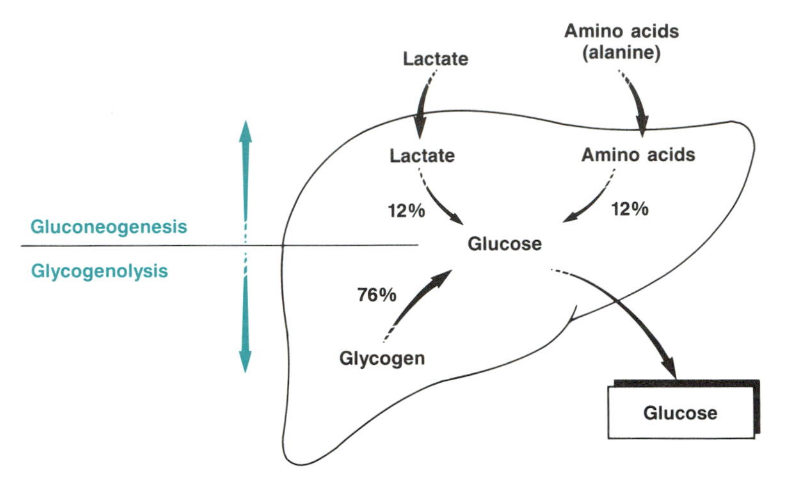

Energy Sources During Starvation

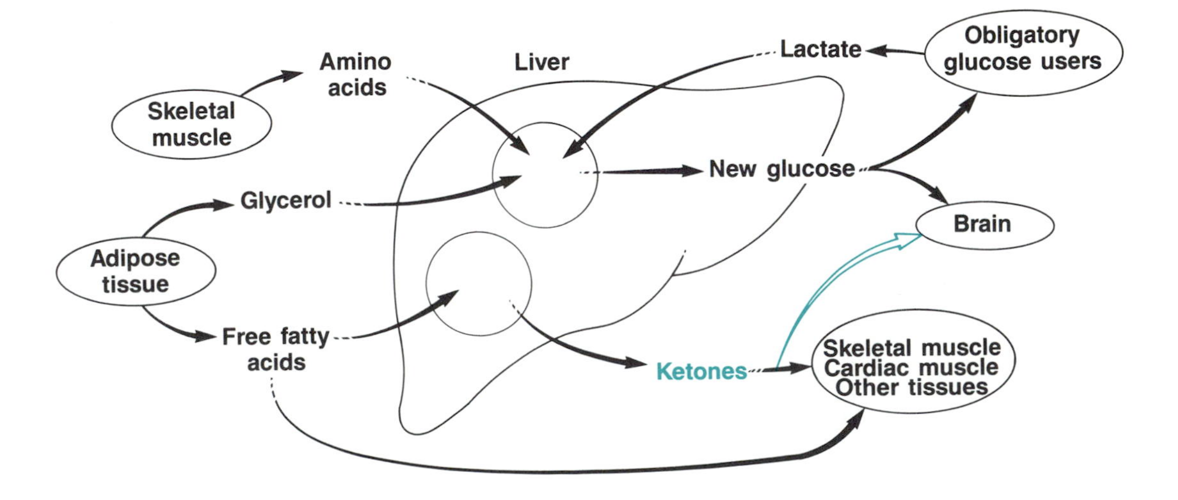

EXERCISE

Cardiorespiratory Responses

Anticipation of exercise

This phase is initiated in the cerebral cortex. The resulting autonomic pattern that influences cardiovascular function is mild **vagal inhibition** and **sympathetic activation.** They cause

- **a small rise in the plasma epinephrine and norepinephrine;**
- **a small increase in cardiac output;**
- a slightly **preferential distribution of cardiac output** toward large skeletal muscle groups. [Two causes have been postulated: (1) general β_2-adrenergic relaxation of vascular smooth muscle in arterioles of skeletal muscle and (2) specific sympathetic cholinergic vasodilatation of selected muscle vascular beds.]

Execution of exercise

Physiologic responses to exercise are driven by (1) the effects of metabolites and (2) neural input from muscle receptors (the spindle and the Golgi tendon organ).

Effects of metabolites

- They change blood gases and H^+. This **stimulates ventilation.**
- They relax vascular smooth muscle. This causes both **capillary recruitment** and **vasodilatation, predominantly in active tissue, where metabolite concentrations are highest.**

Medullary reflex patterns

Skeletal muscle stretch receptors, baroreceptors, and chemoreceptors influence medullary activity during exercise. Medullary efferent patterns have these net cardiovascular effects:

- A **large increase in cardiac output** resulting from

  1. increased heart rate (autonomic nerves, epinephrine);
  2. increased cardiac contractility (sympathetics, epinephrine);
  3. increased preload (respiratory pumping, muscle pumping)

- **Distribution of cardiac output toward active muscle** because inactive regions are constricted (sympathetics; the α effect of epinephrine) and active muscle regions are dilated (metabolites; the β_2 effect of epinephrine)
- **Specific vasodilatation in blood vessels of the skin** if exercise continues long enough to raise the body core temperature

EXERCISE

Cardiorespiratory Responses

Anticipation

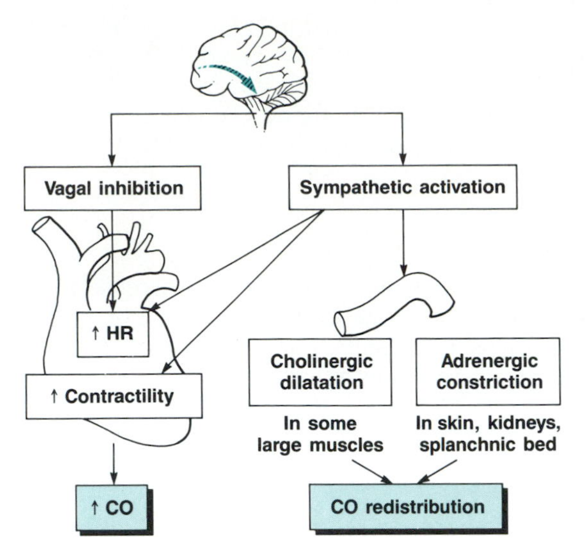

*CO = Cardiac output

Execution

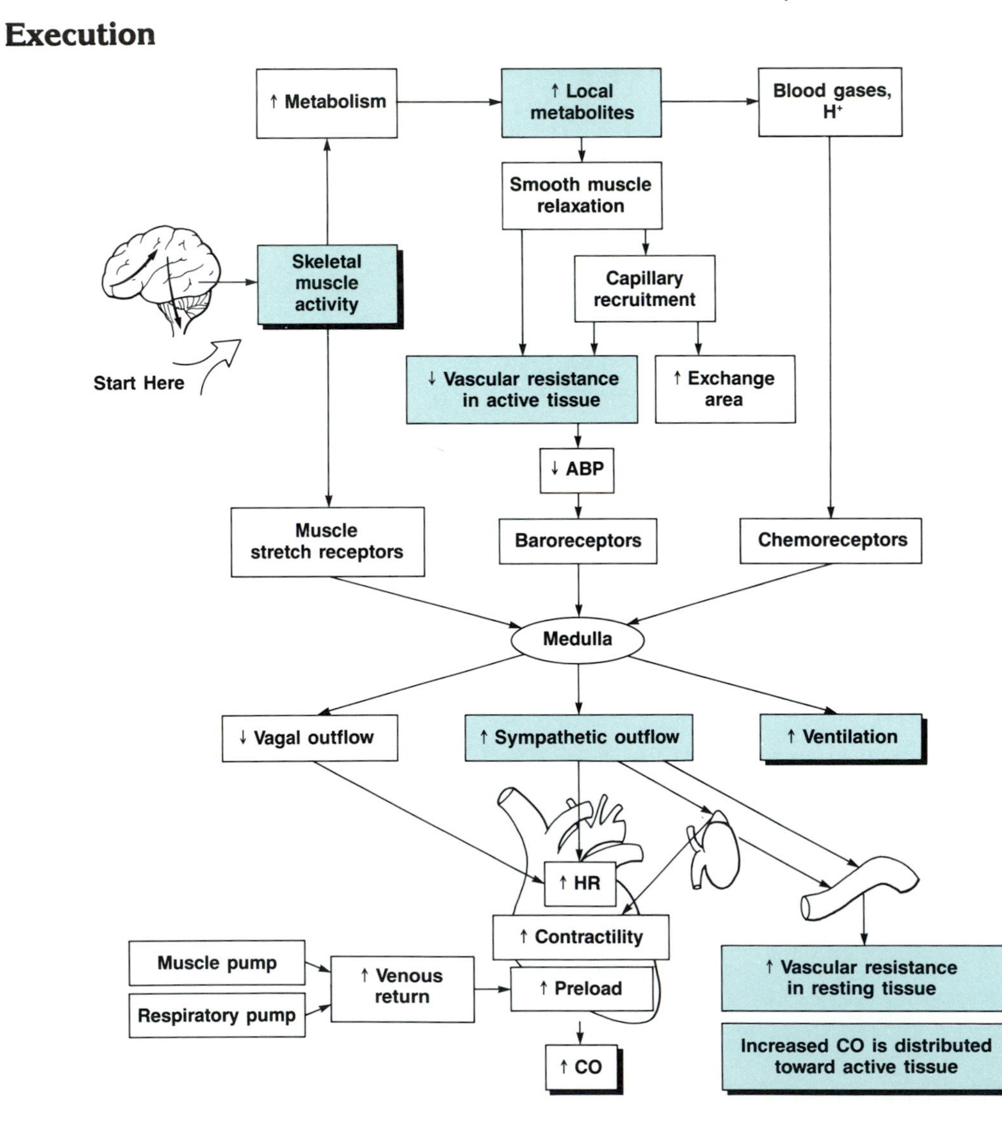

Metabolic Responses

During exercise, **increased utilization of glucose in muscle is matched by increased glucose production in the liver.**

Glucose production

The high level of sympathetic activity during exercise depresses insulin release and promotes glucagon release. As a result, hepatic glucose production increases and tends to hold plasma glucose levels near normal.

Glucose utilization

Catecholamines and lack of insulin both tend to decrease fractional glucose uptake in skeletal muscle. However, changes in flow, increased muscle glycogenolysis (local, direct catecholamine effect), and, perhaps, increased affinity of muscle insulin receptors all contribute to increased glucose utilization in exercising muscle.

EXERCISE—CONT'D.

Metabolic Responses

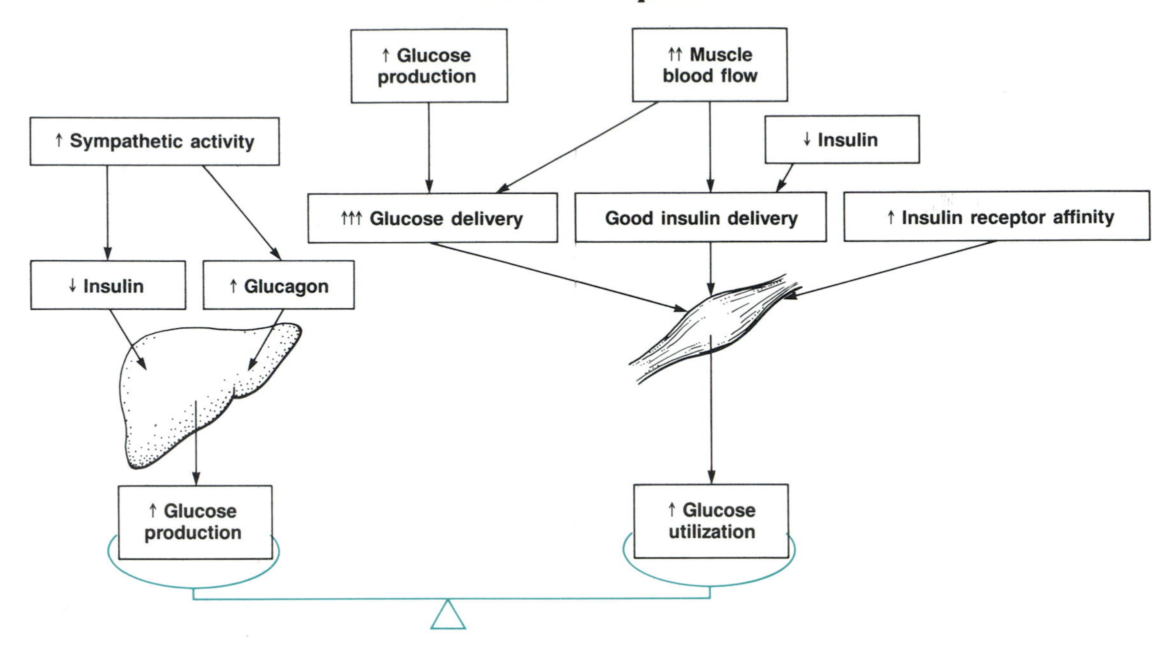

BODY TEMPERATURE REGULATION

Maintenance of constant body temperature in the face of changing ambient temperature requires **regulation of both heat production and heat loss.**

Heat Production

Mammals can respond to a cool environment by producing extra heat to supplement basal metabolic heat production. This occurs via three avenues: (1) **voluntary muscle activity,** (2) **nonshivering thermogenesis,** and (3) **involuntary muscle activity** (e.g., shivering).

Nonshivering thermogenesis (NST)

- This is a cold-induced increase in oxygen consumption.
- It is mediated by the sympathetic nervous system.
- It is a common feature in neonates, but is found in adults chiefly as a long-term adaptive response to cold environments.
- **It depends directly on the amount of brown fat that is present.**

Shivering

- Shivering occurs when NST is not capable of generating sufficient heat to maintain body temperature.
- **Quantitatively it is the most important short-term involuntary thermogenic mechanism,** although it is less efficient than NST because the rhythmic body oscillations associated with shivering increase convective heat loss.

Heat Loss

Heat loss is regulated by control of skin blood flow, the major avenue of heat transport to the body surface. Loss occurs from the surface by four paths: **conduction, convection, radiation,** and **water evaporation.** The relative importance of each depends on ambient conditions such as temperature, wind velocity, humidity, and so on.

Conduction

This is defined as transfer to an object (or air layer) that is in physical contact with the skin.

Convective heat loss

This is removal via mass transport, such as by a stream of air.

Radiation loss

This means loss via electromagnetic waves in the infrared region.

Evaporative heat loss

This occurs when water is evaporated and the vapor is removed from the body surface. In humans, the source of the water is sweat; in animals it is mostly saliva (lost through panting).

BODY TEMPERATURE REGULATION

Heat Production

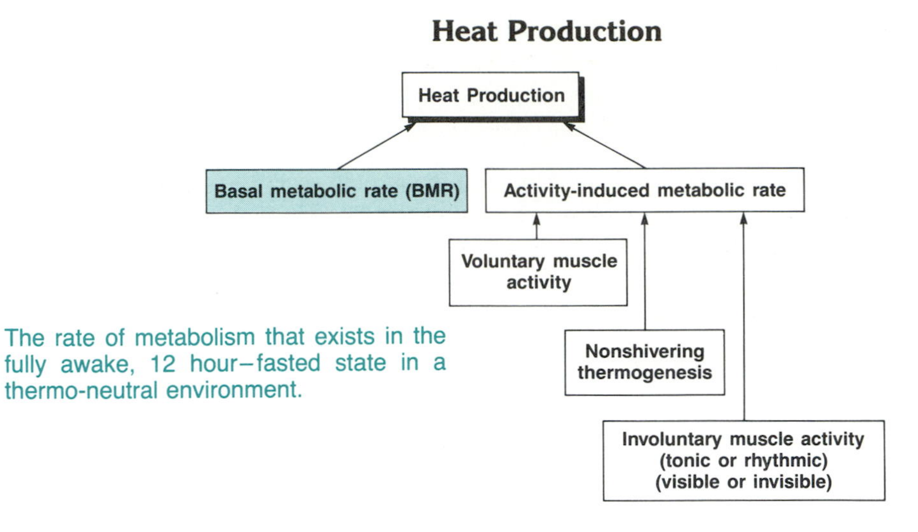

Brown Fat vs. White Fat

Brown Fat	White Fat
Reddish brown in color	White in color
Smaller cell; large number of small fat droplets; centrally located nucleus	Large cell; one large fat droplet; peripherally located nucleus
Many mitochondria	Few mitochondria
High oxygen consumption	Low oxygen consumption
Synaptic contact between cell membrane and sympathetic nerves	Cells not innervated
Selective distribution over the torso	General subcutaneous internal distribution
Total amount decreases with age, but can be improved from any given adult state by the deposition of brown fat in response to prolonged cold exposure	Total amount tends to sneak up with age
Extensive blood supply and blood flow increase with cold	Large blood supply and blood flow decrease with cold
Function is to produce heat	Function is to store fat

Heat Loss

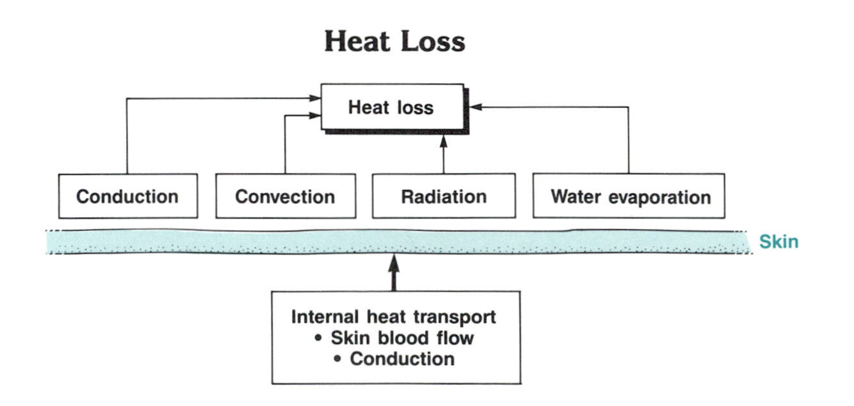

BODY TEMPERATURE REGULATION—CONT'D.

Temperature Homeostasis

Hypothalamic nuclei coordinate reflex responses to a change in the **integrated body temperature.**

- **Defenses against cold are dominated by cutaneous cold receptors.**
- **Defenses against heat are dominated by internal warm receptors.**

The regulation set point can be altered by agents such as leukocyte pyrogens. However, they can cause fever only if they also activate heat-conserving mechanisms (the vasomotor system).

The vigor of individual thermoeffector mechanisms can be modulated by conditions near the receptors that are reporting a temperature change. For example, regional sweat gland activity depends on the local temperature and is modulated by the integrated temperature.

BODY TEMPERATURE REGULATION—CONT'D.

Temperature Homeostasis

Hypothalamic nuclei regulate an integrated temperature detected by sensors in different regions.

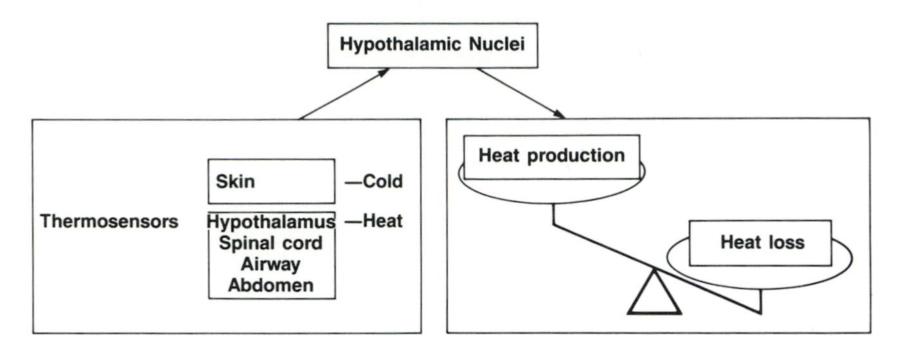

ACID-BASE REGULATION

OVERVIEW

If the volatile acid that is produced each day by metabolism of foodstuffs were dissolved in the body water space, then the resulting H^+ concentration would be 325,000 to 500,000 nM/L.

If only the nonvolatile acid that is produced each day were dissolved in the body water space, then the resulting H^+ concentration would be 1,000 to 1,500 nM/L.

- **Extracellular H^+ concentration is normally maintained in the range of 38 to 43 nM/L (pH 7.42 to 7.37).**
- **Values outside the range of 16 to 160 nM/L (pH 7.8 to 6.8) are *not* compatible with life.**

Disposal of the Daily Acid Load

Volatile acid

CO_2 is excreted by the lungs, the chemoreceptor-ventilation feedback mechanism maintaining extracellular P_{CO_2} near 40 mm Hg.

Nonvolatile acids

These are initially buffered by the body buffer systems and are later excreted by the kidneys.

ACID-BASE REGULATION

OVERVIEW

SOURCES OF H^+

1. Metabolism of carbohydrate and fat produces CO_2. CO_2 combines with body water to generate 13,000 to 20,000 mM of H^+ each day:

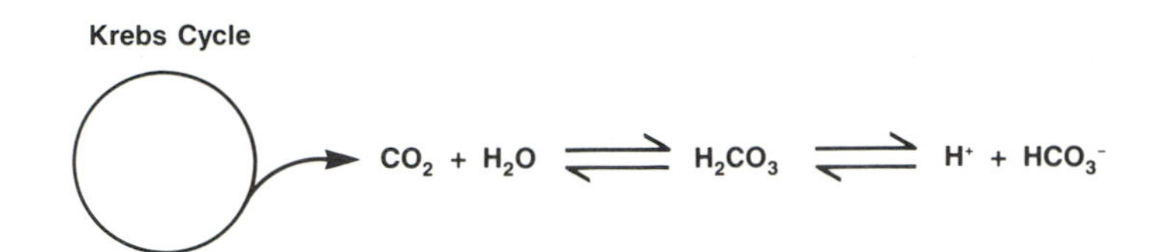

2. Incomplete oxidation of carbohydrates produces lactic acid and pyruvic acid.
3. Incomplete oxidation of fats produces ketoacids.
4. Oxidation of dibasic amino acids (lysine, arginine) produces HCl and other acids.
5. Oxidation of sulfur-containing amino acids (methionine, cystine) produces H_2SO_4.

On a typical North American diet, 2, 3, 4, and 5 account for 40 to 60 mM of H^+ each day.

Buffering of Nonvolatile Acid by the Body Buffering Systems

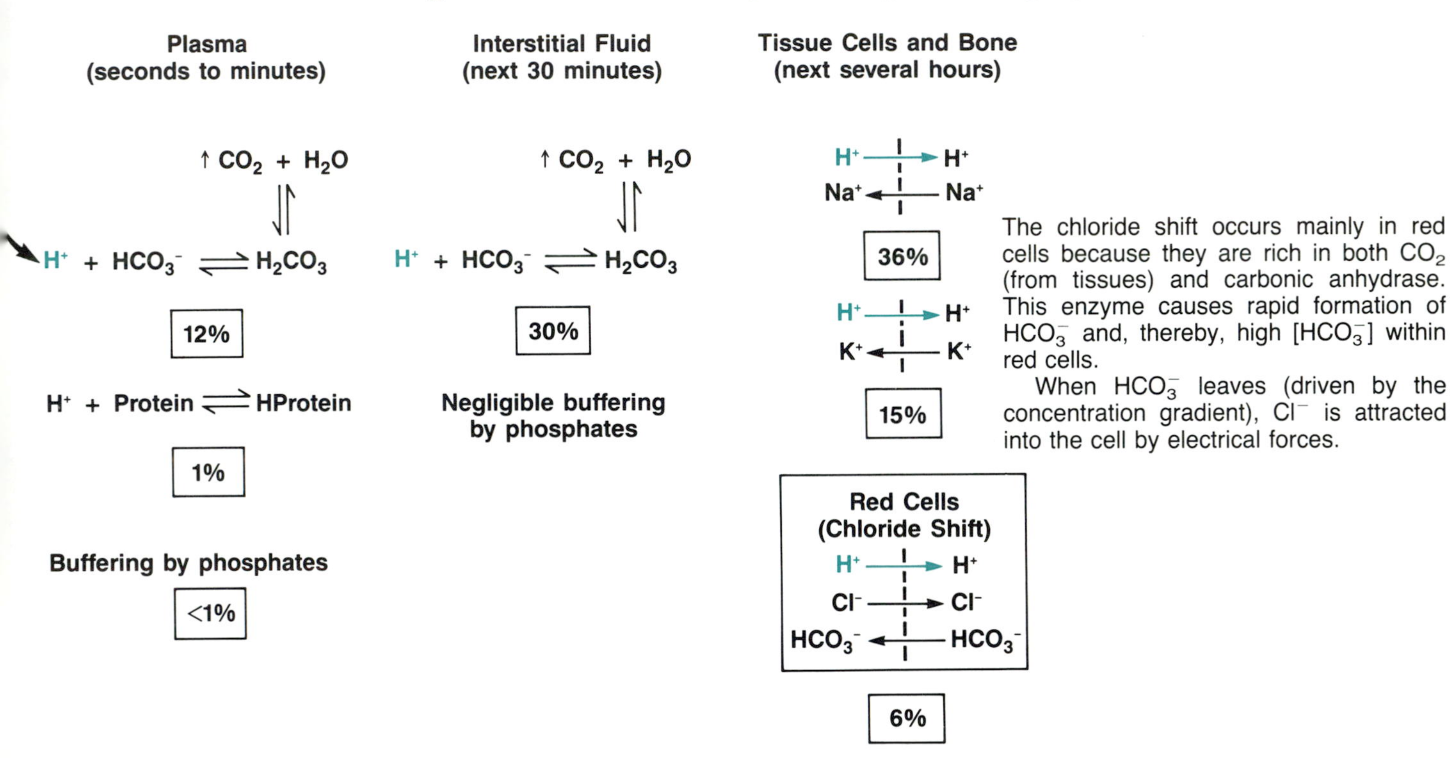

The chloride shift occurs mainly in red cells because they are rich in both CO_2 (from tissues) and carbonic anhydrase. This enzyme causes rapid formation of HCO_3^- and, thereby, high $[HCO_3^-]$ within red cells.

When HCO_3^- leaves (driven by the concentration gradient), Cl^- is attracted into the cell by electrical forces.

For clarity, the anion accompanying the H^+ is not shown.
Percentage figures refer to fractions of total acid load.

Adapted from Valtin H. Renal function: Mechanisms preserving fluid and solute balance in health. Boston: Little, Brown, 1973.

The total capacity of the body buffers is 200 mM of nonvolatile acids per day.

The Role of the Kidneys in Acid-Base Balance

Most renal Na^+ reabsorption occurs in concert with Cl^- reabsorption, but a fraction of it occurs in concert with HCO_3^- or with the anions of fixed acids.

- **Regulated absorption of Na^+ and Cl^- occurs for the purpose of extracellular fluid volume regulation.**
- **Regulated absorption of Na^+, along with either HCO_3^- or the anions of fixed acids, occurs for the purpose of acid-base regulation.**

The kidney has two functions in acid-base regulation: (1) reclaiming HCO_3^- that entered the nephron via glomerular filtration, and (2) excreting buffered, nonvolatile acid and forming new HCO_3^- in the process.

Renal excretion of free H^+ is limited because the distal tubular epithelium in the nephron is able to maintain a concentration difference of only 1,000-fold. That is, at a plasma $[H^+]$ of 40 nmol/L, the maximum $[H^+]$ in urine is near 0.04 nmol/L (pH 4.4), permitting the excretion of only 0.08 nmol of H^+ per day unless exceedingly large volumes of urine are excreted.

Reclaiming of filtered HCO_3^-

The first step is exchange of Na^+ for H^+ across the luminal membrane of tubular cells. **The presence of carbonic anhydrase in the lumen of the proximal tubule is crucial because it allows rapid formation of CO_2, which then diffuses into the cells.** Note that the process does not result in net elimination of H^+ even though a very large number of them are shuttled across the epithelium.

Excretion of buffered, nonvolatile acids

As titratable acid

The first step is, again, exchange of Na^+ for H^+ across the luminal membrane of tubular cells.

- Intracellular formation of H_2CO_3 from body CO_2 generates both H^+ and HCO_3^-.
- Buffering of H^+ by luminal $NaHPO_4^-$ allows net excretion of H^+ as titratable acid.

As ammonium

Intracellular H^+ and HCO_3^- are formed from CO_2 and H_2O. NH_3 is formed mostly during the metabolism of glutamine.

- NH_3 diffuses freely into the tubular lumen, where it combines with H^+ to form NH_4^+.
- Reabsorption of NH_4^+ is prevented by the potential difference that exists at the cell-lumen interface.

ACID-BASE REGULATION—CONT'D.

The Role of the Kidneys in Acid-Base Balance

Reclaiming of filtered HCO_3^-*

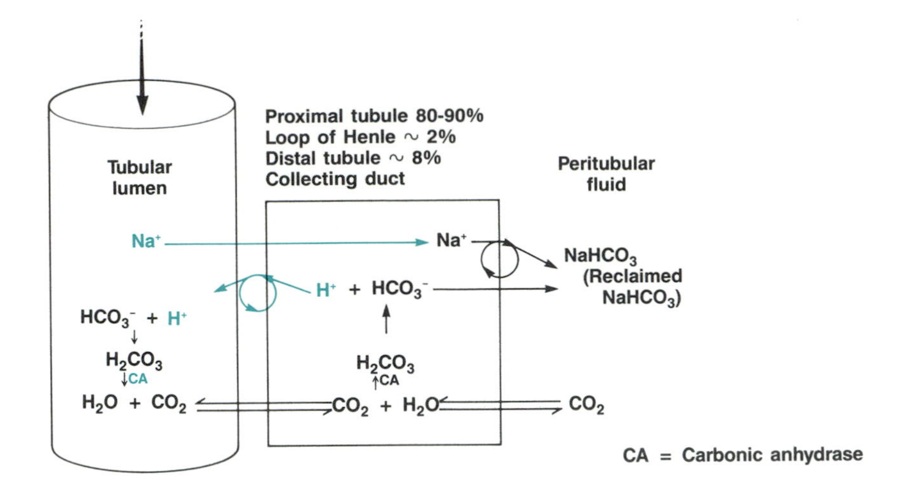

Excretion of buffered, nonvolatile acids

As titratable acid*

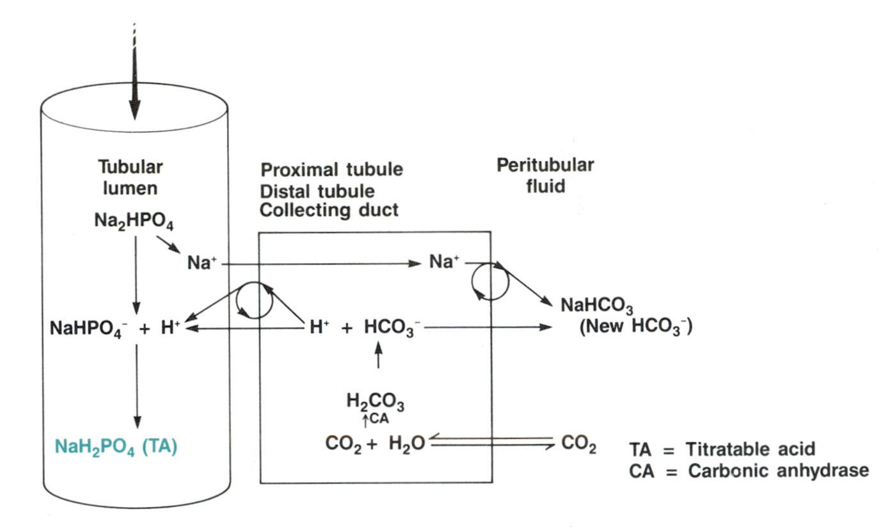

As ammonium (NH_4^+)*

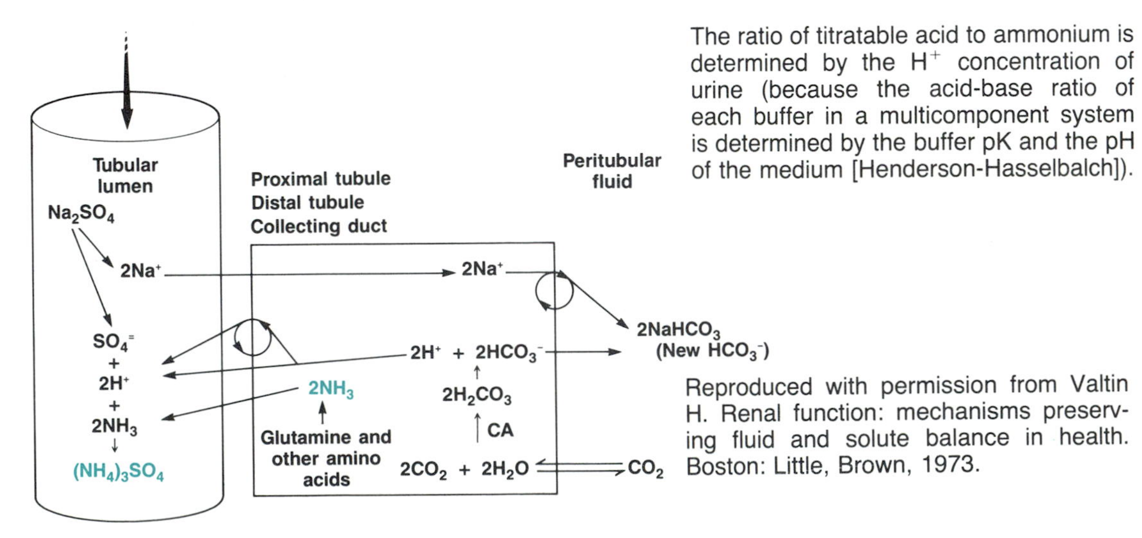

The ratio of titratable acid to ammonium is determined by the H^+ concentration of urine (because the acid-base ratio of each buffer in a multicomponent system is determined by the buffer pK and the pH of the medium [Henderson-Hasselbalch]).

Reproduced with permission from Valtin H. Renal function: mechanisms preserving fluid and solute balance in health. Boston: Little, Brown, 1973.

PHYSIOLOGIC COMPENSATIONS DURING ACID-BASE IMBALANCE

Respiratory Acidosis

Impaired respiration increases arterial P_{CO_2} and raises extracellular $[H^+]$:

$$CO_2 + H_2O \rightleftharpoons H_2CO_3 \rightleftharpoons H^+ + HCO_3^-$$

Note that this reaction is concerned with concentration changes in the nM/L range and, therefore, makes negligible direct contributions to the changes in extracellular $[HCO_3^-]$.

Disposal of H^+

The short-term response is intracellular buffering. Hence,

- extracellular $[HCO_3^-]$ rises slightly (owing to the chloride shift in red cells).

Long-term compensation (chronic respiratory acidosis) is by increased renal NH_4^+ excretion. Hence,

- new HCO_3^- is generated and extracellular $[HCO_3^-]$ rises more.

Acute respiratory acidosis

This exists if $\Delta[H^+] = 0.77\Delta P_{CO_2}$. If metabolic compensation is appropriate, then

- **every ΔP_{CO_2} of +10 mm Hg results in a $\Delta[HCO_3^-]$ of +1 mM/L.** Otherwise, a mixed acid-base disturbance is present.

Chronic respiratory acidosis

This exists if $\Delta[H^+] = 0.32\Delta P_{CO_2}$. If metabolic compensation is appropriate, then

- **every ΔP_{CO_2} of +10 mm Hg results in a $\Delta[HCO_3^-]$ of +3.5 mM/L.**

Metabolic Acidosis

Causes

The possible causes are (1) **loss of HCO_3^-** (from the GI tract or the kidneys), accompanied by **elevation of extracellular $[Cl^-]$** and (2) **gain of a metabolic acid** (diabetic ketoacidosis; lactic acidosis in ischemia; retention of organic, sulfuric, or phosphoric acid in renal failure). This will be accompanied by elevation of the unmeasured anion gap.

Role of the kidney

1. To excrete H^+ (buffered and free)
2. To generate new HCO_3^- for the body buffer stores

If respiratory compensation is appropriate in this disturbance, then $\Delta P_{CO_2} = 1.2\Delta[HCO_3^-]$

Metabolic Alkalosis

Causes

An initial event (e.g., vomiting) **causes HCO_3^- gain;** this is followed by **a set of circumstances that prevents the kidney from excreting the excess HCO_3^-** (e.g., a drive for Na^+ reabsorption that is so intense that it requires accompanying reabsorption of anions in excess of available Cl^-).

If respiratory compensation is appropriate, then $\Delta P_{CO_2} = 0.6\Delta[HCO_3^-]$

PHYSIOLOGIC COMPENSATIONS DURING ACID-BASE IMBALANCE

Analysis of Acid-Base Disturbances

Determine the $[H^+]$, $[HCO_3^-]$ and PCO_2 and ensure that there is no laboratory error by verifying that

$$PCO_2 \text{ (mm Hg)} = \frac{[H^+]\,[HCO_3^-]}{23.9}$$

where the units of $[H^+]$ are nM/L and the units of $[HCO_3^-]$ are mM/L

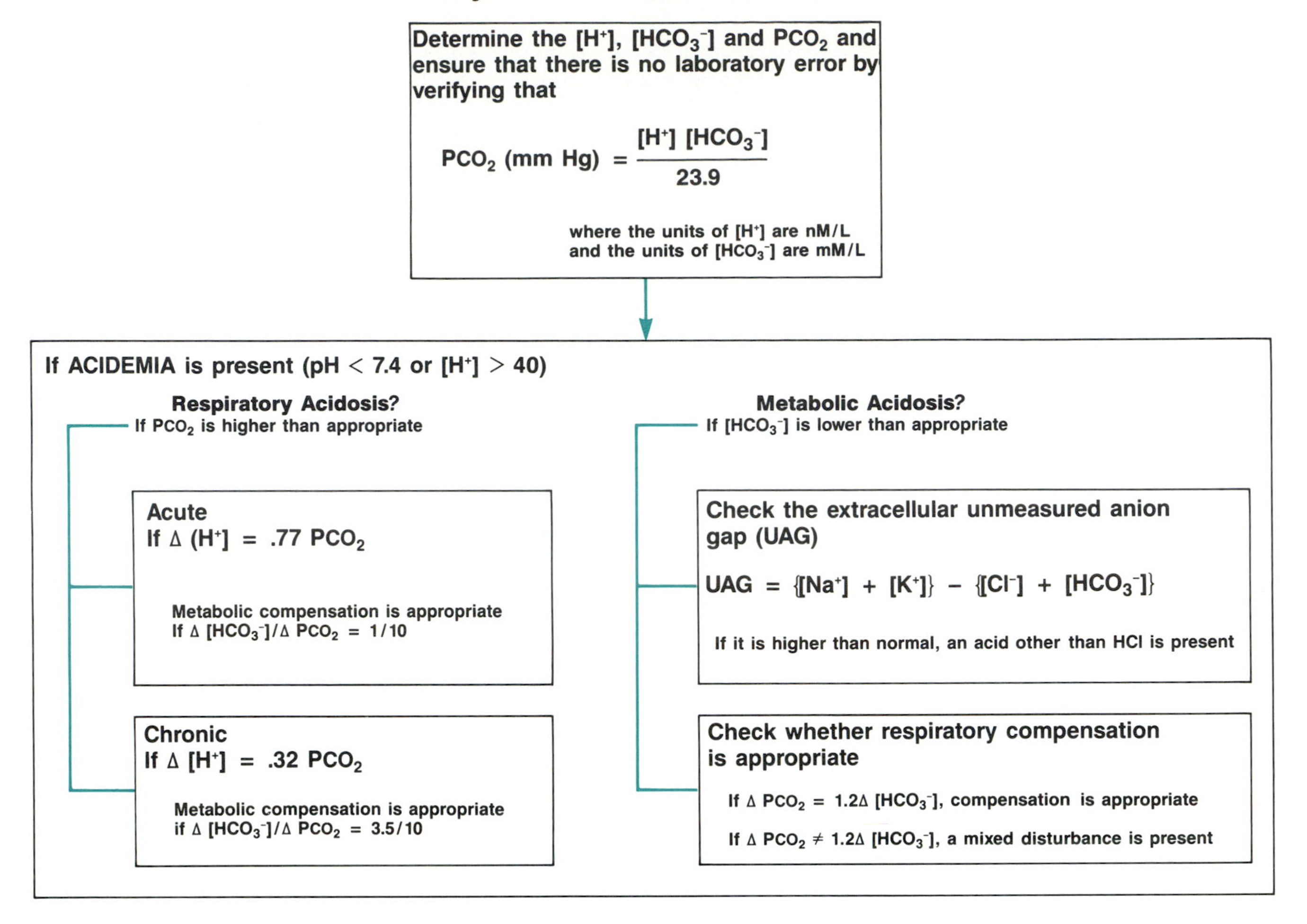

If ACIDEMIA is present (pH < 7.4 or $[H^+]$ > 40)

Respiratory Acidosis?
If PCO_2 is higher than appropriate

- **Acute**
  If $\Delta (H^+) = .77\, PCO_2$
  Metabolic compensation is appropriate if $\Delta [HCO_3^-]/\Delta PCO_2 = 1/10$
- **Chronic**
  If $\Delta [H^+] = .32\, PCO_2$
  Metabolic compensation is appropriate if $\Delta [HCO_3^-]/\Delta PCO_2 = 3.5/10$

Metabolic Acidosis?
If $[HCO_3^-]$ is lower than appropriate

- Check the extracellular unmeasured anion gap (UAG)
  $UAG = \{[Na^+] + [K^+]\} - \{[Cl^-] + [HCO_3^-]\}$
  If it is higher than normal, an acid other than HCl is present
- Check whether respiratory compensation is appropriate
  If $\Delta PCO_2 = 1.2\Delta [HCO_3^-]$, compensation is appropriate
  If $\Delta PCO_2 \neq 1.2\Delta [HCO_3^-]$, a mixed disturbance is present

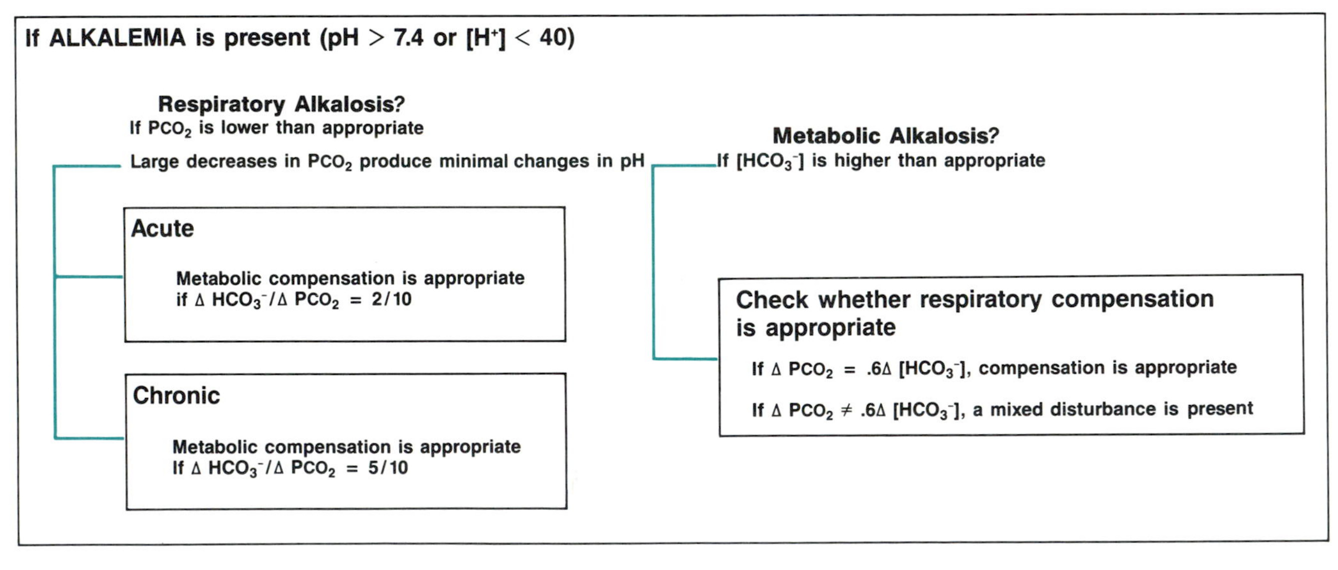

If ALKALEMIA is present (pH > 7.4 or $[H^+]$ < 40)

Respiratory Alkalosis?
If PCO_2 is lower than appropriate
Large decreases in PCO_2 produce minimal changes in pH

- **Acute**
  Metabolic compensation is appropriate if $\Delta HCO_3^-/\Delta PCO_2 = 2/10$
- **Chronic**
  Metabolic compensation is appropriate if $\Delta HCO_3^-/\Delta PCO_2 = 5/10$

Metabolic Alkalosis?
If $[HCO_3^-]$ is higher than appropriate

- Check whether respiratory compensation is appropriate
  If $\Delta PCO_2 = .6\Delta [HCO_3^-]$, compensation is appropriate
  If $\Delta PCO_2 \neq .6\Delta [HCO_3^-]$, a mixed disturbance is present

If $[H^+]$ is normal, check PCO_2, $[HCO_3^-]$ and UAG to ensure that there is no mixed acid-base disturbance.

MICTURITION

Anatomy

- The body of the bladder is a smooth-muscle chamber formed by the **detrusor muscle.**
- The ureters and the urethra connect to the bladder through the **trigone,** a triangular area of fine smooth muscle fibers, located near the bladder neck. This arrangement creates a functional internal sphincter that can be relaxed only by contraction of the detrusor muscle (mostly parasympathetic control).
- The urethra is surrounded near its origin by the **external sphincter** (striated muscle; somatic control).

Bladder Filling

During this phase **the external sphincter is constricted** (somatic control) and **the bladder fills passively** from the ureters, the urine being propelled by **peristaltic waves in the ureters.**

When about 300 mL of urine has been collected and the **chamber pressure** reaches about 20 mm Hg, a tension threshold is reached and the **urge to micturate** is perceived. This urge originates in **stretch receptors** that project to the anterior pons. It can be suppressed **(continence)** by inhibitory activity from hypothalamic and cortical centers to prevent emptying at unsuitable times. If it is not suppressed, then the **voiding reflex** is initiated.

Bladder Emptying (The Voiding Reflex)

The first steps are

- **contraction of the detrusor muscle** (parasympathetic control)—this permits passive opening of the internal sphincter;
- **relaxation of the external sphincter** (somatic control).

Once the bladder has begun to empty, the process accelerates explosively until emptying is completed. The acceleration is mediated by several factors:

- Contractions of the detrusor muscle while the bladder is still full will further excite bladder afferents.
- The appearance of urine flow in the urethra may excite afferent receptors.
- Central inhibitory processes may be blocked.

Central Nervous Influence

Transection of the cord above the sacral level may remove sympathetic influences and does remove higher nervous contributions to bladder control. **Individuals** with such injuries **can learn to control bladder evacuation on the basis of local, spinal reflex paths.** This involves initiation of detrusor contractions at suitable intervals by momentary elevation of bladder pressure above the tension threshold (e.g., by slight tapping of the abdomen).

MICTURITION

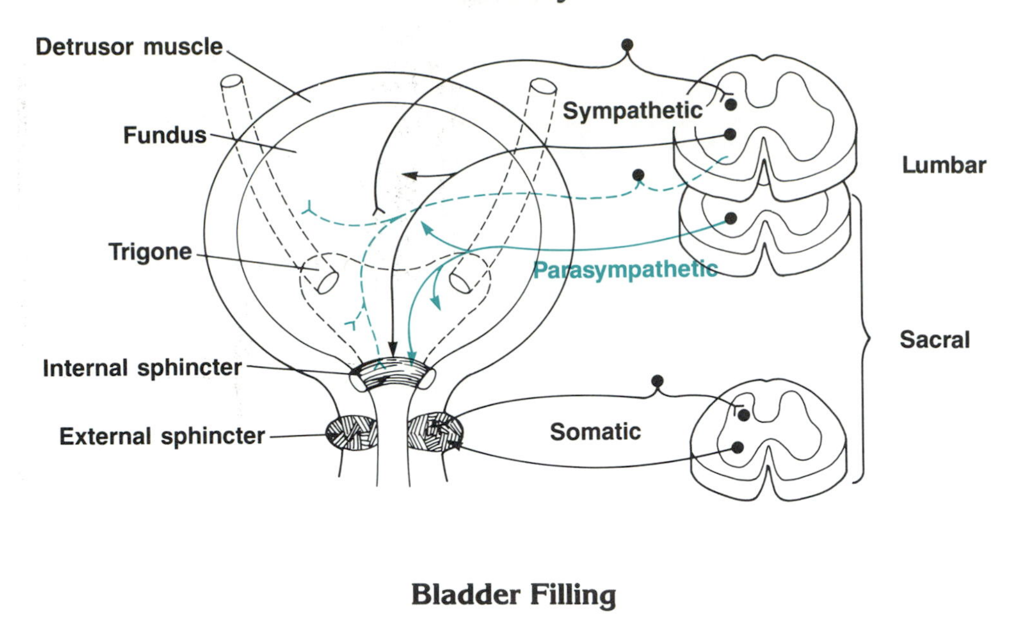

Bladder Filling

Urge

Continence

Stretch receptors

External Sphincter

Somatic

Bladder Emptying

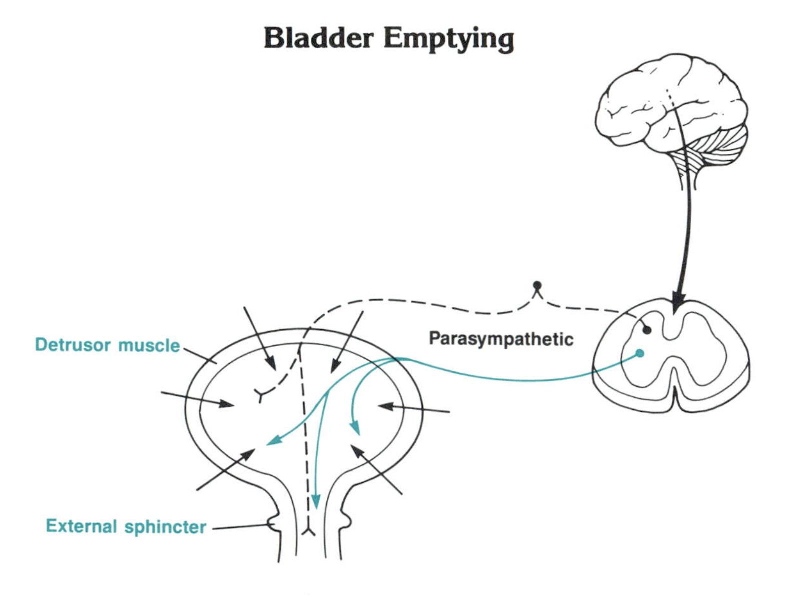

BOWEL EVACUATION

Filling of the Rectum

The rectum fills by peristaltic contractions of the descending colon. Each wave initiates **stretch receptor activity** in afferents of the pudendal nerve and pelvic nerve afferents. Once the rectum contains about 2 L of feces, such stretch receptor activity leads to conscious perception of an **urge to defecate.**

Anomalously, the arrival of each bolus in the rectum is also accompanied by transient relaxation of the internal sphincter (a reflex governed by the enteric nervous system). As a result, **continence** must be **maintained by tonic contraction of the external sphincter** (resulting from a somatic, spinal reflex with afferents and efferents in the pudendal nerve). Soon after the arrival of each bolus, the rectum accommodates the increased volume, wall stretch is reduced, and the internal sphincter contracts.

Emptying of the Rectum

At a suitable time, evacuation of the rectum is initiated by **voluntary effort.** This involves four components:

1. **Intra-abdominal pressure and intrarectal pressure are increased** by voluntary contraction of abdominal muscles.
2. Rectal contents are moved distally by **reflex contraction of longitudinal musculature** in the colon and rectum. This further increases intrarectal pressure.
3. The **internal sphincter relaxes** as each pressure wave arrives in the rectum.
4. The **external sphincter is relaxed** by voluntary effort.

Central Nervous Influence

Transection of the spinal cord above the sacral level abolishes the voluntary motor patterns that assist defecation. As a result, paraplegics must learn special techniques for relaxing the external anal sphincter.

BOWEL EVACUATION

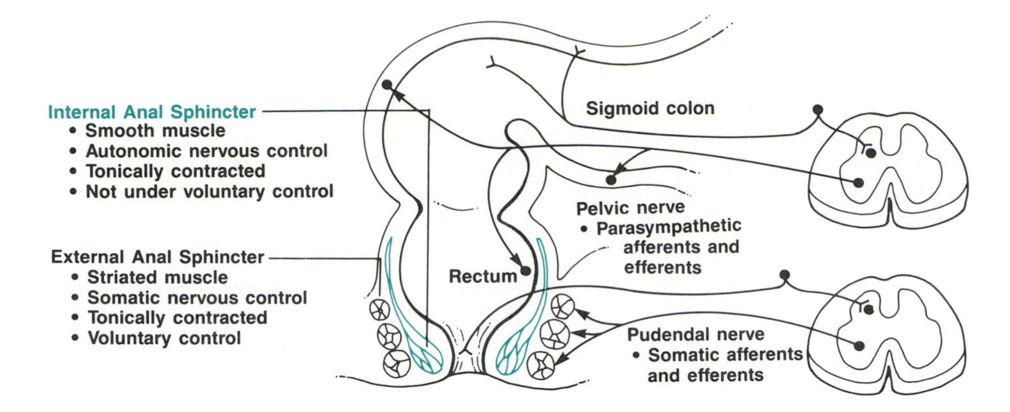
Internal Anal Sphincter
• Smooth muscle
• Autonomic nervous control
• Tonically contracted
• Not under voluntary control
External Anal Sphincter
• Striated muscle
• Somatic nervous control
• Tonically contracted
• Voluntary control
Sigmoid colon
Pelvic nerve
• Parasympathetic afferents and efferents
Rectum
Pudendal nerve
• Somatic afferents and efferents

OVARIAN FUNCTION

Anatomy

The **cortex** is the dominant anatomical feature of the ovaries. It is composed of three areas: the interstitium, the **follicles,** and the **corpus luteum.**
FOLLICLES: These are the sites of **oocyte nurture.**
CORPUS LUTEUM: This is the final stage of the follicle from which the oocyte was expelled during ovulation.

The Follicle

- Almost all follicles are **primary follicles.** They are only **protective spherical structures** in which an oocyte is surrounded by a layer of granulosa cells, a basement membrane, and a layer of theca cells.
- **Some follicles grow, for a brief period each month,** into concentrated hormone sources that profoundly influence the emotional, physical, and reproductive potentials of a woman. These are the **Graafian follicles.**

Primary follicles

These follicles dominate in the mature woman.

- They were **present at birth,** and at maturity they are only double or triple their original size.
- They have a centrally placed oocyte, and the most advanced among them show small pockets of **antral fluid** separating some granulosa cells.
- Some follicles are clearly undergoing **atresia,** a process of deterioration that steadily reduces the number of oocytes.

Graafian follicles

Although most of the 2 million oocytes present at birth are lost to attrition over the next 50 years,

- **Each month before menopause, 6 to 12 follicles develop beyond the primary stage.** Within the month these follicles grow to be up to 30 times larger.
- Their granulosa and theca cells grow so rapidly that the **interior structure** becomes **asymmetrical;** they are then called Graafian follicles.

Regional membrane receptor activity and hormone synthesis

While the follicle was in the primary stage,

- **theca interna cells** (the layer of theca cells closest to the basement membrane) had few, if any, receptors;
- **granulosa cells** contained only receptors for follicle stimulating hormone (FSH).

As the follicle grows to the preovulatory stage,

- **Theca interna cells** develop a large number of membrane **luteinizing hormone** (LH) **receptors.** As a result, these cells, which contain the enzymes required for the full synthetic pathway from acetate-cholesterol to testosterone, produce ample androstenedione and testosterone.
- **Granulosa cells** use androstenedione and testosterone as substrates for the production of **estradiol.** They also **develop receptors** for **LH** and for **prolactin.**

OVARIAN FUNCTION

Anatomy

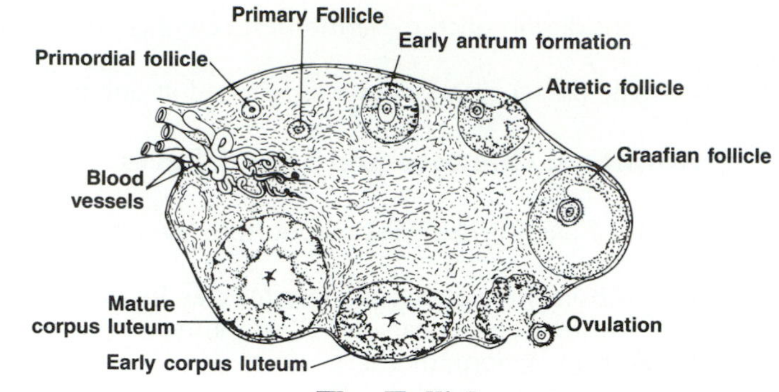

The Follicle

Primary Follicle

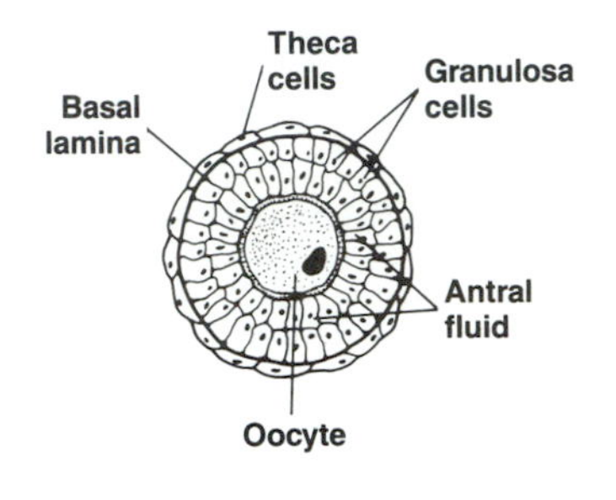

Graafian Follicle

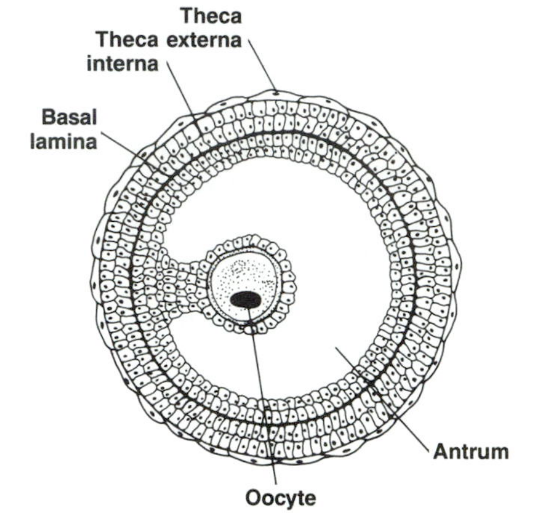

Steroid Synthesis

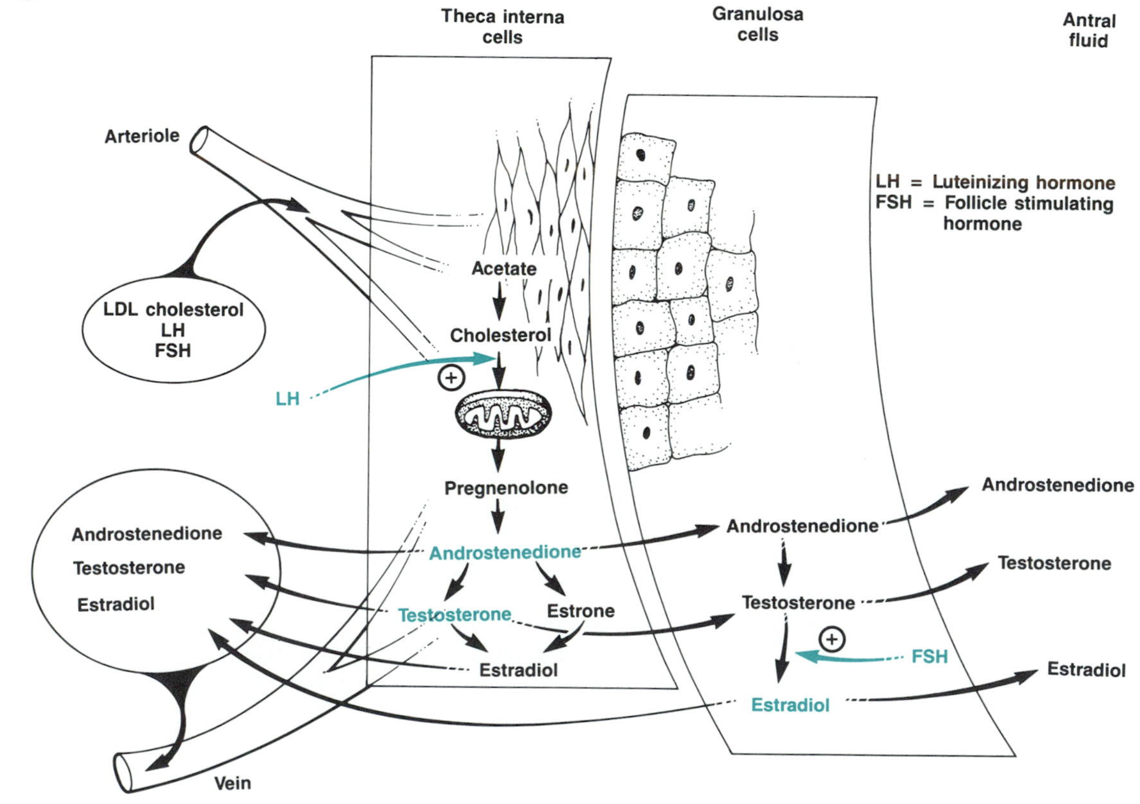

WOMEN'S MONTHLY RHYTHM

Purpose

1. To create at some point during each cycle the best possible conditions for reproduction and nurture
2. To ensure that normally only a single oocyte is fertilized at one time

Hormonal Changes and Their Effects

1. At about the 24th day of each cycle, **FSH** output from the pituitary begins to rise.

- This rise **induces growth in three to six primary follicles in each ovary.** They reach a peak size of about 5 mm diameter near day 10 of the next cycle, and a single dominant follicle is selected from among them.
- The rise in FSH also **stimulates estradiol formation in granulosa cells.** Not enough estradiol is produced to increase its level in plasma at this time, but its local concentration rises enough to induce more FSH receptors in granulosa cells, generating even more estradiol production. Eventually:

2. At days 10 to 12 of the next cycle, there is a steep rise in **plasma estradiol** concentration. The increased estradiol

- exerts **positive feedback on pituitary LH release.** The **LH surge triggers ovulation from the dominant follicle.**

A surge in FSH release occurs as well.

- This induces growth in a further group of follicles. The function of this latter group is to contribute to estradiol production in the luteal phase of the cycle.

3. Following the LH and FSH surges there is a rapid down-regulation of LH and FSH receptors. This causes

- **decreased output of androgens from theca interna cells and of estradiol from granulosa cells.**

The released oocyte is not under any immediate further hormonal influence. On the other hand, the remaining granulosa cells, set free from an apparent inhibition exerted either by the oocyte or by the follicular fluid, proliferate over the next 3 days and develop into the **corpus luteum.**

Corpus luteum

The formation, growth, and secretory behavior of the corpus are under hormonal control.

- The corpus is initially maintained by LH.
- In the later luteal phase it is maintained by human chorionic gonadotropin (hCG), an LH-equivalent of placental origin.
- Its major secretory product is **progesterone.** Progesterone **inhibits the hypothalamus and the anterior pituitary.** The resultant progressive decrease in LH and FSH levels has 2 consequences: It leads to **atresia** in the post-ovulatory follicles and it causes regression in the corpus luteum (provided that fertilization has not occurred).

4. By the 26th day of the cycle,

- the levels of estradiol and progesterone are so low that
  a. the pituitary can begin the FSH rise that starts the next cycle and
  b. the endometrium is without adequate steroid support and disintegrates, to be sloughed in the menstrual flow.

WOMEN'S MONTHLY RHYTHM

How Do Estradiol and Progesterone Act to Achieve the Purposes of the Rhythm?

By creating **optimum conditions** for reproduction and nurture.

Estradiol

- Promotes endometrial growth and vascularization
- Increases vaginal secretions that provide lubrication as well as a capacitating medium for sperm
- Stimulates growth and development of breasts

Progesterone

- Facilitates sperm transport by decreasing viscosity of vaginal secretions
- Slows endometrial growth, but induces supportive secretory activity instead
- Stimulates mammary growth and development so that lactation may occur
- Inhibits uterine motility

By ensuring that **only a single oocyte is fertilized** at one time.

- A pregnant uterus maintains a functioning corpus luteum and develops a placenta.

↓

Progesterone

↓

- Progesterone blocks the positive feedback effect of estradiol on the pituitary.

↓

LH surge is prevented.

↓

No further ovulation.

Hormonal, Follicular, and Endometrial Phases

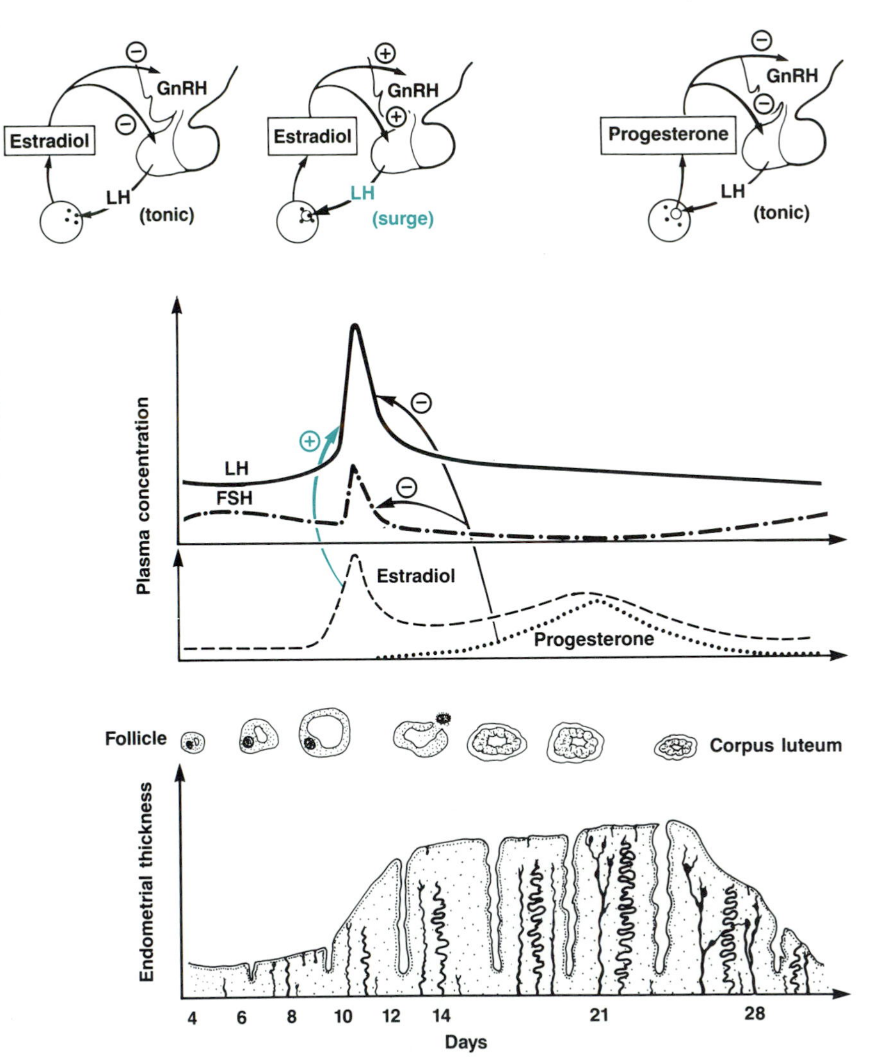

At normal plasma levels (50 to 60 pg/mL), estradiol inhibits LH synthesis by negative feedback on pituitary gonadotrophs as well as on hypothalamic GnRH-producing neurons. However, when plasma estrogen levels have been maintained at a level above 150 pg/mL for more than 36 hours, then estrogen exerts a transient positive feedback and produces a surge in the release of LH from the anterior pituitary.

FERTILIZATION AND PREGNANCY

Fertilization

Fertilization occurs when **the genetic material of a sperm combines with that of an ovum.**

Sperm transport

Of the several million sperm that are deposited during ejaculation near the cervix,

- many leak out through the vagina;
- many are incapacitated by the ambient acidity;
- many are immobilized by the viscous cervical mucus;
- many are phagotized by intrauterine leukocytes;
- only a few experience full capacitation (i.e., they undergo sufficient membrane deterioration to permit release of hydrolytic enzymes from their acrosomes).

Penetration of the ovum

Hydrolytic enzymes released from the acrosomes of several sperm clear a path by dissolving the ground substance of corona radiata and zona pellucida. (This is called an **acrosomal reaction.**)

As soon as one sperm reaches the oocyte cytoplasm,

- **entry of further sperm is prevented**
  a. by depolarization of the oocyte membrane (inward Na^+ current) and
  b. by an alteration of the space between the oocyte membrane and zona pellucida (exteriorization of cortical granules after the sperm-mediated increase in intracellular $[Ca^{++}]$ is involved);
- **the first cell division is initiated by a surge in intracellular $[Ca^{++}]$.**

FERTILIZATION AND PREGNANCY

Fertilization

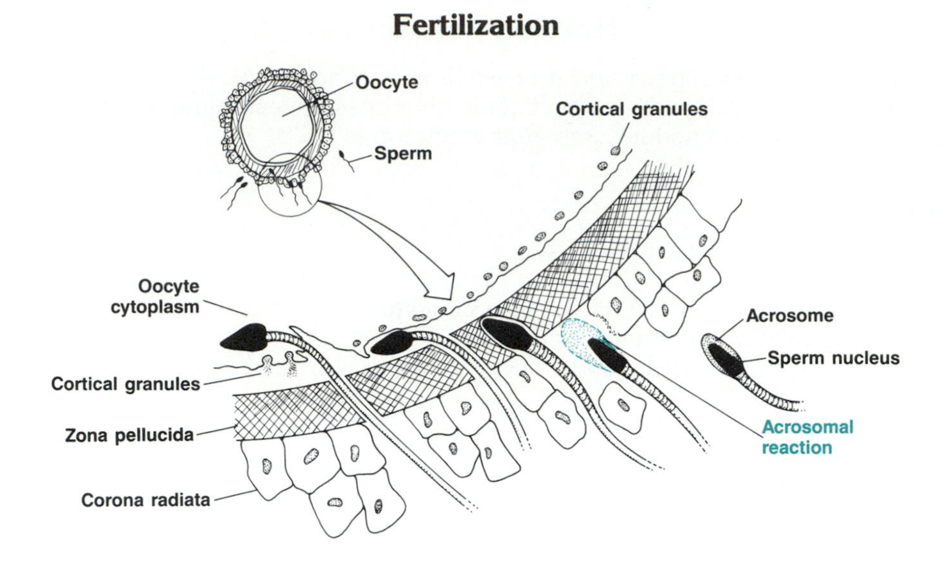

Hormone Synthesis

After fertilization, **estrogens** and **progesterone** continue to provide important hormonal support for the endometrium. Two new trophic hormones appear: **human chorionic gonadotropin** (hCG) and **human chorionic somatomammotropin** (hCS).

Estrogens

During pregnancy, these hormones

- **foster continuous growth of the myometrium;**
- **foster growth of the mammary ducts;**
- act to soften and reshape abdominal and genital structures so as to accommodate the expanding uterus.

Progesterone

- **ensures a quiescent uterus** by blocking spread of contractile activity;
- promotes nutritive secretions from fallopian tubes and endometrium.

Human chorionic gonadotropin (hCG)

As the levels of estrogens and progesterone increase, secretion of LH and FSH is progressively suppressed. Instead, hCG replaces pituitary LH as the controlling hormone.

- It appears early after fertilization.
- It is synthesized by a layer of cells (trophoblasts) whose major function is attachment of the blastocyst to the endometrium.
- It acts to maintain corpus luteum function beyond its usual life span and promotes steroid synthesis in the placental-fetal unit.

Human chorionic somatomammotropin (hCS)

- A peptide hormone that is unique to pregnancy
- Appears by the fifth week, and its plasma concentration rises continuously until birth
- Adapts maternal metabolism to the needs of the growing fetus

Hormone Synthesis

Role of the placenta

Over a period of 4 to 8 weeks after fertilization and implantation, synthesis of estrogens and progesterone is shifted from the corpus luteum to the placenta and the fetus. (The placenta lacks sufficient enzymes to generate dehydroepiandrosterone [DHEA] for subsequent conversion to androgens and estradiol. It obtains DHEA by extraction from fetal and maternal blood.)

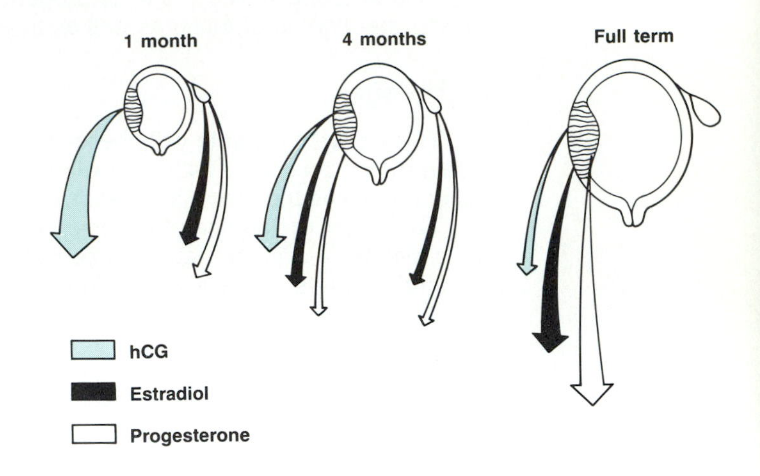

Control of hormone synthesis

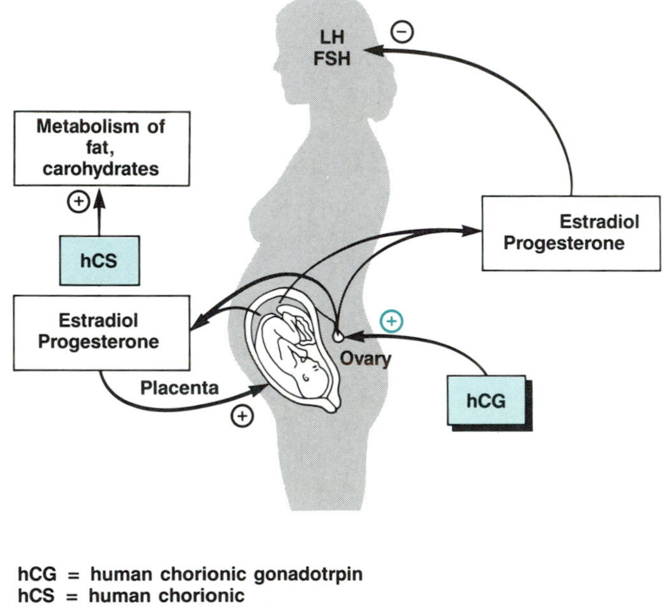

hCG = human chorionic gonadotrpin
hCS = human chorionic somatomammotropin

TESTICULAR FUNCTION

Androgens

Testosterone is the major androgen of the testes. It is carried in blood by a specific testosterone-binding globulin as well as by albumin and other plasma proteins.

Synthesis

This occurs in **Leydig cells** because (1) they contain LH receptors; (2) their mitochondria contain the cytochrome P-450 enzyme that converts **cholesterol to pregnenelone;** (3) their cytoplasm contains the enzyme necessary for conversion of **androstenedione to testosterone.**

The major metabolite of testosterone, estradiol, is produced in several tissues. Testosterone is synthesized in the fetus (while hCG is present) and after puberty (when LH levels are sufficient), but it is not synthesized in childhood.

Actions

Testosterone has six major functions:

- Regulation of gonadotropins (GnRH and LH [adult] or hCG [fetus])
- Formation of the male phenotype during sexual differentiation
- Induction of sexual maturation and of secondary sexual characteristics at puberty
- Modulation of sexual desire
- Modulation of bone formation
- Promotion of protein deposition, especially contractile proteins in skeletal muscle

TESTICULAR FUNCTION

Anatomy

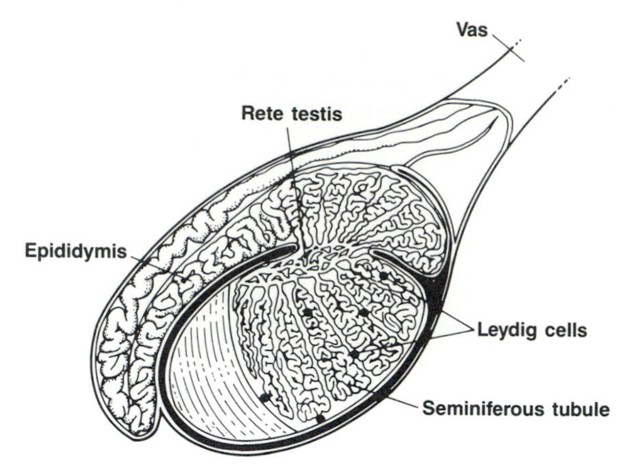

The two major functions of the testes are **androgen production** and **sperm formation.** They take place in different cells.

Leydig cells

- Located in the connective tissue that lies between the seminiferous tubules
- Site of androgen synthesis
- Open into the rete testis

Seminiferous tubules

- Make up the bulk of the testes
- Site of sperm production

Rete testis

- A sperm-collecting chamber that is connected to the **epididymis** by efferent ductules

Androgen Synthesis and Regulation

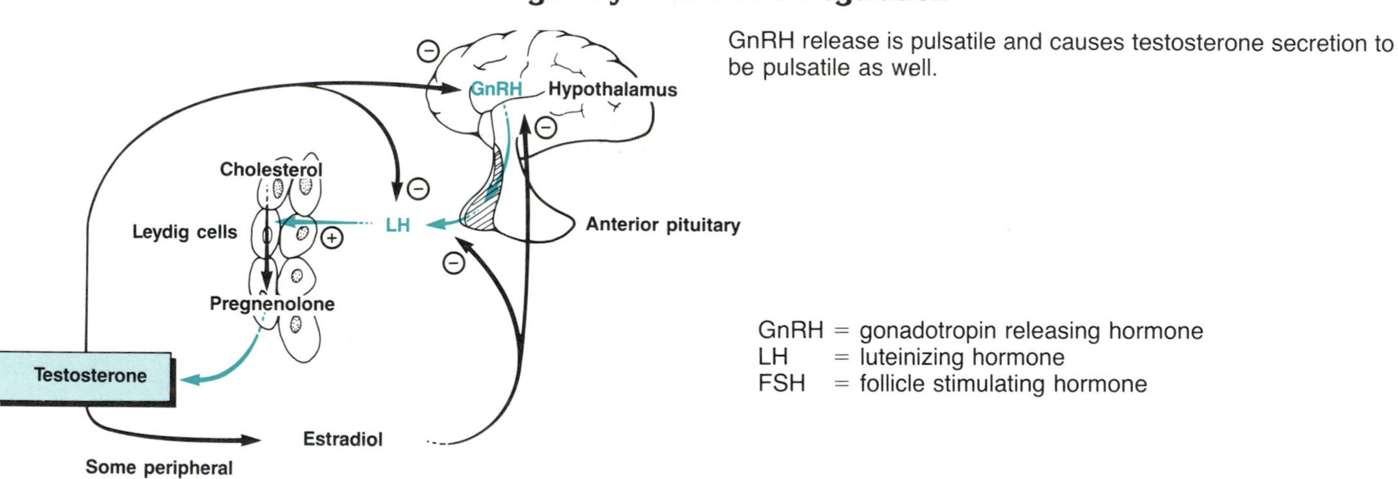

GnRH release is pulsatile and causes testosterone secretion to be pulsatile as well.

GnRH = gonadotropin releasing hormone
LH = luteinizing hormone
FSH = follicle stimulating hormone

TESTICULAR FUNCTION—CONT'D.

Sperm Production

Over a period of 60 to 80 days, undifferentiated germ cells **(spermatogonia)** become highly differentiated **spermatozoa** that are independently mobile and carry in their heads **23** ***unpaired*** **chromosomes** plus the enzymes required for penetration of the ovum.

Hormonal control

Spermatogenesis requires LH, testosterone, and FSH.

- **LH** exerts all of its spermatogenic effects via high intratesticular levels of testosterone.
- **Testosterone** promotes termination of one of the intermediate stages of sperm formation.
- **FSH** stimulates secretory activity in **Sertoli cells.**

Sertoli cells

These specialized cells line the inside of the seminiferous tubules and create within them a unique environment. Three aspects of Sertoli cell physiology contribute to this:

1. Sertoli cells form **tight junctional complexes** among themselves and thereby control the flow of molecules between interstitial space and tubular lumen.
2. Their secretions create the **chemical milieu** for sperm formation.
3. They act as mothers to spermatids throughout spermatid formation, maintaining a **constant physical contact, supplying substrates,** and **absorbing redundant cytoplasm** from the transforming spermatids.

In addition, **Sertoli cells secrete a variety of proteins:**

- Transport proteins for Cu^{++}, Fe^{3+}, and androgens
- Proteins that have hormone-like activity (including **inhibin,** a feedback suppressor of pituitary FSH synthesis)
- Proteins that have enzymatic activities (e.g., plasminogen activator)
- Proteins that help maintain the basement membrane

Epididymis

The end product of Sertoli cell nurture is nonmotile (and, therefore, impotent) spermatozoa that are washed out of the tubule into the rete testis. Their final processing occurs in the epididymis.

Sperm Production

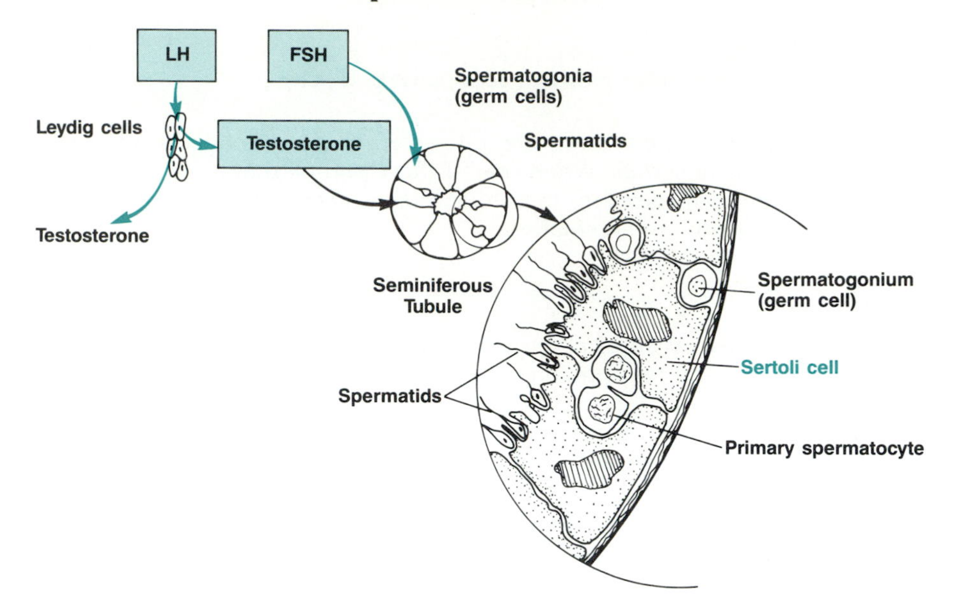

THE HUMAN SEXUAL RESPONSE

This area of human physiology is dominated by misinformation, psychological characteristics, and cultural mythology.

The Sexual Response Cycle

Most often we are in a nonsexual state, characterized by attention to other matters and by a receptiveness to sexual stimuli. When one of these stimuli draws our attention and a full sexual response is allowed to develop, then four phases can be recognized in the response: **excitement, plateau, orgasm,** and **resolution.**

Excitement

This phase is initiated by physical (visual or tactile) or psychogenic factors. Its continuation and development depend on situational reinforcement. Excitement **begins with the onset of erotic feelings.** It is marked by

- progressive increase in autonomic nervous activity, leading to regionally specific changes in blood flow as well as increased ventilation, heart rate, and arterial blood pressure;
- progressive increase in vasocongestion and muscular tension in the erogenous areas (causing skin mottling, erection of nipples and penis, clitoral enlargement, vaginal lubrication, or penile erection).

Plateau

This is an advanced state of arousal in which vasocongestion is at its maximum, causing the formation of the orgasmic platform in the woman or the attainment of maximum penis size in the man. Additional physical changes include

- lifting and tilting of the uterus to form a seminal receptacle;
- raising of the testes to the perineum;
- discharge of a few mucoid drops from the penis.

The duration of the plateau phase varies with the effectiveness of the erotic stimuli and the effectiveness of the situational reinforcement as well as with the desire, training, and ages of the individuals involved. The plateau phase can lead to orgasm.

Orgasm

This phase is brief and marked by **rapid release of the developed vasocongestion and muscular tension.** It involves the whole body.

- Reactions in the genitalia include rhythmic contractions, emission and ejaculation of semen, and increased liberation of vaginal secretions.
- Reactions in skeletal muscles include voluntary and involuntary contractions such as nearly spastic contractions of facial, abdominal, and intercostal muscles.
- Reactions in whole-body sympathetic activity include changes in respiration, heart rate, and blood pressure.
- Central nervous effects operate to exclude all other sensory perceptions.

Orgasm is followed by a refractory period in the man. During this period he will not be able to respond physically to further sexual stimuli. **The length of the refractory period increases with age.**

Although the woman may not wish to be further aroused after orgasm, she does not have a refractory period during which she cannot be aroused.

Resolution

All physiologic variables return to their resting states within 30 minutes after orgasm, and **frequently there is a desire to sleep. Often a desire to urinate** is felt as well. This is most likely due to activation during orgasm of sympathetic efferents with bladder branches.

THE HUMAN SEXUAL RESPONSE

Myths and Realities

Myth

- Individuals with stereotypically attractive physique are capable of better sexual performance and greater sexual enjoyment.
- Large breasts are more satisfying to a male sexual partner.
- A large penis is more satisfying to a sexual partner.
- Penis size is correlated with body build, race, and frequency of sexual experience.
- Only vaginal penetration can lead to a full orgasm in the female sexual partner.

Reality

- Body configuration and sexual function are not anatomically or physiologically correlated. They are, however, linked by the psychological effects of a perceived mismatch between an individual's body configuration and the culturally fostered stereotype.
- A large penis may cause a partner to be apprehensive about inability to accommodate.
- There is great variation in the size of the flaccid penis, but when it is erect it is 15 ± 2.5 cm long and 3 to 4 cm in diameter. Its size is not predictable by race or influenced by sexual experience.
- Sensory nerve endings are highly concentrated in the clitoris. In addition, there may be particularly responsive areas along the vaginal walls (e.g., the Grafenberg [G] spot). As a result, a woman experiences different kinds of orgasm, varying in intensity of feeling and depending on the degree of stimulation of each of her centers of sexual sensory perception.

Innervation of Genitalia

Women

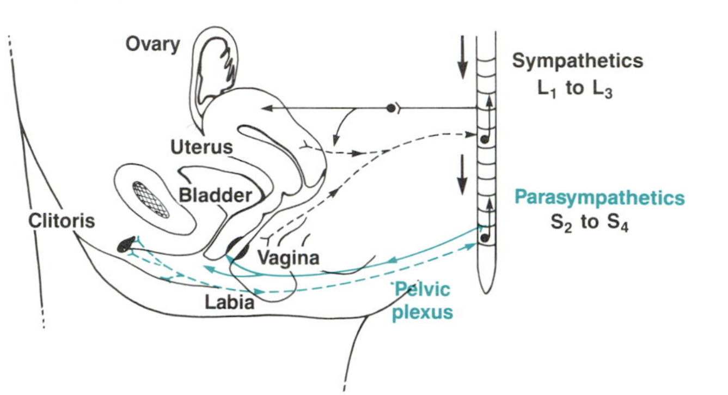

The major innervation is via the **pelvic plexus.** Its fibers distribute to the clitoris or glans penis, the perineal nerve, and the inferior rectal nerve. They arise from and terminate in sacral segments S_2 to S_4. As a result of this distribution, **sensations received from the clitoris, penis, labia, scrotum, or anus may be perceived as similarly pleasurable.**

Erection is almost totally **a local parasympathetic reflex** and does not require higher nervous centers. If the local spinal cord area (S_2 and below) is injured, then erection is possible via psychogenic factors in about 25% of patients.

Emission is a **centrally coordinated** reflex. **Ejaculation** is under **local parasympathetic control,** requiring integrity of cord segments S_2 to S_4. If the cord were injured above S_2, then there would be no emission, there would be dry "ejaculation," and there would be no sensation of orgasm.

Men

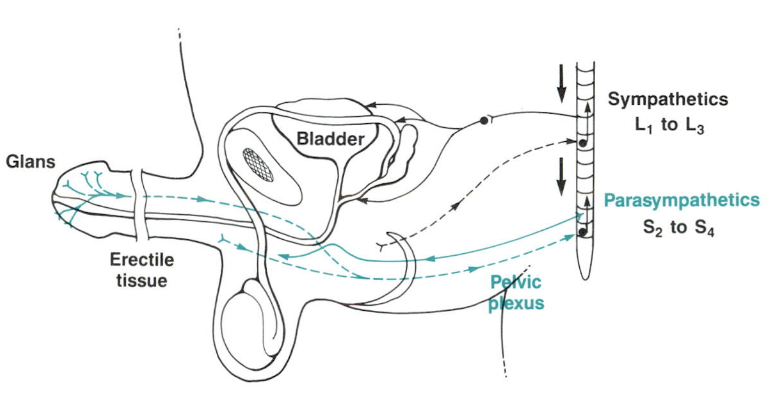

The Sexual Response Cycle

Women

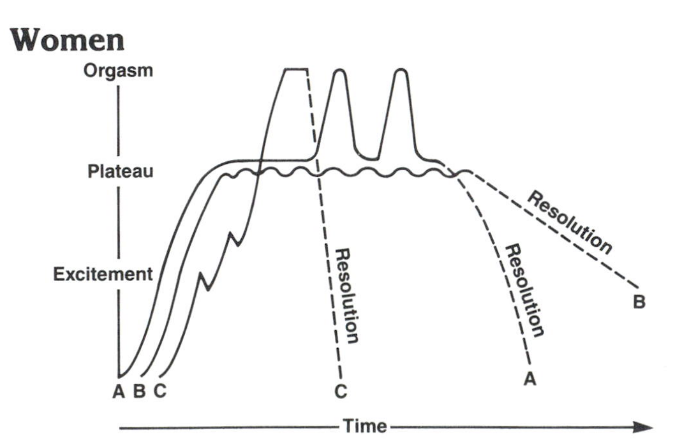

Men

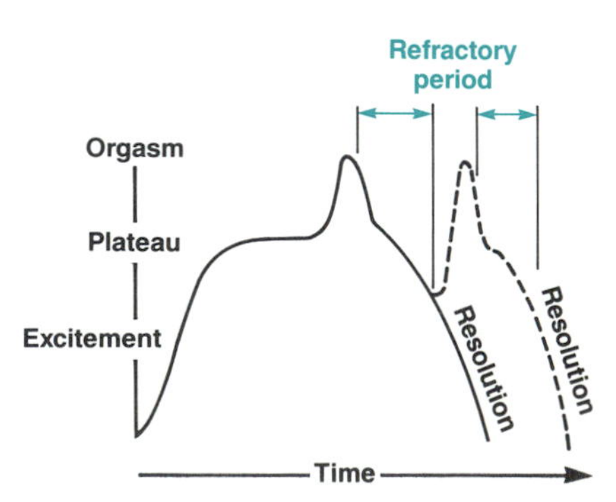

A, B, and C represent responses observed in different women.

13

PHYSIOLOGY OF AGING

AGING OF ORGANISMS

The process of physiologic aging has effects quite different from those of the physical and mental pathologies that afflict the aged, particularly after age 75. Aging encompasses changes in many biochemical and physiologic variables. Despite these, **total body function is impressively well maintained, under nonstressful conditions,** in optimally healthy 70-year-olds.

Theories of Aging

The causes of aging have not been identified apart from a **strong genetic link to the age of the mother.** Six classes of theories have been offered:

1. LATE-ACTING DELETERIOUS GENE: Aging is the result of **random accumulation of deleterious mutations.**
2. WEAR AND TEAR: Aging is caused by the **gradual wearing out of irreplaceable body constituents.**
3. EXHAUSTION: **When an irreplaceable substance has been exhausted,** aging follows.
4. ACCUMULATION OF TOXICANTS: Aging results from the **accumulation of autointoxicants.**
5. ENDOCRINE CHANGES: Aging results from **variations in the levels of certain hormones.**
6. IMMUNE REACTIONS: Aging is a consequence of **declining immune competence and increasing autoimmunity.**

Immune competence

- Immune competence declines with age.
- **Diminishing immune protection exposes the elderly to higher incidence of disease.**

Autoimmunity

- The ability to distinguish self from nonself decreases with age.
- **There is increasing production of antibodies that react with normal cells and destroy them.**

AGING OF ORGANISMS

Definitions of Aging

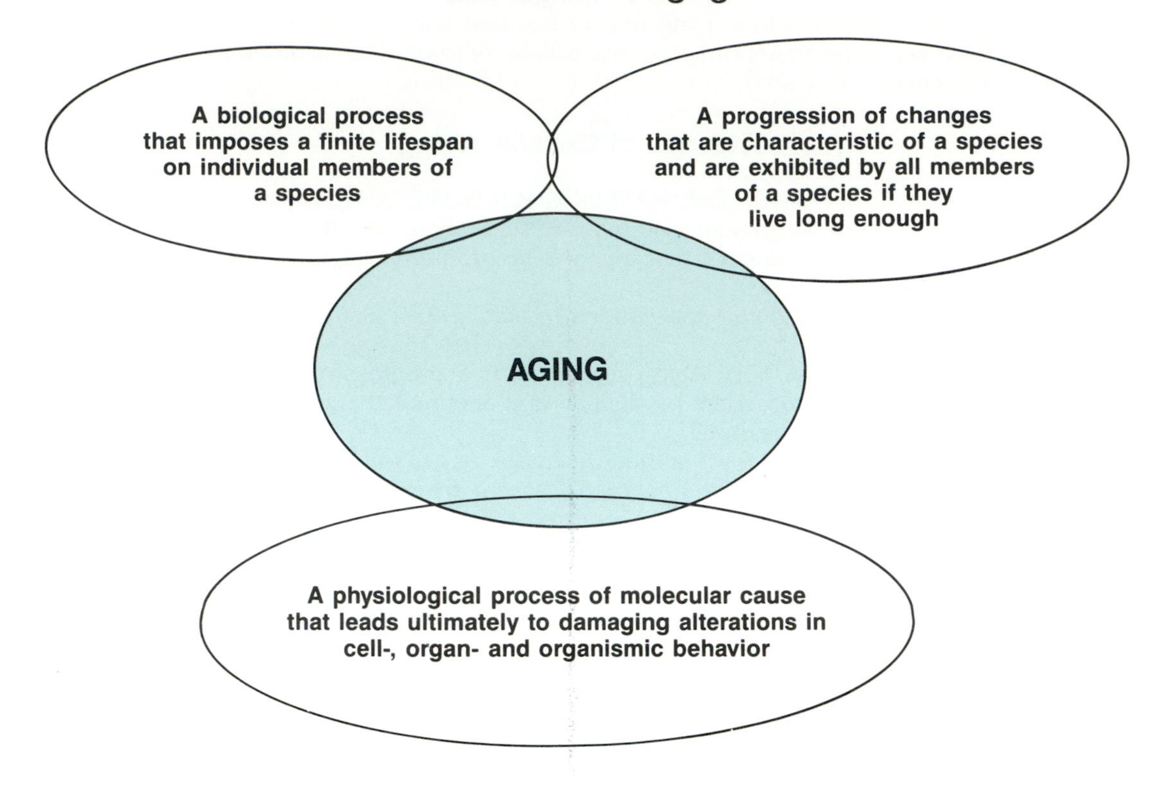

A biological process that imposes a finite lifespan on individual members of a species
A progression of changes that are characteristic of a species and are exhibited by all members of a species if they live long enough
AGING
A physiological process of molecular cause that leads ultimately to damaging alterations in cell-, organ- and organismic behavior

Genetic Links in Aging

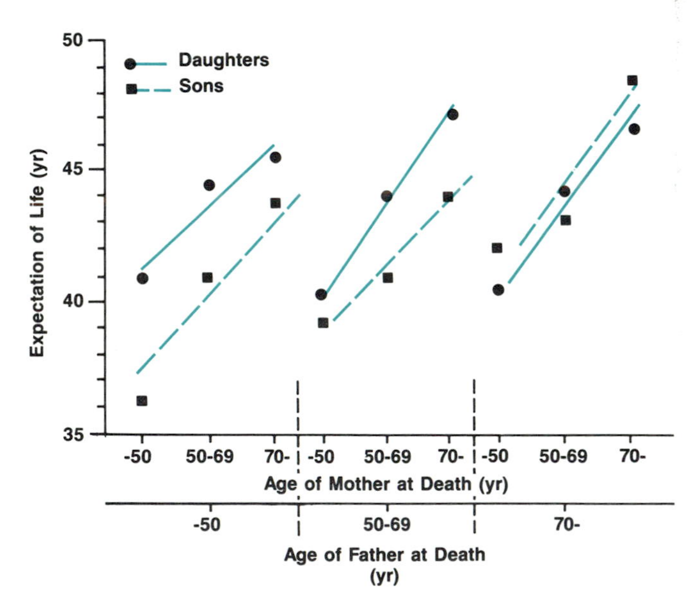

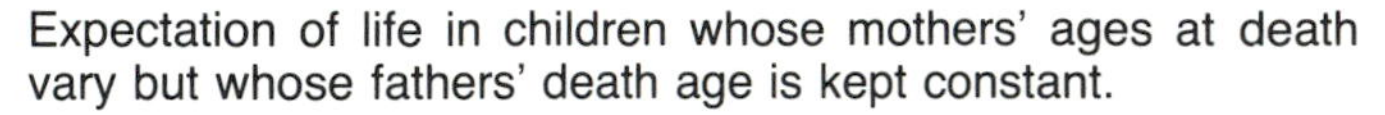

Expectation of life in children whose mothers' ages at death vary but whose fathers' death age is kept constant.

Reproduced with permission from Jalavisto. Ann Med Int Fenniac 1951; 40:267.

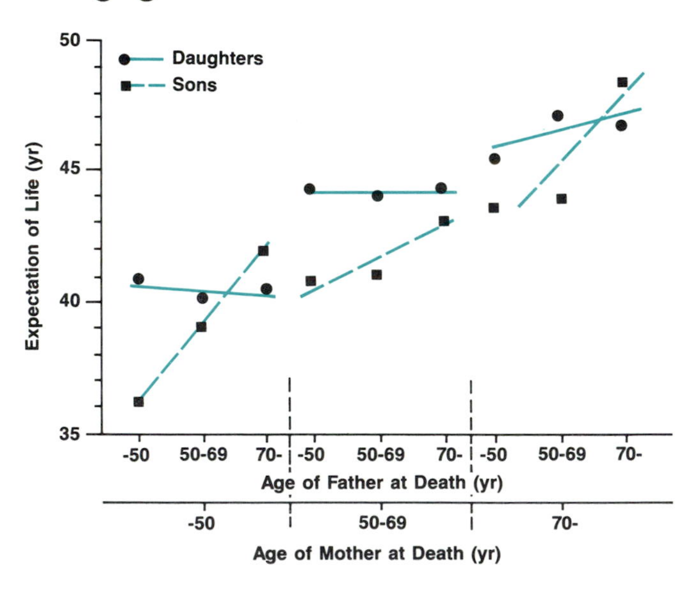

Expectation of life in children whose fathers' ages at death vary but whose mothers' death ages are kept constant.

Reproduced with permission from Jalavisto. Ann Med Int Fenniac 1951; 40:268.

AGING OF CELLS

The most striking age-related cellular changes occur in the cell membrane. Nevertheless, it is commonly believed that aging occurs because some intracellular process prevents complete molecular replacement or successful cellular renewal. The following six theories have enjoyed prominence.

Theories of Cellular Aging

1. PROGRAMMED AGING (THE HAYFLICK PHENOMENON): Human fetal lung fibroblasts, in vitro, undergo a finite number of replications and then die. This observation was interpreted as evidence that aging is preprogrammed and controlled by one or more **aging genes.**
   (However, replicating cell populations in vivo survive well beyond the life span of the organism.)
2. ERROR ACCUMULATION: Aging results from a **progressive increase in erroneous gene replication,** subsequent production and accumulation of defective proteins, and an eventual error catastrophe.
   (Experimental support for this theory is weak because (1) accuracy of protein synthesis does not decline with age, and (2) there are organisms in which no errors in the fidelity of protein synthesis occur [e.g., drosophila], but aging takes place nevertheless.)
3. SOMATIC MUTATION: In the absence of evidence for a decline in either DNA synthesis or the capacity for DNA repair, it has been suggested that aging is caused by a **mutational alteration in cellular DNA stores.**
   (This is unlikely because, without ionizing radiation, the occurrence of mutations is too infrequent to account for overall age changes.)
4. PROTEIN LINKAGE: Aging results from **changes in protein properties** arising from progressive and excessive cross-linking between adjacent molecules.
5. FREE RADICALS: Aging is caused by an **accumulation of toxic byproducts of oxygen metabolism.**
   (This hypothesis has been reconciled neither with the fact that enzymatic systems [superoxide dismutase, catalase, and glutathione peroxidase] and scavenger systems [vitamins E and C, cysteine, glutathione, uric acid, and β-carotene] are plentiful nor with the fact that experiments with antioxidant compounds have not yielded conclusive results.)
6. LIPOFUSCIN ACCUMULATION: The rate of **intracellular lipofuscin deposition has been inversely correlated with life span in many animals.**
   (However, at least two observations speak against a direct causal link: (1) myocardial cells show no loss of ability to hypertrophy despite large lipofuscin depositions; (2) antioxidant therapy in animals has significant effects on reducing intracellular lipofuscin accumulation, but has no significant effect on extending maximum life span.)

AGING OF CELLS

Cell Membrane

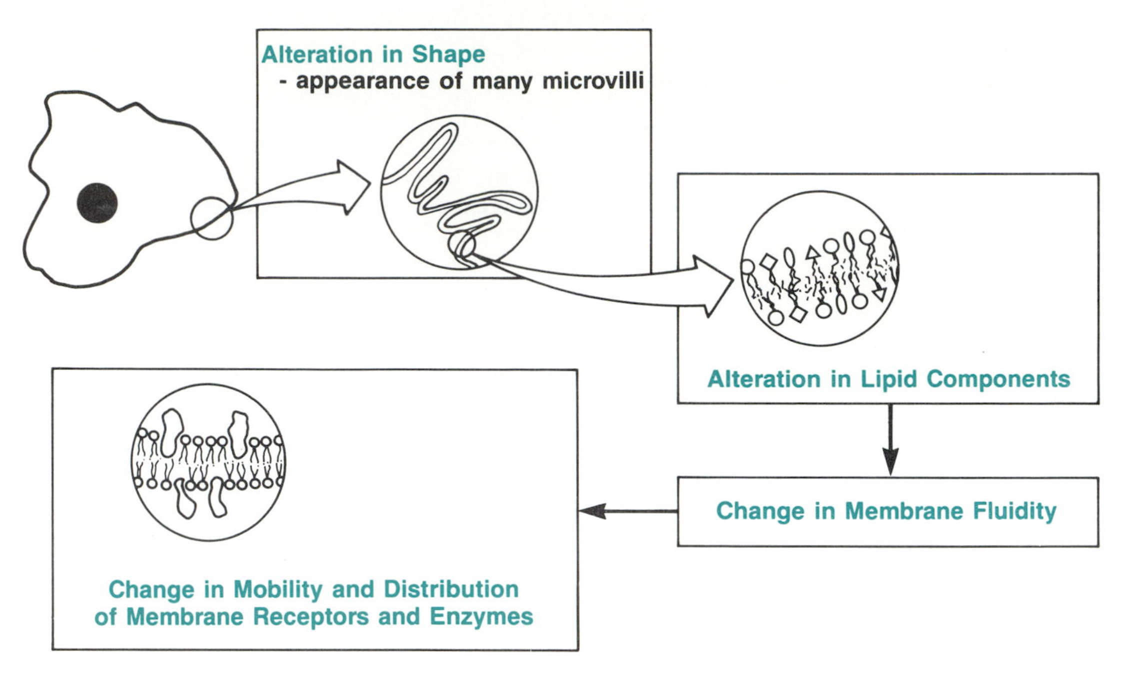

Cell Interior

Lipofuscin* accumulation is the most prominent change in many cells from many organisms. It accumulates particularly in nondividing cells (e.g., neurons and cardiomyocytes).

---

*Lipofuscin (age pigment) is a byproduct of lipid peroxidation during self-digestion of cells.

MODULATION OF AGING PROCESSES

Aging involves a variety of mechanisms, and although many attempts have been made to alter its progress, no single global intervention is likely to succeed—there is no elixir of youth.

Pharmaceutical interventions

Gerovital-H_3

Studies of its effect on longevity in animals and humans have been conflicting, and **its only consistent effect in humans is a mild antidepressant action** (presumably related to the monoamine oxidase [MAO] inhibitor action of procaine).

L-Dopa

Experiments were prompted by the finding that deficiencies in brain dopaminergic systems occur with age. **Long-term therapy in patients with Parkinson's disease has not increased their life spans beyond those of the general population.**

Dehydroepiandrosterone (DHEA)

DHEA is a precursor for other steroids, but many additional effects have been attributed to it. **Its effectiveness as an anti-aging factor in humans is not yet known.**

Antioxidants and free-radical scavengers

The severe weight loss that results from antioxidant administration is a confounding influence in the application of these agents.

In humans, only vitamin E has been systematically investigated. **Population studies have not shown a dose-related beneficial effect of vitamin E on mortality.** On the contrary, decreased life span was observed in a group of individuals over age 65 who were taking more than 1,000 IU per day.

Body temperature

So far, there is **no clear evidence** from human studies on the long-term effects of body cooling.

Nutrition

Human studies have shown **no clear relationship** between body weight and longevity. Studies on the effects of diet composition have focused on disease processes such as GI cancer and atheroma, not on aging itself.

Immunologic interventions

Only animal studies have been performed. They were based on the observations that (1) thymectomy accelerates the immune deficiencies observed with aging and (2) coenzymes Q, a group of compounds that participate in the mitochondrial electron transport chain, are deficient in old age. **No information on effects in humans is available yet.**

Exercise

Regular exercise reduces mortality in individuals at risk of coronary artery disease, presumably by increasing the ability of heart muscle to withstand ischemic insult.

MODULATION OF AGING PROCESSES

Prolongation of life is one of the oldest and most persistent dreams of humankind.

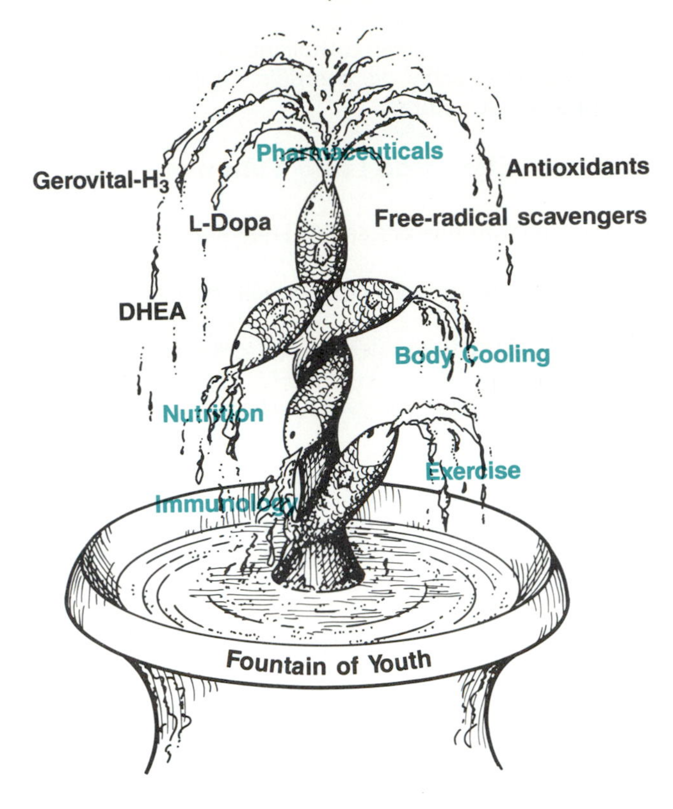

Pharmaceutical Interventions

Gerovital-H_3

A preparation of procaine, benzoic acid, and metabisulfite. Promoted since the 1950s as a treatment for age and age-related disorders.

L-Dopa

Long-term L-dopa therapy in patients with Parkinson's disease yielded increased survival compared with untreated patients.

Dehydroepiandrosterone (DHEA)

DHEA administration in mice has resulted in prolongation of normal immune function, reduced incidence of mammary cancer, and increased survival.

Antioxidants and free-radical scavengers

A variety of antioxidants have been used in animal studies (mostly in mice) to delay aging.
A life-prolonging effect was demonstrated in some studies.

Body Cooling

In hibernating animals and in poikilotherms, lifespan is prolonged by decreased body temperature.

Nutrition

Calorically deficient diets leading to weight reduction on the order of 20 to 25% have consistently prolonged lifespan in rats.

Immunological Interventions

Attempts have focused on replacement of thymus-derived factors or on reversing deficiencies in coenzymes Q.

Exercise

Current evidence suggests that regular exercise affects lifespan in groups that are susceptible to coronary artery disease.

AGING AND THE RESPIRATORY SYSTEM

Changes in Structure

With age, lung tissue becomes more compliant and loses elastic recoil. The thoracic cage, on the other hand, becomes stiffer. Its overriding influence causes an **overall decrease in the compliance of respiratory structures.**

Effects on Lung Volumes

Changes in lung volumes with age are small. They are dominated by

- **enlargement of the airways** (leading to a small increase in tidal volume that has no effect on alveolar ventilation);
- **reduced elastic recoil of the lungs** (leading to reduced intrapleural pressure).

Diminution of intrapleural pressure has two functional consequences:

1. It causes **higher residual volume** (RV) because the outward pull of the chest wall is less forcefully opposed by the inward pull of the lungs.
2. It causes **higher "closing volume"*** during forced expiration because the surface-tension forces that tend to collapse small airways are less strongly opposed by the intrapleural elastic forces in neighboring tissue.

Effects on Blood Gases

Parts of the aging lungs are underventilated during portions of the breathing cycle. This is caused by accumulation of particulate matter and greater alveolar collapse (i.e., higher closing volume).

The resulting mismatch between ventilation and perfusion leads to **significant reduction in arterial Po_2** with age, falling to about 75 mm Hg by age 75. **Arterial Pco_2 remains constant.**

Changes in the Control of Respiration

Ventilatory responses to hypercapnia and hypoxia are reduced. It has not yet been determined whether this is due to reduced chemoreceptor sensitivity, to impaired central nervous responses, or to compromised neuromuscular mechanisms.

*During forced expiration to residual volume, sufficient intrapleural pressure is developed to collapse some small airways. Closing volume is the volume of air trapped distal to the closure. It is larger than residual volume, but smaller than functional residual capacity.

AGING AND THE RESPIRATORY SYSTEM

Changes in Structure

Structural Change	Functional Consequence
Thoracic Cage	
Wasting of respiratory muscles	Decreased maximal ventilation
Calcification of cartilage	More rigid chest wall —Greater use of diaphragm —Preference for increasing rate, rather than depth, of breathing in hyperpnea
Lung Tissue	
Enlarge alveolar ducts, but shallower, flatter alveoli	Altered air distribution —More in ducts —Less in alveoli Diminished transport surface area
Less (and differently distributed) elastic tissue, but more fibrous tissue Cross-linking of elastin	Decreased elastic recoil —Reduced vital capacity —Increased residual volume —Decreased ventilatory flow rate —Increased flow resistance
Vascular Tissue	
Decreased capillarity Altered capillary-alveolar interface	Ventilation-perfusion mismatch

Effects on Lung Volumes

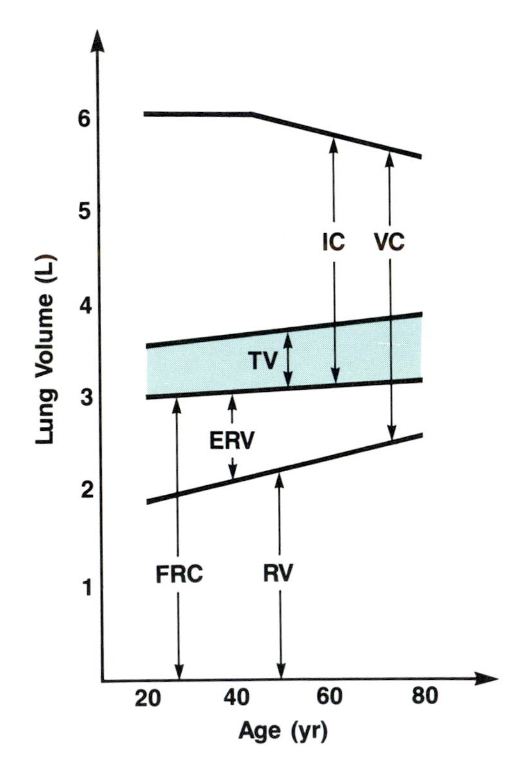

TV = Tidal volume
ERV = Expiratory reserve volume
IC = Inspiratory capacity
VC = Vital capacity
FRC = Functional residual capacity
RV = Residual volume

AGING AND THE CARDIOVASCULAR SYSTEM

Cardiovascular Function at Rest

Aging of the cardiovascular system is a focus of study because the most common diseases among the elderly are those of the heart and blood vessels (atherosclerosis). Their occurrence rises exponentially with age, and they have come to be regarded as inevitable manifestations of aging. There is little evidence to support this view. **In the aged who are without symptoms of cardiovascular disease, most indices of cardiac function remain normal at rest.**

Cardiovascular Function During Exercise

Exercise studies, restricted to individuals who are free of occult cardiovascular disease, also fail to support the common belief in severely deteriorating function with age alone. **It is chiefly the maximum attainable exercise level that declines with age.**

The following observations are made when indices of cardiac performance in 80-year-olds are compared with those in 20-year-olds at any one level of exercise:

- The decrease in total peripheral resistance is not different.
- The increase in cardiac output above rest is not different.
- The increase in heart rate above rest is smaller. (The cause is thought to be diminished myocardial responsiveness to β-adrenergic stimulation.)
- **The increase in stroke volume is larger.**
- The change in ejection fraction from rest is smaller.
- **Left ventricular end-diastolic filling volume (preload) and filling pressure are increased.** Nevertheless,
- early diastolic filling rates are reduced.

Reliance on increased preload to meet cardiac demands results in two additional age-related changes:

1. Increased left ventricular diastolic filling pressure causes **elevated pulmonary venous pressure.** (This predisposes the elderly to pulmonary congestion and shortness of breath during exercise.)
2. The force required to develop a given intraventricular pressure increases with preload (Laplace's Law). As a result, the elderly have **mild cardiac hypertrophy.** An additional cause for age-related ventricular hypertrophy may be diminished target organ responsiveness to β-adrenergic stimulation (most likely due to diminished phosphorylation of cellular proteins or to impairment of phosphorylation-dependent ion transport mechanisms).

AGING AND THE CARDIOVASCULAR SYSTEM

Cardiovascular Function at Rest

Studies of the general population, unscreened for coronary artery disease, are skewed by the prevalence of cardiovascular pathology among the elderly and give the impression that cardiovascular function deteriorates with age. This view is not supported by studies restricted to healthy, physically active individuals (the Baltimore Longitudinal Study).

	Institutionalized and Unscreened for Coronary Artery Disease	Active in Community Life; No Evidence of Coronary Artery Disease (Stress Test)
	Change from Age 20 to Age 90	Change from Age 30 to Age 80
Heart rate	Slight decrease	Slight decrease
Stroke volume	Decrease	Slight increase
End-diastolic volume	Increase	Increase
Ejection fraction	Decrease	No change
Cardiac output	Decrease	No change
Total peripheral resistance	Increase	No change
Systolic blood pressure	Increase	Increase
Diastolic blood pressure	No change	No change

Indices of Cardiac Function

- No deterioration in cardiac function with age per se.
- No left ventricular dilatation.
- No decline in the velocity of left ventricular fiber shortening (contractility).
- Early (rapid) diastolic filling rate decreases progressively, reaching about 50 percent of normal at age 80. However, this decline does not compromise left ventricular end-diastolic volume.
- There is an increase in arterial systolic pressure (in response to age-related increases in aortic stiffness), and this increase is associated with a mild left ventricular hypertrophy.

Cardiovascular Function During Exercise

The aging cardiovascular system adapts to exercise by drawing on the intrinsic myocardial response to preload (the Frank-Starling Law of the heart) in the face of an age-related decline in maximum heart rate.

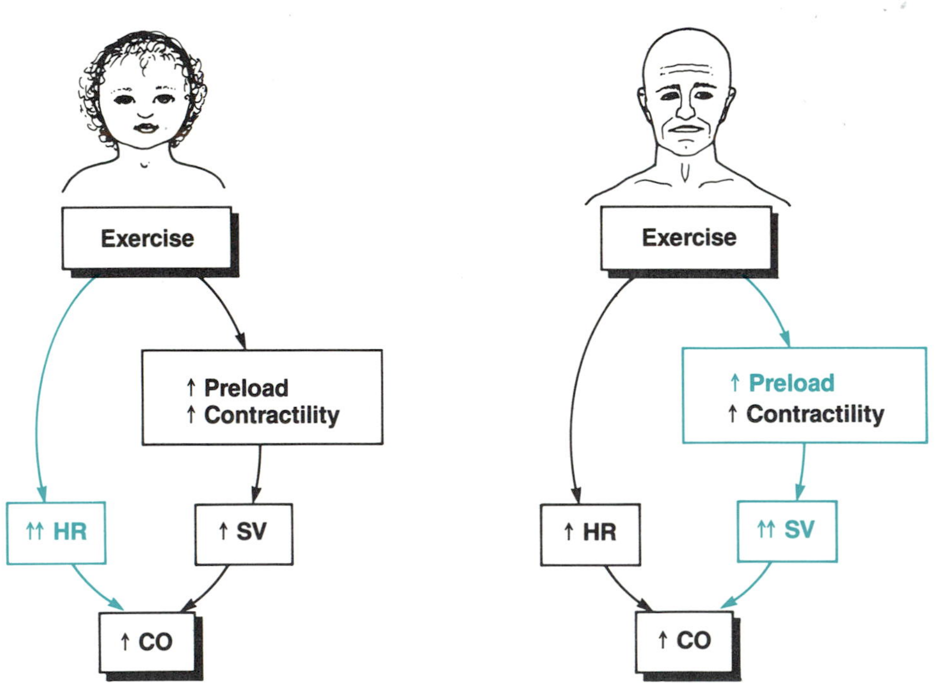

AGING AND THE RENAL SYSTEM

Changes in Body Composition

Cell mass decreases progressively with age (muscle shrinkage) and body fat content increases (in the mesenteric and perirenal areas). **Intracellular fluid volume (percentage of body weight) decreases because adipose tissue has a low water content.** Changes in extracellular fluid volume (percentage of body weight) are small.

Changes in Kidney Function

Structure

Organ level

Total kidney mass is stable at about 350 g from maturity to age 40, and thereafter it declines steadily, reaching about 300 g in the mid-eighties (owing to **loss of nephrons** plus **diminished aptitude for compensatory hypertrophy**).

Histologic level

- The basement membrane thickens.
- The amount of connective tissue increases (especially in the medulla).
- The medullary interstitium becomes increasingly dehydrated.
- The glomerular capillary loops degenerate (especially in the juxtamedullary region), resulting in loss of glomeruli and loss of peritubular capillary networks.
- Arteriolar bypasses appear around nonfunctional glomeruli.
- Arcuate and interlobar arteries become increasingly tortuous.

Function

In addition to the altered renal morphology, hormonal changes occur in the elderly:

- Aldosterone levels are decreased (because of decreased juxtaglomerular function).
- The vasopressin system is changed (in the form of diminished ability to suppress vasopressin secretion and diminished tubular epithelial responsiveness to vasopressin).

The net functional effects of age-related changes in renal morphology and endocrinology are

1. **enhanced urinary excretion of protein;**
2. **impaired ability to cope with fluctuations in fluid intake** (expressing itself as a decline in the ability to concentrate urine and the ability to produce more dilute urine during osmotic and volume stimuli);
3. **impaired excretion of metabolites and drugs.**

Although the loss of renal mass with age reduces the ability to form NH_3, **the elderly are well able to maintain acid-base balance.**

AGING AND THE RENAL SYSTEM

Changes in Body Composition

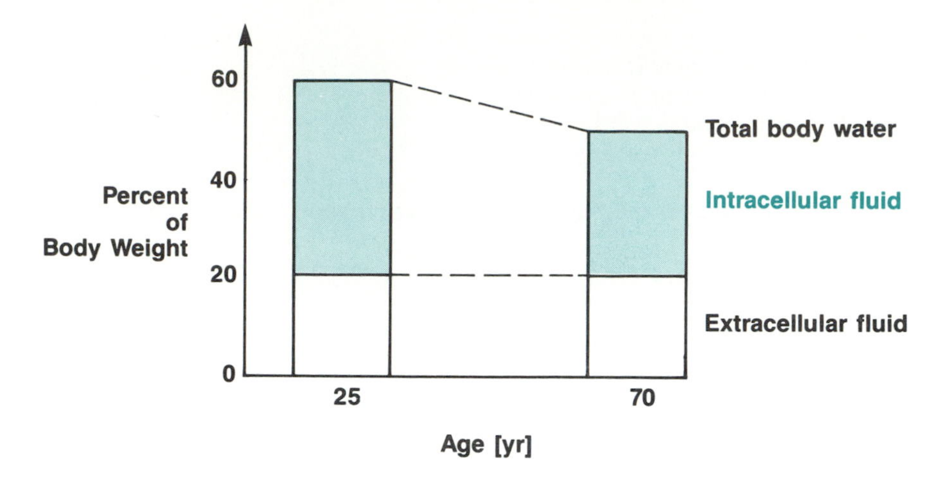

Changes in Kidney Function

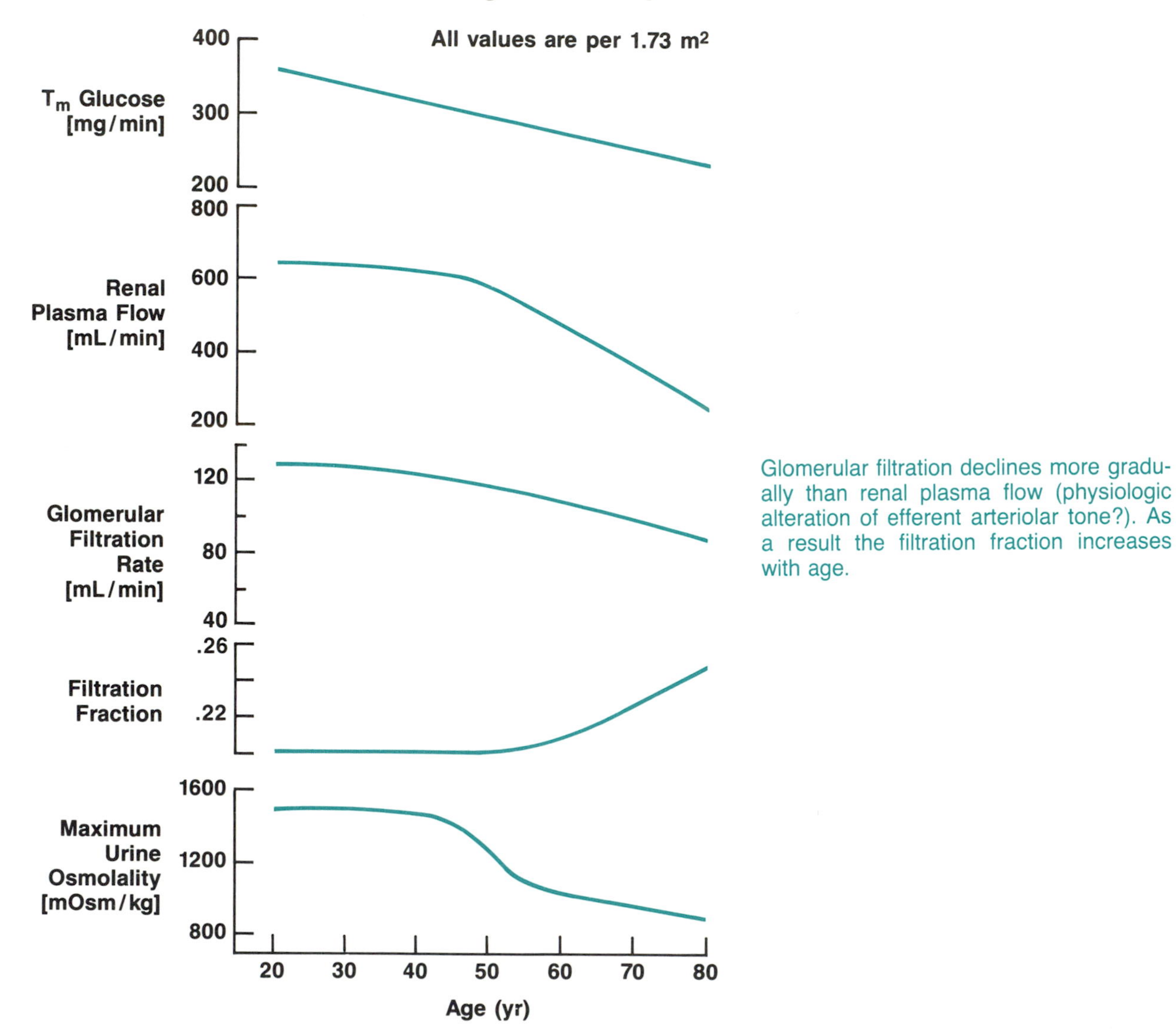

Glomerular filtration declines more gradually than renal plasma flow (physiologic alteration of efferent arteriolar tone?). As a result the filtration fraction increases with age.

AGING AND MICTURITION

Urinary incontinence is a problem in many elderly. Its components include

- motivation,
- level of cognitive function,
- neurologic competence,
- level of mobility, and
- dexterity.

It also reflects **changes in the physiology of lower urinary tract function.**

Men

In men, function of the lower urinary tract is often impaired because the **enlarged prostate creates a physical outflow obstruction.** This leads to a decline in urine outflow rate through the urethra, retention of urine within the bladder, bladder expansion, and **instability of the detrusor muscle.**

Women

In women, especially those who have had several children, the **musculature of the pelvic floor, including the external bladder sphincter, is often weak.** This impairs the ability to remain continent at moments of increased intra-abdominal pressure such as during sneezing, coughing, or laughing.

AGING AND MICTURITION

Control of Lower Urinary Tract Function

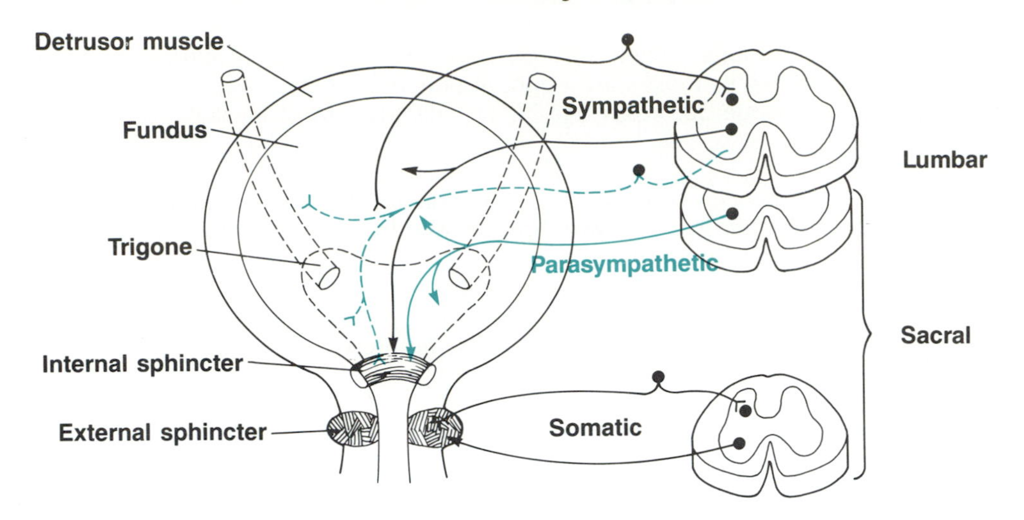

Sympathetic efferents supply mostly the blood vessels in the bladder wall and the portion of the detrusor muscle that surrounds the bladder neck. They have little influence on normal bladder function. Their chief role may be closure of the internal sphincter during orgasm.

Parasympathetic efferents innervate the bladder wall and the urethra.

Afferent fibers, subserving the sensations of stretch and pain, travel mostly with the parasympathetic nerves.

The Voiding Reflex

Urine is transported from the renal pelvis into the bladder by peristaltic waves in the ureters. It is held in the bladder by two mechanisms:

1. Reflux into the ureters is prevented by the nature of the ureter-trigone junction (oblique entry of the ureters into the smooth muscle layer forms a sphincter-like junction).
2. Escape into the urethra is prevented by tonic constriction of the external bladder sphincter (by somatic nerves).

As the bladder fills, wall stretch increases and the distension excites stretch receptor afferents projecting to the brainstem. When the filling volume reaches 200 to 300 mL, the activity of these afferents is such that a conscious desire to void is perceived. Voiding is suppressed until conducive circumstances exist. At a filling volume of 400 to 500 mL, the desire to void becomes very strong.

Voiding begins with contraction of the detrusor muscle (parasympathetic activation) and of the abdominal muscles. The resulting increase of pressure within the bladder further excites stretch receptor afferents. When the pressure within the bladder reaches about 40 cm H_2O, the external sphincter is relaxed by inhibition of its somatic innervation, and urine enters the urethra.

AGING AND THE IMMUNE SYSTEM

Lymphocytes in the Elderly

Changes in the immune system with age have been related to **involution of the thymus gland.** This gland reaches its maximum weight at puberty and decreases to near 5 percent of its maximum weight at age 50. From age 30 on, this process of involution is accompanied by the following:

- **Decline in the plasma concentration of thymic polypeptide hormones**
- **Decrease in the number of T-lymphocytes** (the total number of circulating B-lymphocytes remains constant with age)
- Shifts in the proportion of T-cell subpopulations
- Decrease in the ratio of mature T-lymphocytes to B-lymphocytes in the blood
- Increase in the proportion of circulating **T-lymphocytes displaying surface marker deficiencies**

Lymphocytes from elderly individuals are more susceptible to damage by ultraviolet light, ionizing radiation, and other mutagenic influences.

Cell-Mediated Immune Responses in the Elderly

The ability of T-cells to proliferate in response to a stimulus such as skin testing antigens or tissue grafts declines with age. There are three reasons:

1. There is a **decreased number of responsive T-cells.**
2. The **vigor** of these responsive cells **is reduced.**
3. **Humoral suppressor activity** of unknown origin is present.

Humoral Immune Responses in the Elderly

Age brings changes in the plasma concentrations of immunoglobulins.

- IgM decreases; IgA and IgG both increase.

The ability to mount a humoral response to antigen is reduced (mostly because of decreased helper T-cell activity). The impairment does not express itself as an overt decline in either the total number of antibody-producing cells or the total amount of antibody formed after antigenic stimulation, but it does express itself as a diminished rate of response to novel antibodies.

There is an increase in the production of autoimmune antibodies as well as in the frequency and severity of autoimmune responses (rheumatoid arthritis, lupus, Addison's disease, rheumatic fever, multiple sclerosis, and so on).

AGING AND THE IMMUNE SYSTEM

Involution of the Thymus with Age

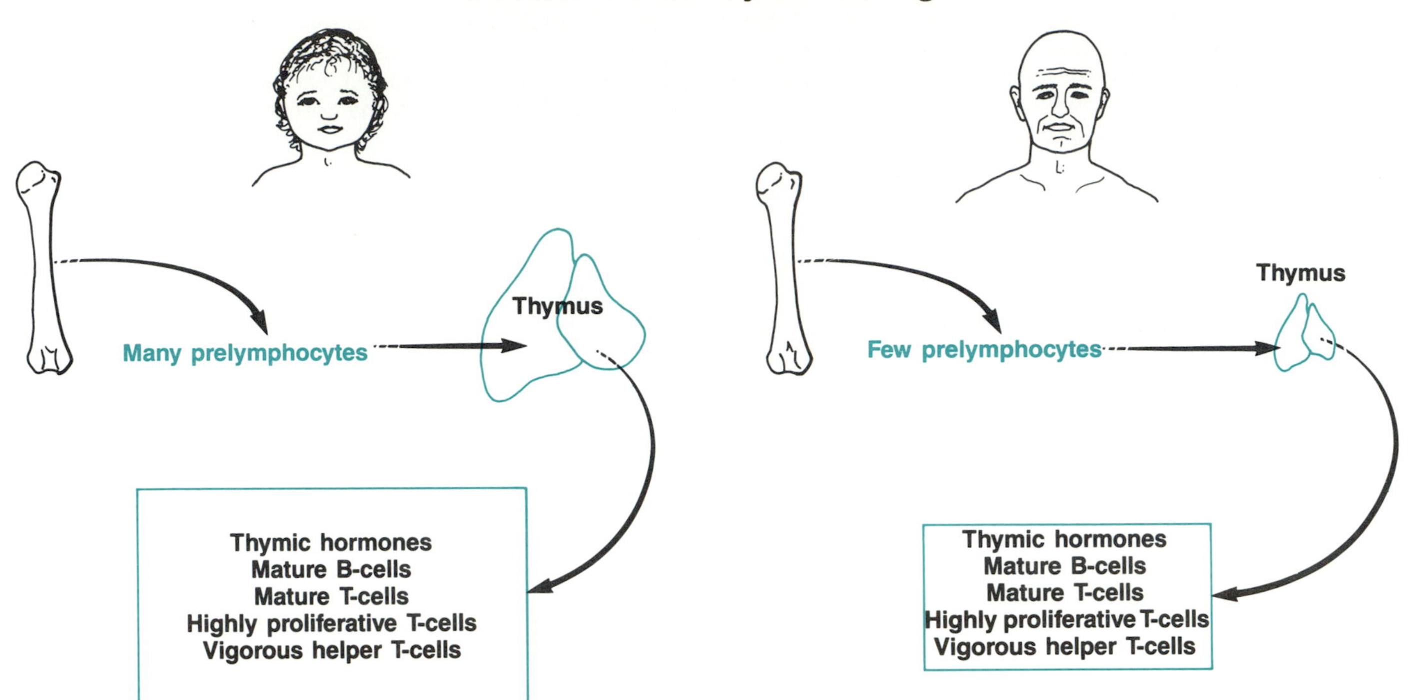

The thymus contains both endocrine and lymphoid tissue.

As an endocrine organ it secretes a family of thymic polypeptide hormones that play vital physiologic roles in stem cell differentiation, T-cell maturation, lymphokine production, and interleukin secretion.

Its lymphoid tissue receives immature prelymphocytes from the bone marrow, processes a small number of them (about 5 percent), and releases them into the blood as mature T-lymphocytes. (By regulating stem cell differentiation, the thymus also indirectly participates in the maturation of B-cells.)

Aging as an Immune Phenomenon

Although it has been clearly demonstrated that changes occur in the immune system with age, and although there is a clear age-related increase in life-threatening diseases, **it has not been possible to demonstrate that there is a correlation between immune competence measured in early life and subsequent life span.** Human studies have suggested that individuals with severely impaired reactions to foreign antigens or with high titers of autoantibodies have higher mortality, but these studies were not able to demonstrate that altered immune responses were the cause of increased mortality rather than just another effect of those factors that were the true cause of increased mortality.

AGING AND ENDOCRINE SYSTEMS

OVERVIEW

Immunologically measured basal plasma levels of most hormones are not affected by age because the ability to adjust secretion rates to clearance rates is maintained. However, the portion of total immunoreactive product that represents the active hormone form declines in some cases. Among these are renin and tri-iodothyronine (T_3) as well as the adrenal cortical hormones, aldosterone and dehydroepiandrosterone (DHEA).

Changes in the Mechanisms of Hormone Action

The finding of **decreased cellular responses to hormones with advancing age** has been consistent. However, there is no consistent pattern of cell surface receptor changes with age. This suggests that **intracellular mechanisms are responsible.** Intracellular changes are also suggested by the observation that several **intracellular steroid receptors are down-regulated with advancing age.**

AGING AND ENDOCRINE SYSTEMS

OVERVIEW

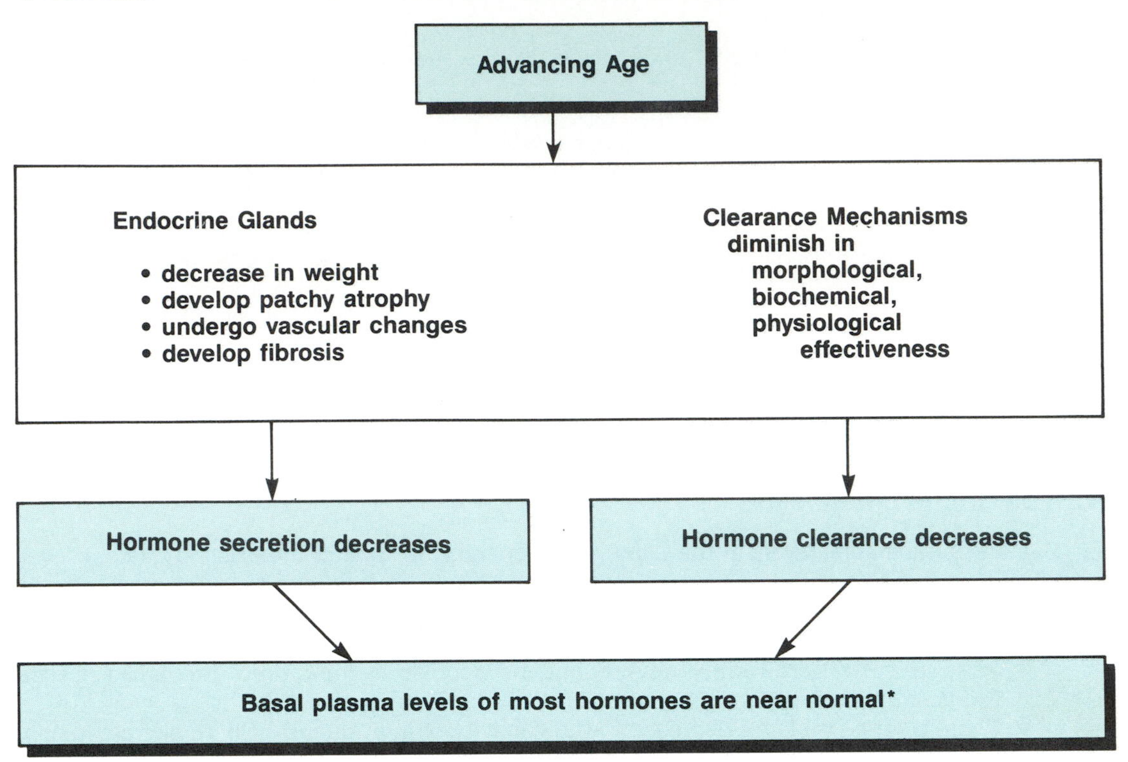

*Some hormones show *decreased* levels of their *active form*

The Endocrine Pancreas and Aging

Few morphologic changes take place in the healthy, aging endocrine pancreas. Nevertheless,

- the ability to regulate plasma glucose levels after a glucose load diminishes because
- there is a decline in the rate at which plasma glucose levels decrease after insulin injection.

Changes in the physiology of insulin, glucagon, or somatostatin may be responsible.

Insulin

Diabetes mellitus (lack of insulin) is associated with many symptoms (cataracts, neuropathy, atherosclerosis, skin changes) that are also seen in aging. As a result, **linkages between pancreatic function and aging have been suggested.**

Circulating insulin levels

◆ **Basal plasma insulin levels in the elderly are generally normal** for the existing glucose level.

Release of insulin

- Most studies report a **small age-dependent increase in the ratio of proinsulin release to insulin release after a glucose load.**
- There is an age-related **decrease in the capacity of individual B-cells to secrete insulin** (related to a decline in intracellular cAMP levels).
- With age there is also an **increase in the number of secreting B-cells,** but it is often insufficient to restore full secretory capacity to the pancreas as a whole.

Insulin receptors

- **Number and affinity of receptors** in liver and muscle **decrease.**
- Receptor density on adipocytes also decreases, because of increasing adipocyte size. (However, adipose tissue removes less than 5 percent of injected glucose.)

Tissue sensitivity to insulin

◆ **Tissue sensitivity to insulin decreases with age.** (Both a decline in receptor number or affinity and a decline in cytoplasmic, postreceptor mediators may be involved.)

Glucagon and somatostatin

Plasma glucagon levels and glucagon kinetics in response to stimulants (e.g., alanine) and suppressors (e.g., glucose) **do not appear to be influenced by age.**

Basal plasma somatostatin levels increase slightly with age.

The Endocrine Pancreas and Aging

Response to a 75-Gram Oral Glucose Load

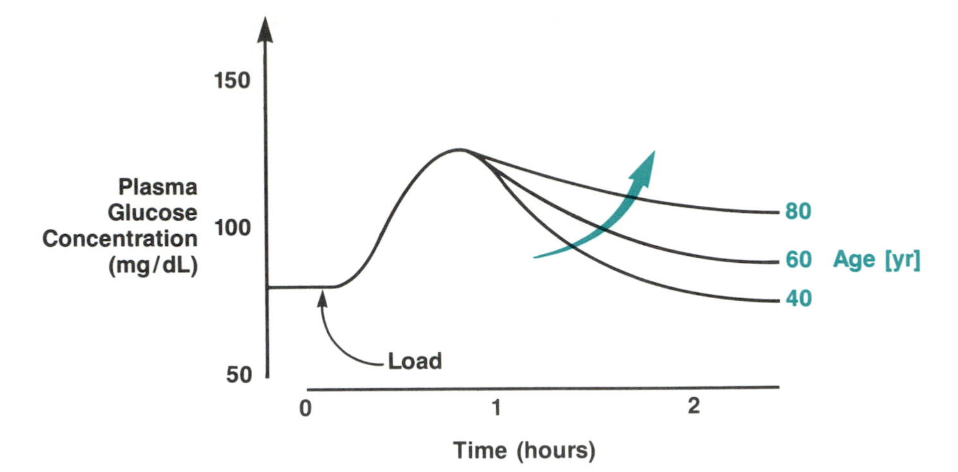

Age and Insulin Actions

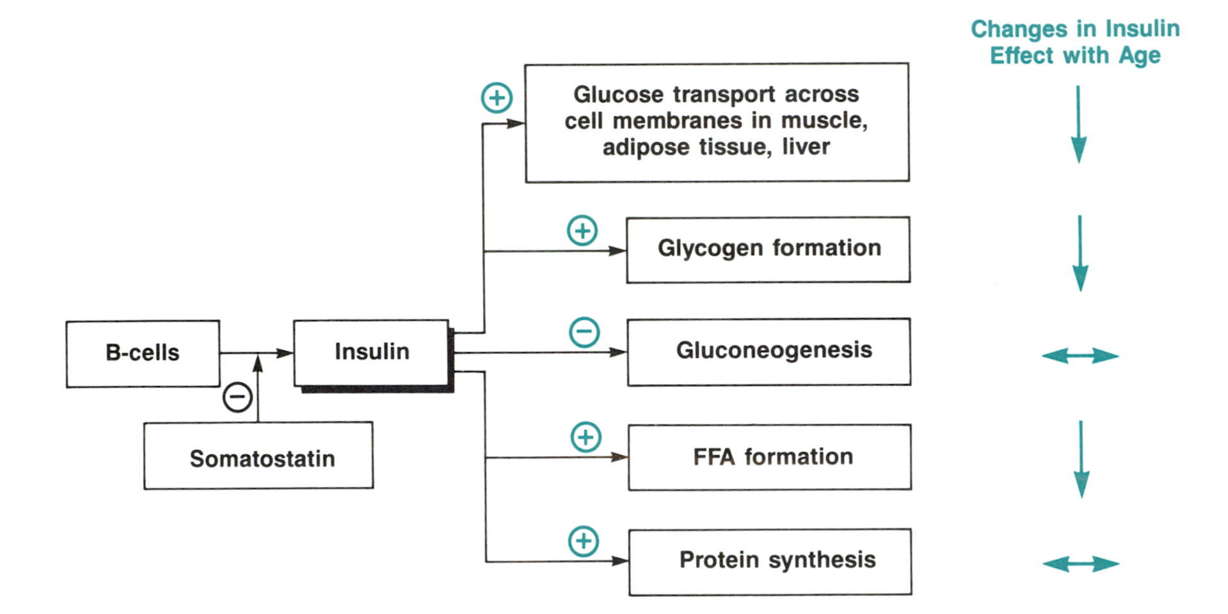

Age and Glucagon Actions

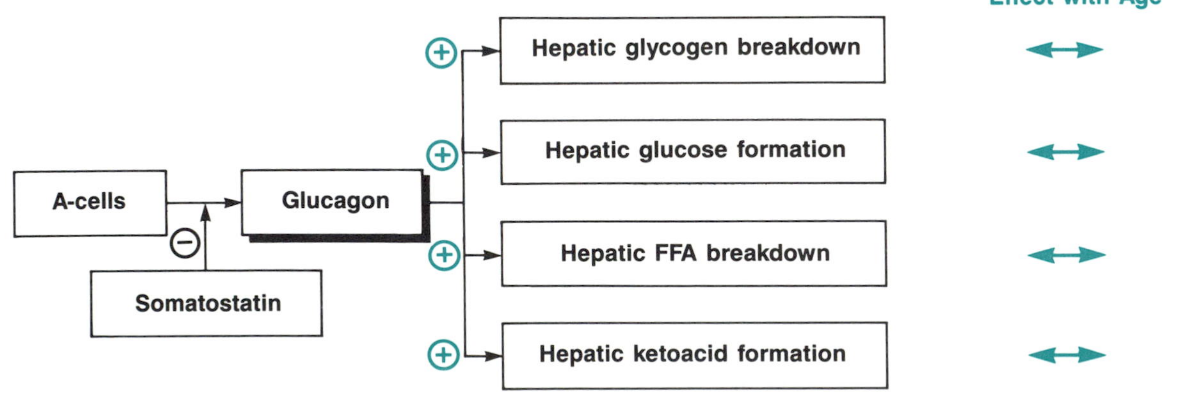

The Thyroid and Aging

Thyroid Deficiency as a Cause of Aging

Many symptoms of thyroid insufficiency resemble signs of normal aging. As a result, attention has been focused on thyroid involution as a hormonal cause of aging and on thyroid hormone supplementation as a vehicle for rejuvenation.

Structure

Age-related morphologic changes are small at the organ level but are noticeable at the cellular level and in the vascular supply:

- The absolute size of the thyroid decreases from adulthood on, but its **size relative to body weight remains constant.**
- **Cellular changes consistent with decreasing secretory activity** occur (reduced mitosis, flattened epithelium, more fibrous tissue).
- Vascular atherosclerotic changes suggest that **hormone transport should be impaired.**

Hormone secretion

- **T_4 release from follicular cells in response to stress or direct stimulation by TSH is unchanged with age.**
- **Production of reverse T_3 increases, and production of T_3 decreases,** with advancing age.

Plasma levels of hormones

Although dissenting findings are numerous, it is generally agreed that, with advancing age,

- plasma TSH levels are unchanged;
- the circadian periodicity of TSH release is unchanged;
- the ratio of protein-bound T_3 and T_4 to freely circulating hormone remains constant at about 99.8 percent;
- **basal plasma levels of T_3 decrease slightly;**
- **basal plasma levels of reverse T_3 increase;**
- metabolic clearance rates for T_3 and T_4 decrease slightly.

Hormone receptors

- There is an inverse relationship between T_3 availability and the number of nuclear T_3 receptors.
- Receptor up-regulation may not be sufficient in all target cells to compensate for the age-related decrease in T_3 production within peripheral cells.

The Thyroid and Aging

Thyroid Deficiency as a Cause of Aging

Symptoms of Thyroid Insufficiency in Normal Humans

- Decreased metabolic rate
- Hyperlipidemia
- Increased serum cholesterol
- Increased atherosclerosis
- Decreased mental performance
- Decreased reaction time
- Aging of skin and hair

Because all of these are associated with advancing age, the question arises: is thyroid involution a hormonal cause of aging?

Changes in Hormone Secretion and Deiodination

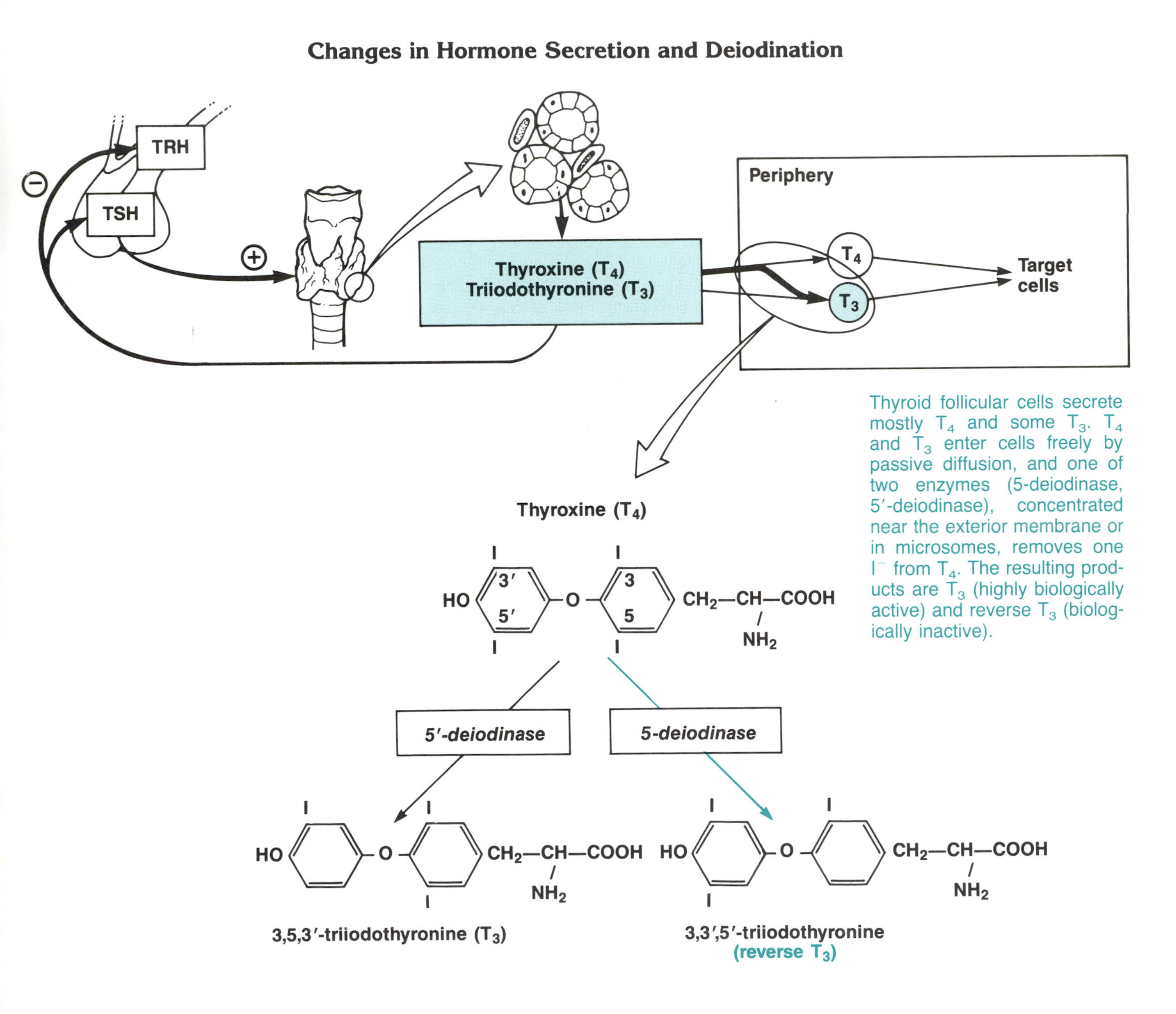

Thyroid follicular cells secrete mostly T_4 and some T_3. T_4 and T_3 enter cells freely by passive diffusion, and one of two enzymes (5-deiodinase, 5′-deiodinase), concentrated near the exterior membrane or in microsomes, removes one I^- from T_4. The resulting products are T_3 (highly biologically active) and reverse T_3 (biologically inactive).

Energy Metabolism and Age

Basal Metabolic Rate (BMR) and Daily Metabolic Rate

When allowance is made for the general decrease in muscle mass and increase in body fat with age, then this **composition-corrected basal metabolic rate changes little with age.** Nevertheless, some tissues may show a decrease in the number of mitochondria along with decreased metabolic capacity.

- **Daily metabolic rate changes in parallel with physical activity.**

Carbohydrate metabolism

Age-related depression of glucose tolerance has been suggested as evidence for a link between age and carbohydrate metabolism.

Carbohydrate stores

- The glycogen stores of muscle and liver are not changed with age.
- There is also no change in the ability to mobilize these stores (e.g., during exercise).

Carbohydrate utilization

- Although there is no global change, glucose utilization may be altered within particular tissues.

Fat metabolism

Although food restriction in rats has been shown to increase their life span dramatically, **there is no general correlation between life span and fat content in persons of average adiposity.** Moreover, very thin individuals behave contrarily to what would be expected from rat experiments: their life span increases with body fat content.

Protein metabolism

Body protein stores decline with age, but **it is not yet known whether protein mobilization during fasting or during trauma is affected by age.**

Thermoregulation

With advancing age, the hormonal, enzymatic, muscle, and sweat responses to both cold and heat exposure are altered.

Hormone responses to cold exposure

There is diminished TSH elevation after cold exposure.

Enzyme responses to cold exposure

Metabolic conversion of tyrosine (to epinephrine or to thyroid hormones) is depressed.

Skeletal muscle responses to cold exposure

Shivering is less intense.

Sweat gland responses to heat exposure

- The temperature threshold for sweating is increased.
- The volume of sweat is reduced even though there is no change in the total number of sweat glands.

Energy Metabolism and Age

Metabolism of Proteins, Carbohydrates, and Fats

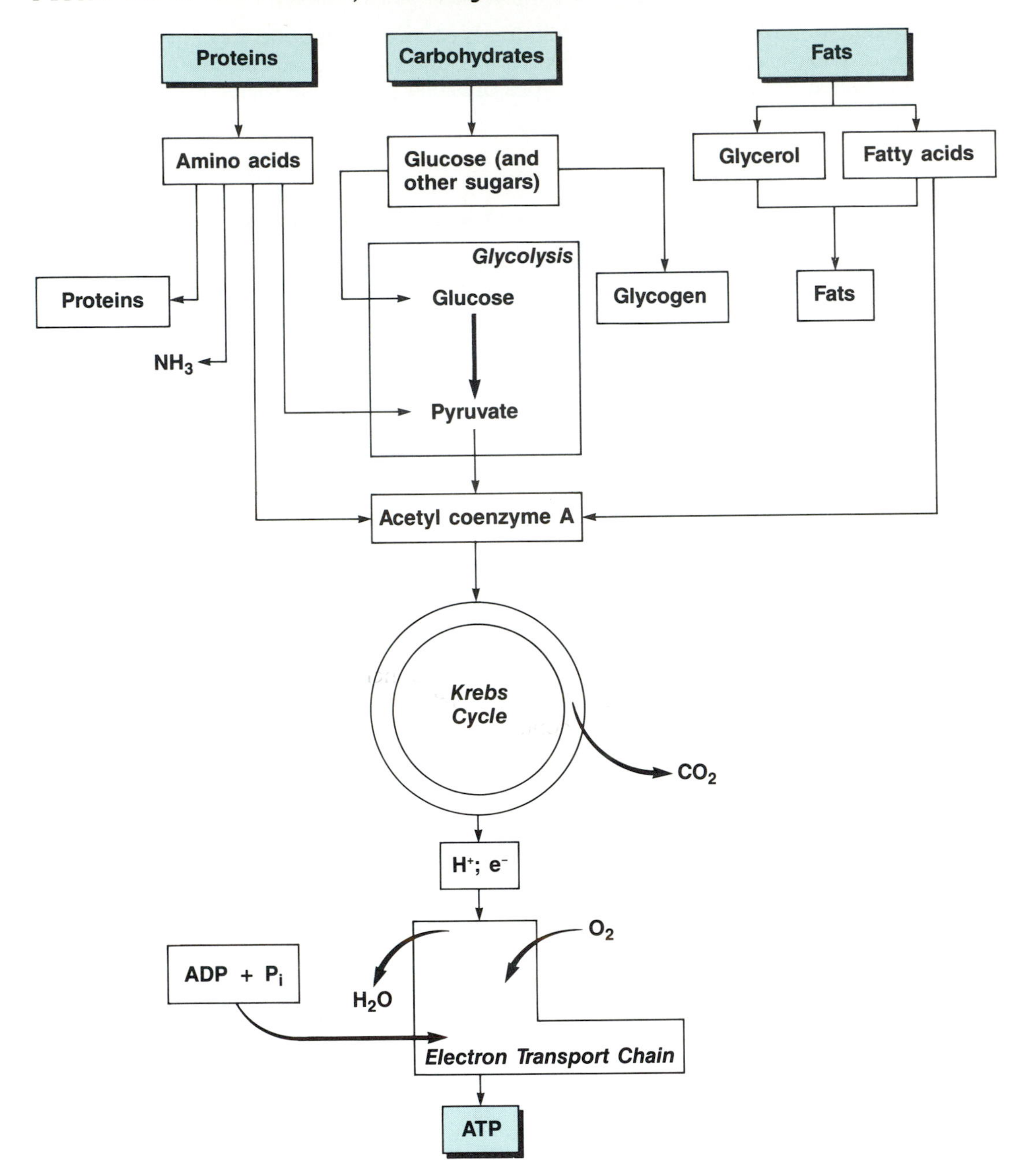

Adapted from Marieb EN. Human anatomy and physiology. Redwood City, CA: Benjamin Cummings, 1989.

No conclusive evidence has emerged to demonstrate that aging is related to altered enzymatic activities in the metabolic pathways.

Aging and the Anterior Pituitary

Structure

Age-related morphologic changes in pituitary cells are nonspecific only (they include accumulation of lipofuscin, diminished number of dendrites and synapses, and altered neurotransmitter balance). The numbers of most types of pituitary hormone-secreting cells are maintained.

Hormones

Follicle-stimulating hormone (FSH) and luteinizing hormone (LH)

- These change more in aging women than in aging men.
- The dominant influence after menopause is **diminished feedback inhibition from ovarian steroids.** As a result,
- **plasma levels of FSH and LH are increased.**
  - Elevated LH pulses are closely coordinated with the "hot flushes" that are a common complaint of women undergoing menopause. This observation, combined with anatomic proximity of hypothalamic thermoregulatory neurons to gonadotropin releasing hormone (GnRH)–secreting cells, has led to the hypothesis that hot flushes arise from augmented GnRH pulses.

Prolactin (PL)

- Pituitary content and **mean plasma levels of PL increase with age** (a result of depressed hypothalamic dopamine levels); however,
- daily circadian fluctuations in plasma PL decrease in amplitude.

Thyroid stimulating hormone (TSH)

- Plasma TSH levels are **unchanged.**
- The circadian periodicity of TSH release is unchanged.

Adrenocorticotropic hormone (ACTH)

ACTH levels remain **constant,** but the responsiveness of some adrenal pathways to ACTH declines.

Growth hormone (GH)

- Total pituitary GH content, the number of GH-containing pituitary cells, and GH clearance remain constant.
- Whether or not GH release and plasma GH levels remain constant with age is a matter of some controversy, centered around the influence of obesity. **The present consensus is that the age-related decline in GH release exceeds the suppression that might be explained by age-related obesity.**
- Responsiveness of most somatomedin*-secreting cells to GH is preserved. However, the responsiveness of liver cells is depressed in old age.

*The somatomedins are insulin-like growth factors I and II, nerve growth factor, and epidermal growth factor.

AGING AND ENDOCRINE SYSTEMS—CONT'D.

Aging and the Anterior Pituitary

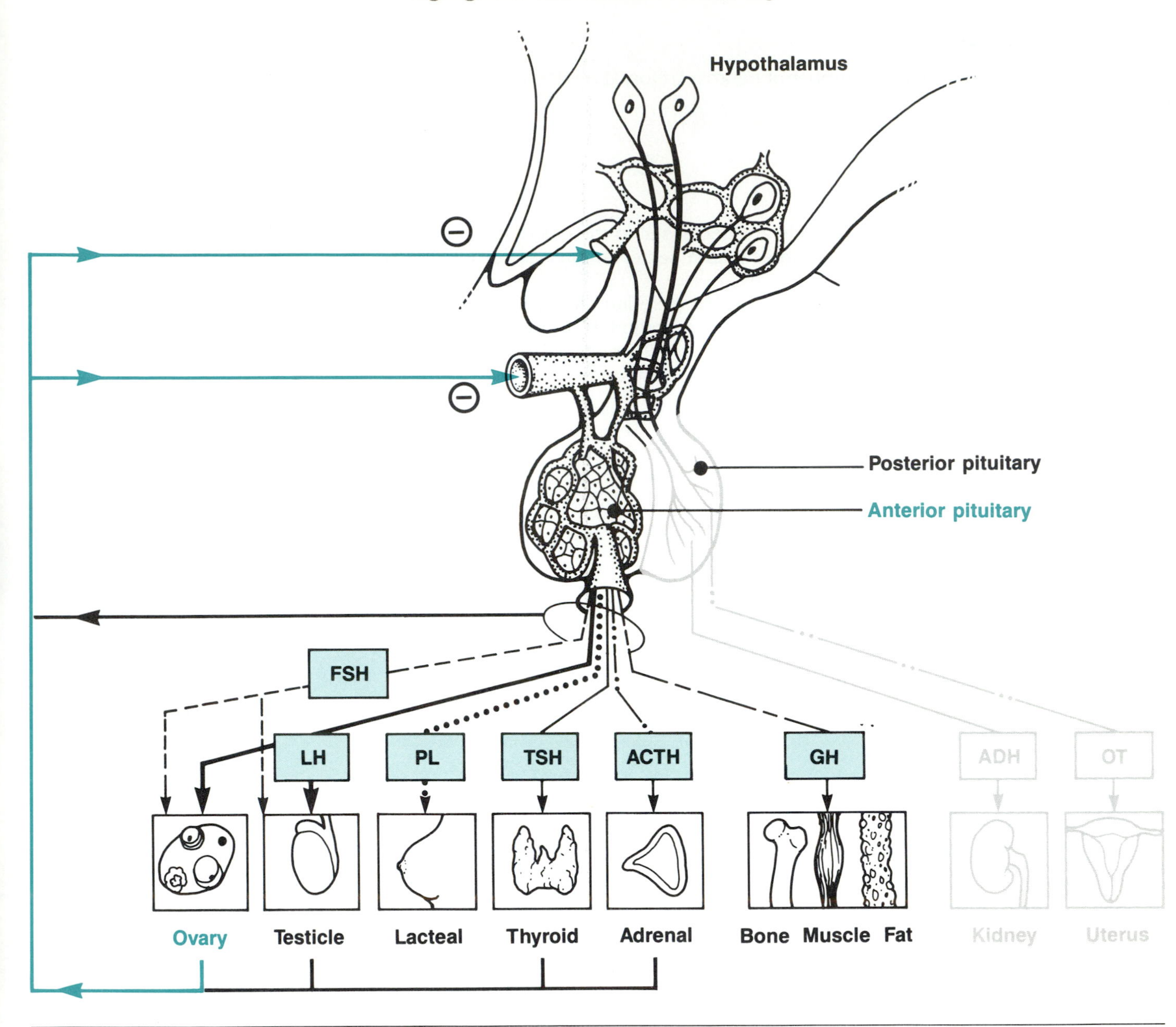

Adapted from Timiras PS, ed. Physiological basis of geriatrics. London: Macmillan, 1988.
FSH = follicle stimulating hormone; **LH** = luteinizing hormone; **PL** = prolactin; **TSH** = thyroid stimulating hormone; **ACTH** = adrenocorticotropic hormone; **GH** = growth hormone.

The dominant influence after menopause is diminished feedback inhibition from ovarian steroids.

Aging and the Posterior Pituitary

Structure

Cell volume, number of neurosecretory granules, and indices of endoplasmic reticular activity are decreased.

Hormones

Vasopressin (ADH)

- **Plasma ADH levels are decreased** in the elderly.
- **The ability to suppress ADH** secretion after all stimuli (hypotonic or hypervolemic) **is attenuated.**
- **The ability to increase ADH secretion after a hypertonic stimulus is enhanced,** but
- **the ability to increase ADH secretion in hypovolemia is reduced.**
- Vasopressin administration in old monkeys has improved memory performance in some animals. Findings in elderly humans have been equivocal.

Aging and the Posterior Pituitary

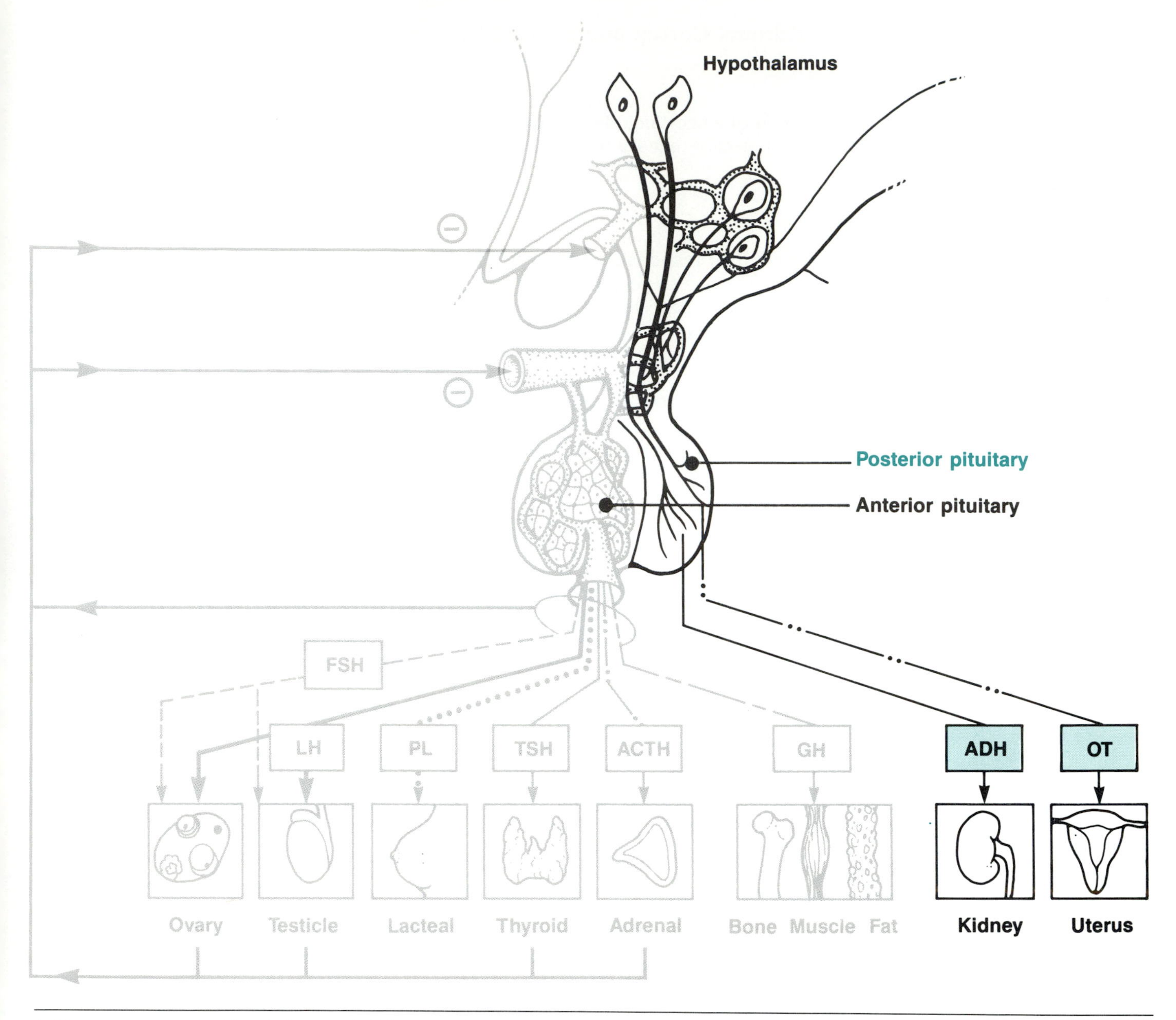

ADH = vasopressin (antidiuretic hormone); **OT** = oxytocin.
Adapted from Timiras PS, ed. Physiological basis of geriatrics. London: Macmillan, 1988.

Aging and the Adrenal Glands

Adrenal Cortex and Steroid Synthesis

Basal levels

There is little change in glucocorticoid levels, but a much greater change in adrenal androgen levels. The greatest decline occurs in the plasma levels of dehydroepiandrosterone (DHEA) and aldosterone.

Stimulated levels

- **Adrenal target cells retain their responsiveness to ACTH,** but some of the consequent enzymatic steps deteriorate. Thus, in the elderly, the following occur after ACTH stimulation.
  - The increase in Δ^4 steroids (progesterone to 17α OH progesterone to androstenedione or cortisol) is preserved; however,
  - **There is a decline in** the synthesis of Δ^5 steroids—pregnenolone to **dehydroepiandrosterone (DHEA).**
  - Aldosterone levels continue to show circadian fluctuations, and they increase during salt restriction, but
  - **Even in salt deficient states, plasma aldosterone levels remain depressed** in the elderly. (The chief cause is changes in renal juxtaglomerular vasculature and their effects on renin secretion.)

Receptors

Animal studies suggest that, in addition to the changes in clearance and synthesis, there are age-related **decreases in hormone receptor number and affinity.** This has been most clearly demonstrated for the corticosteroids, for androgens, and for estrogens. There have been reports of age-related increases in the number of aldosterone receptors in renal target cells.

Adrenal Medulla and Catecholamines

Basal levels

Overall **sympathoadrenal activity,** and therefore **basal plasma levels of epinephrine and norepinephrine, are increased with age.**

Stimulated levels

Epinephrine and norepinephrine often show differential responses in elderly individuals:

- During mental arithmetic, plasma norepinephrine is elevated, but plasma epinephrine is not. It has not been resolved whether such observations represent sympathetic hyperactivity or adrenal suppression.

AGING AND ENDOCRINE SYSTEMS—CONT'D.

Aging and the Adrenal Glands

Adrenal Cortex and Steroid Synthesis

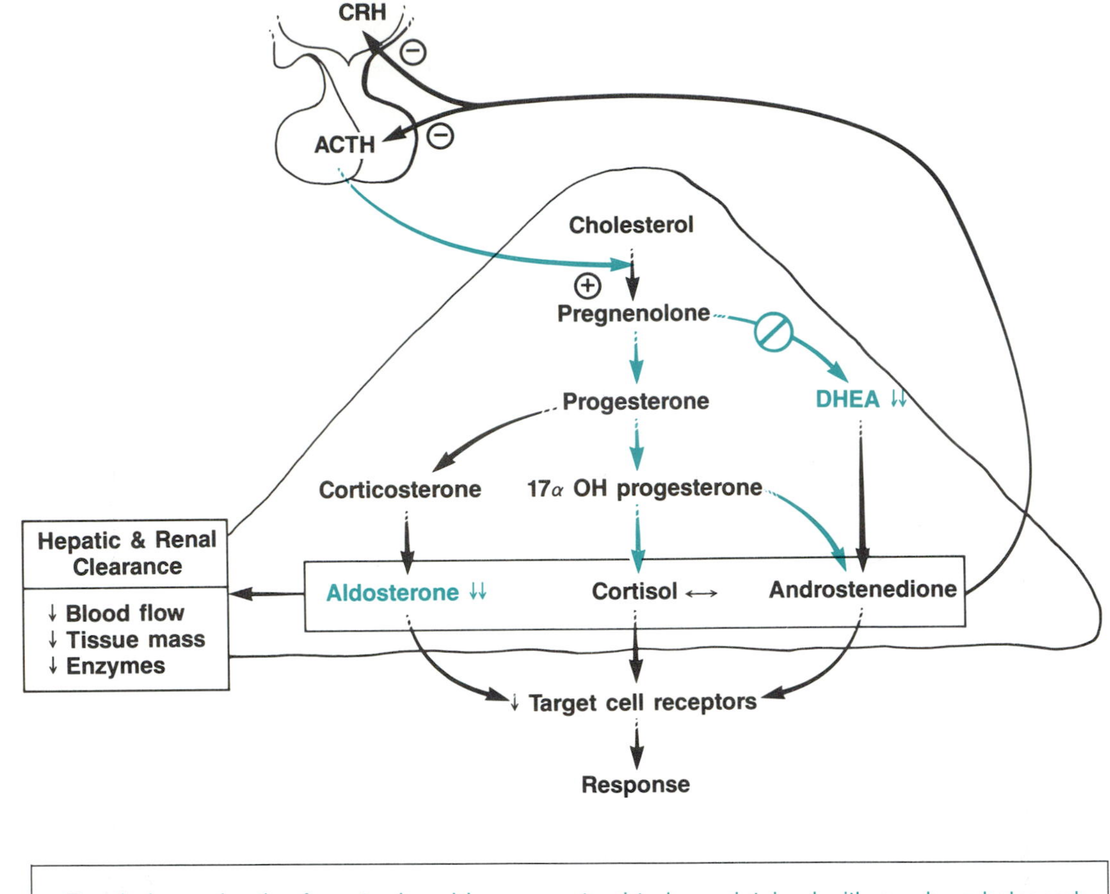

Basal plasma levels of most adrenal hormones tend to be maintained with age by a balanced reduction in secretion and clearance rates.

Adrenal Medulla and Responses to Stress

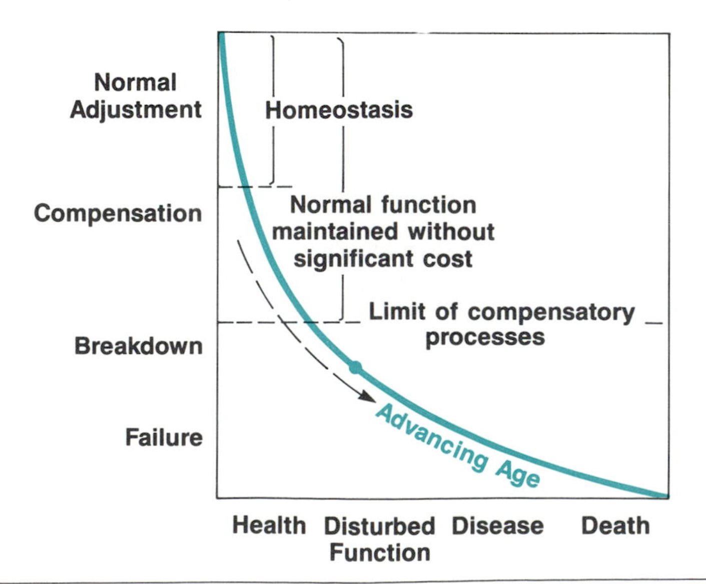

The elderly show delayed and impaired responses to stress. This implies a deterioration in the ability to react to challenges with coordinated adjustments of several overlapping homeostatic control mechanisms. The process of aging can be pictured as a progressive decline along a curve of adaptive capability.

Adapted from Timiras PS, ed. Physiological basis of geriatrics. London: Macmillan, 1988.

AGING AND WOMEN'S SEXUAL AND REPRODUCTIVE PHYSIOLOGY

Ovaries

Changes in structure

- The ratio of primary follicles to growing follicles is 50:1 at puberty and declines to 3:1 by about age 40.
- **By the beginning of menopause** (at an average age of 50 years), the **ovary is virtually without oocytes.** (In contrast, rodents retain large numbers of apparently normal oocytes into extreme old age.)
- Oocytes change in appearance (fewer chiasmata per metaphase).

Changes in function

- The frequency of chromosomal abnormalities increases with age.
- Corpora lutea cease to be formed when ovulatory function ceases; as a result,
- **Synthesis and plasma levels of progesterone, estrone, and estradiol decrease sharply.**
- After menopause the production of estrogens is restricted to nonovarian tissues and varies directly with body fat: the more fat, the greater the potentiation of conversion of androstenedione to estrone.

Hypothalamus-pituitary axis

After menopause, **absence of ovarian feedback inhibition causes increased amplitude of GnRH pulses.**

◆ "Hot flushes" are closely coordinated with GnRH and LH pulses. This has led to the hypothesis that menopausal women show intensified activity in GnRH-secreting hypothalamic neurons and that this activity spreads to adjacent thermoregulatory neurons.

Uterus

When estrogen levels are diminished,

- **uterine epithelium atrophies;**
- **uterine blood vessels gradually narrow or become occluded** and coiled arterioles disappear;
- the rate of these changes is modulated by a shift in estrogen receptor populations: low-affinity receptors disappear, but high-affinity receptors are retained.

Vagina

In the absence of estrogen stimulation, the vagina of the postmenopausal woman **becomes smaller** in length and diameter, **loses pliability, and shows a thinning epithelial wall as well as inadequate glandular secretions.**

Breasts

- The **breasts undergo involution** following the loss of ovarian hormones.
- **Mucoid connective tissue** in the lobules **is converted to** a denser form of collagen **(hyaline collagen).**
- With age **the incidence of breast cancer increases** dramatically, accompanied by a steady increase in the proportion of estrogen receptor-positive tumors (a relationship of breast tumors to the pathophysiology of estrogen and prolactin has been suggested).

AGING AND WOMEN'S SEXUAL AND REPRODUCTIVE PHYSIOLOGY

Many aspects of women's sexual and reproductive physiology are dominated by the age-related attrition of oocytes.

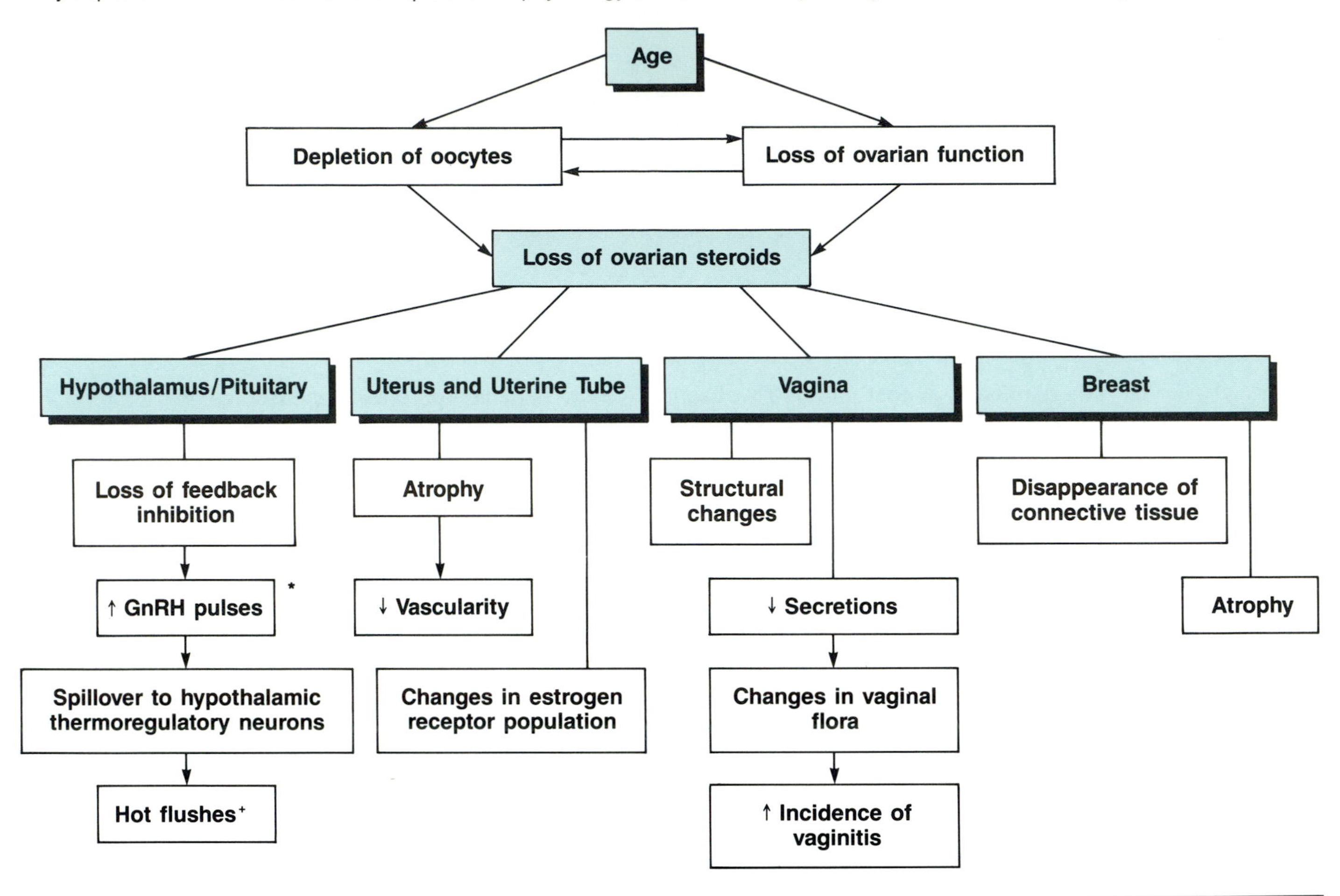

*GnRh = gonadotropin releasing hormone.
†Hot flushes = a symptom complex that involves a feeling of pressure in the head and warmth or flushing that begins in the skin of the face and neck and spreads down the trunk. It may be accompanied by sweating, nausea, tachycardia, and tremor.

Loss of ovarian steroids affects many other tissues and organs:

Tissue	Effects of Ovarian Steroid Loss
Skin	Becomes thinner and less pliable; "liver" spots develop
Bladder and urethra	Epithelial atrophy
Cardiovascular system	Altered serum lipids; increased atherosclerosis; increased incidence of coronary heart disease
Bone	Osteoporosis

AGING AND MEN'S SEXUAL AND REPRODUCTIVE PHYSIOLOGY

Reproductive aging in men is not as clearly associated with diminished function of one key organ as it is in women.

Prostate

After age 40, histologic, biochemical, and functional changes take place in the prostate.

Prostate histology

- There is evidence of atrophy, accumulation of stagnant secretions, and decreased protein synthesis.
- There is also evidence of hypertrophy.

Prostate function

The dominant, almost universally observed feature is benign prostatic hypertrophy. (It is thought to be caused by changes in the steroidal milieu, including LH, FSH, prolactin, estrogen, and androgen.)

Penis

After the age of 40 years, small arteries and veins of the penis undergo **progressive sclerotic changes** that eventually involve the corpora cavernosa. It has been suggested that they might feature in the increasing incidence of impotence seen in older men.

Leydig cells

Leydig cell function determines the plasma level of testosterone.

Basal plasma testosterone levels

In the male population as a whole, average plasma levels of testosterone and its major active metabolites begin to decline continuously by age 50, but there is great individual variability in this phenomenon (due to pulsatile release, health status, medication, obesity, alcohol intake, environmental stress, and status of binding proteins).

Stimulated testosterone release

- Leydig cell testosterone production in response to LH stimulation decreases. As a result,
- there are increases in the average plasma levels of FSH and LH, reflecting disinhibition of hypothalamic secretion.

Seminiferous tubule

The testes of old men commonly show patches of senile degeneration of seminiferous tubules and their supporting interstitium.

- The number of chromosomal abnormalities in spermatogonia may increase with age.

- **The rate of conception decreases with male age** even though
  1. total sperm count and ejaculate volume increase with age (because of age-related decrease in intercourse frequency) and
  2. the ability of sperm to penetrate ova does not deteriorate with age.

- **Male germ cells continue to be replenished to extreme age.**
- **The time required for the maturation of spermatogonia remains constant with age.**

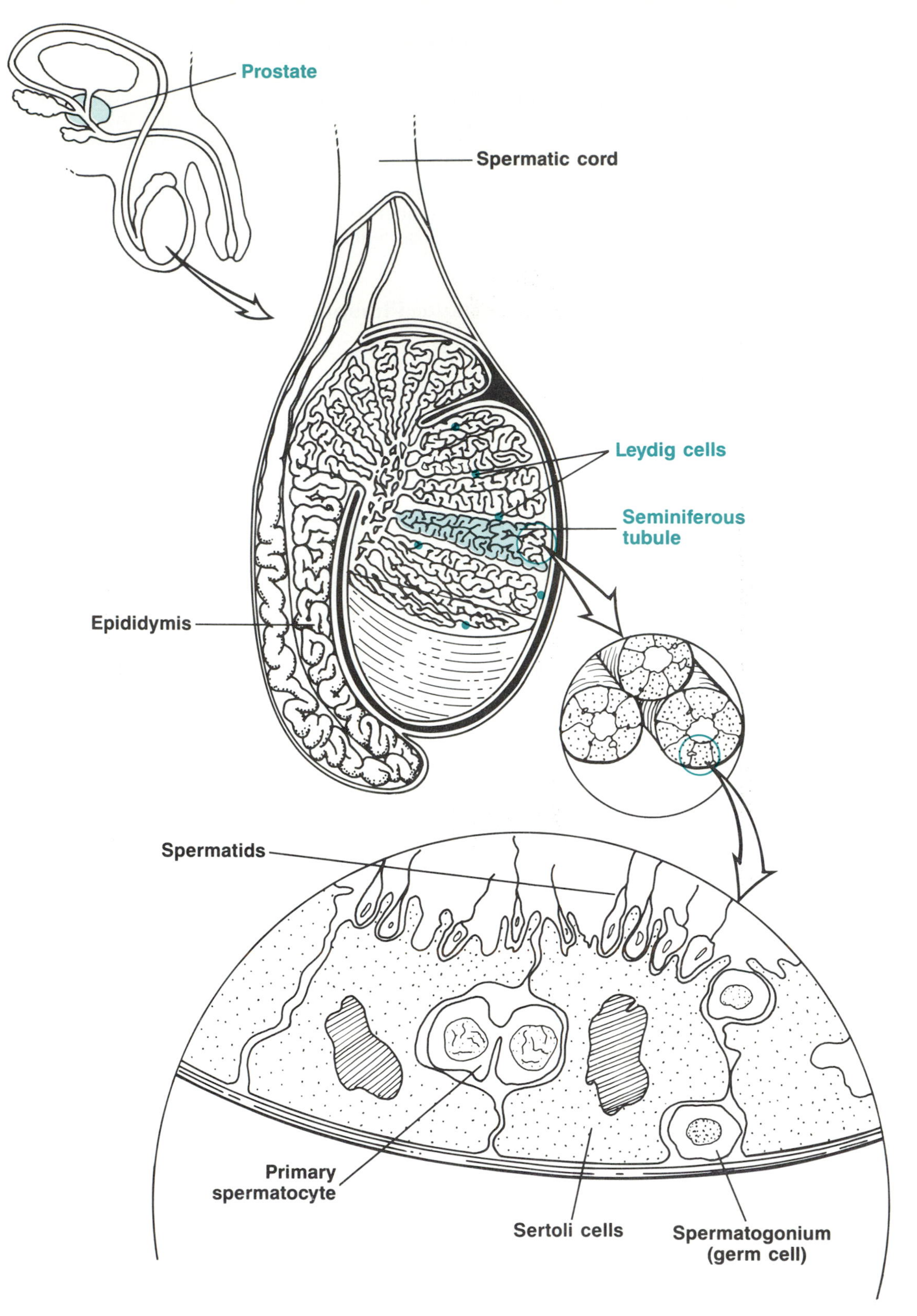
Prostate
Spermatic cord
Leydig cells
Seminiferous tubule
Epididymis
Spermatids
Primary spermatocyte
Sertoli cells
Spermatogonium (germ cell)

AGING AND MINERAL METABOLISM

Mechanisms of Extracellular Calcium Regulation

With age there is a progressive decline in active, vitamin D–dependent Ca^{++} absorption from the GI tract. The decline is more pronounced in women than in men, mostly because postmenopausal decline in estrogen levels makes vitamin D deficiency more frequent in women.

Ca^{++} reabsorption from bone and nephrons, on the other hand, is generally increased to a homeostatically effective level. As a consequence, plasma levels of Ca^{++} show little change with age in healthy individuals when adjustments are made for age-related decline in plasma protein concentration.

Mechanisms of Extracellular Phosphate Regulation

As a result of Ca^{++}-driven increased bone resorption, plasma phosphate levels increase with age in women. Men usually show a slight decline in plasma PO_4^{3-} by their eighth decade.

AGING AND MINERAL METABOLISM

Mechanisms of Extracellular Calcium Regulation

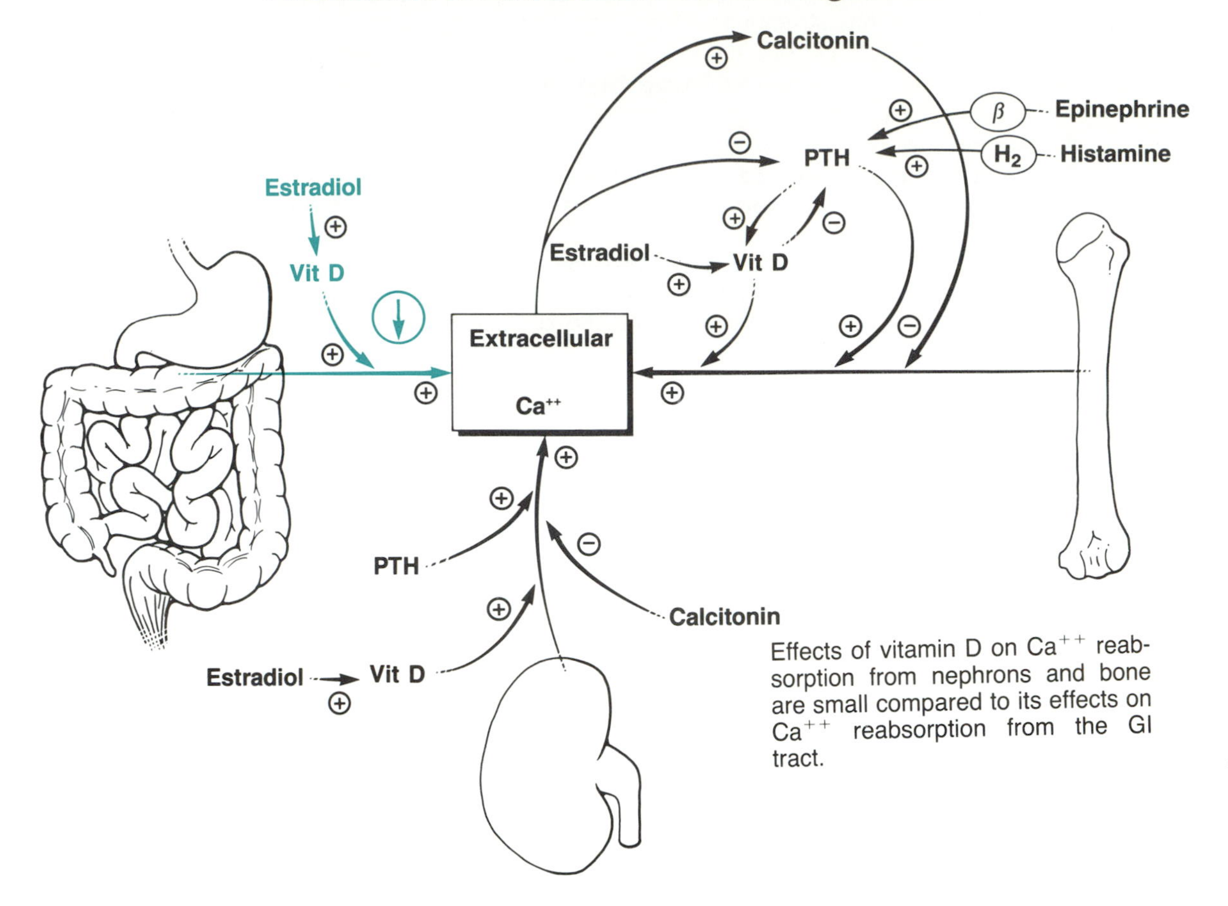

Effects of vitamin D on Ca^{++} reabsorption from nephrons and bone are small compared to its effects on Ca^{++} reabsorption from the GI tract.

Mechanisms of Extracellular Phosphate Regulation

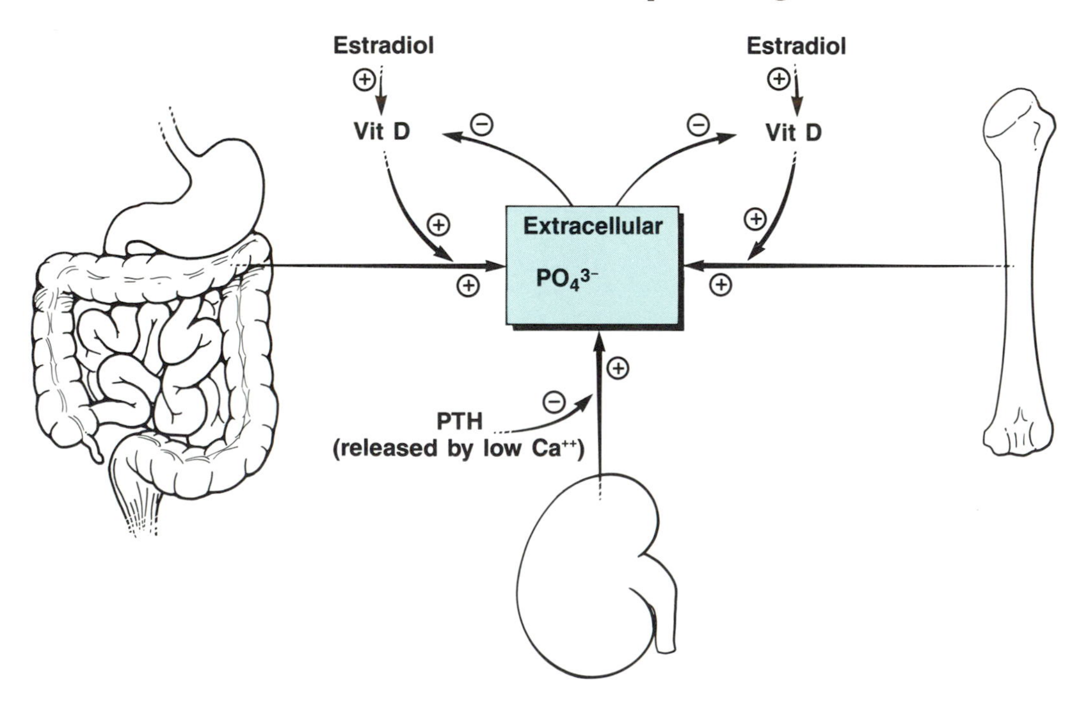

AGING AND BONE

Aging is accompanied by net loss of bone tissue. The rate of loss is greater in women than in men. This loss has structural and functional consequences.

Bone Structure

- Osteons show incomplete closure.
- Bone porosity increases because the Haversian canals erode.
- Canals nearest the marrow cavity become wider and the endosteal surface becomes eroded.
- The number and size of osteocytic lacunae increase.
- In vertebral bones the horizontal trabeculae disappear, followed by the vertical trabeculae at a later age. The result is diminished strength of these bones because the trabeculae acted as supporting "struts."

Bone Remodeling

- Deficits in bone mass accumulate gradually because of a decrease in the speed of bone remodelling around resorption cavities.
- The mineralization rate decreases.
- The number of osteoblasts decreases.
- The rate of bone resorption increases, particularly in postmenopausal women.

Bone Mass

When the loss of bone mass has reached a level where structural failure occurs, then a state of **osteoporosis** is said to exist.

◆ **Low initial bone mass appears to be the most frequent cause of osteoporosis, but hormonal factors, nutrition, and level of physical activity each influence the rate of development of osteoporosis.**

Women are more affected for two reasons:

1. At age 20 women have 20 percent less bone mass per body weight than men.
2. The plasma levels of estrogens and progesterones undergo more profound changes in women (during menopause) than they do in men.

AGING AND BONE

Bone Remodeling

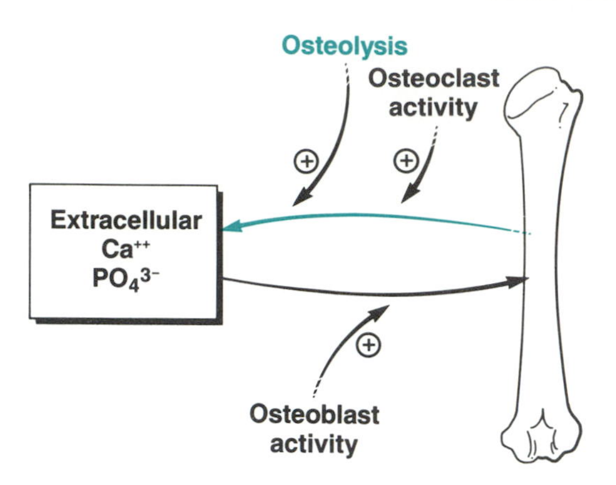

The normal sequence at an active site is a bone resorption phase lasting 25 days, followed by a rest phase of 30 days, followed by a bone deposition phase of 75 days, followed by a resting phase of several years.

Bone Mass

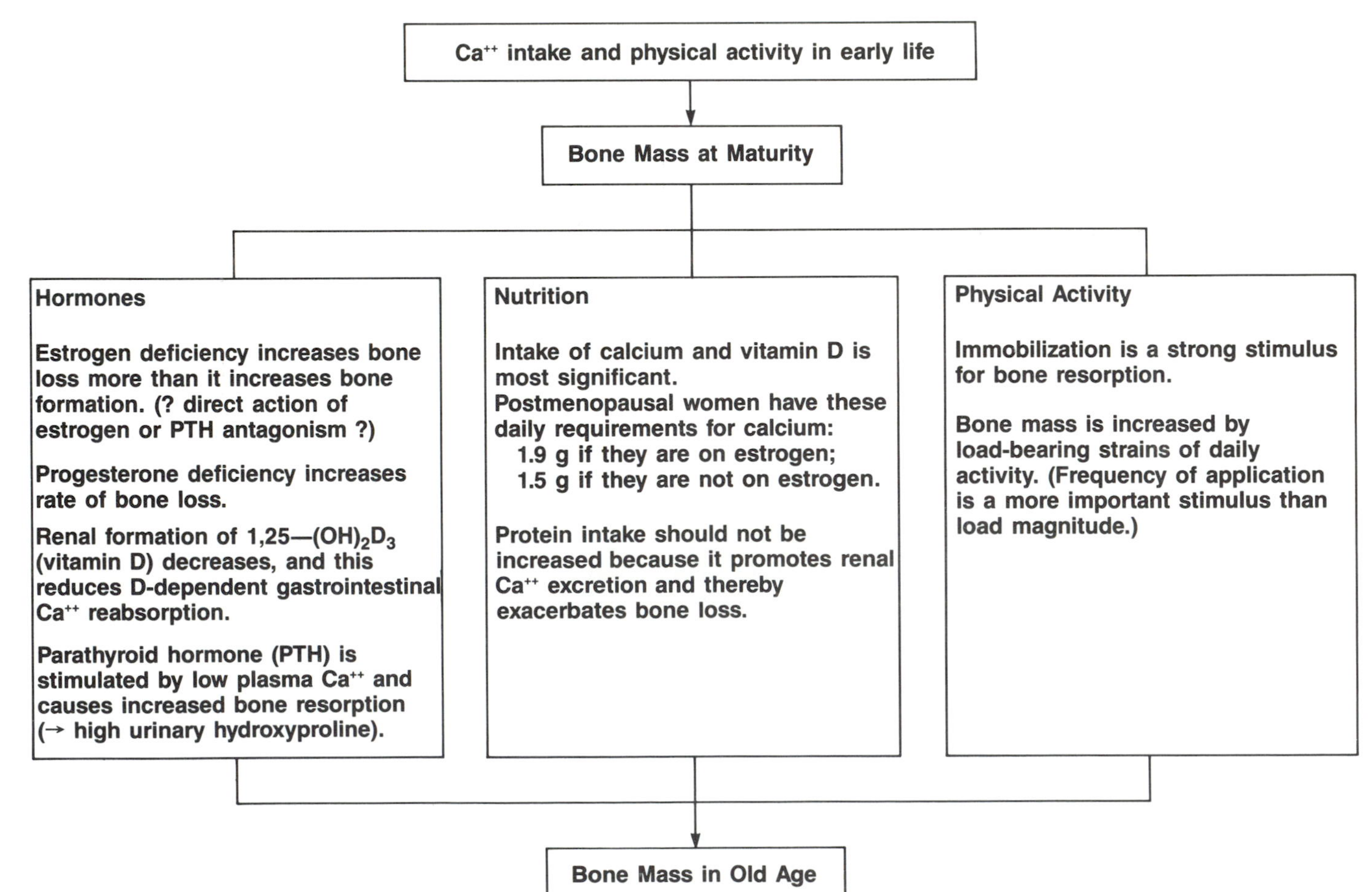

Ca^{++} intake and physical activity in early life
Bone Mass at Maturity
Hormones
Estrogen deficiency increases bone loss more than it increases bone formation. (? direct action of estrogen or PTH antagonism ?)
Progesterone deficiency increases rate of bone loss.
Renal formation of 1,25—$(OH)_2D_3$ (vitamin D) decreases, and this reduces D-dependent gastrointestinal Ca^{++} reabsorption.
Parathyroid hormone (PTH) is stimulated by low plasma Ca^{++} and causes increased bone resorption (→ high urinary hydroxyproline).
Nutrition
Intake of calcium and vitamin D is most significant.
Postmenopausal women have these daily requirements for calcium:
1.9 g if they are on estrogen;
1.5 g if they are not on estrogen.
Protein intake should not be increased because it promotes renal Ca^{++} excretion and thereby exacerbates bone loss.
Physical Activity
Immobilization is a strong stimulus for bone resorption.
Bone mass is increased by load-bearing strains of daily activity. (Frequency of application is a more important stimulus than load magnitude.)
Bone Mass in Old Age

AGING AND CONNECTIVE TISSUE

Connective tissue consists of

1. cells (fibrocytes) that secrete collagen and ground substance (mostly proteoglycans),
2. extracellular matrix,
3. collagen fibers, and
4. elastin fibers.

Fibrocyte function is relatively unaffected by age, but each of the other three constituents undergoes changes.

Collagen

- The **number and size of collagen fibers increase.**
- Collagen **molecules become structurally tougher and more resistant to enzymatic degradation** (as a result of an increase in the number of intra- and intermolecular bonds).
- Collagen **fibers become less flexible, more resistant to stretch, and less able to return to resting length after a stretch** (as a result of an increase in the number of cross linkages between adjacent fibers).

Elastin

- **Elastin fibers develop cross linkages and become more rigid.** They may fray and tear.

Ground Substance

- The ground substance matrix **becomes denser and more aggregated.** This has wide-ranging effects because it hinders intercellular permeation of many substances.

◆ The net result of all these changes is that **connective tissue becomes less pliant with age and offers greater hindrance to the interstitial transport of substances.**

AGING AND CONNECTIVE TISSUE

Secretion of Collagen and Formation of Collagen Fibers

Amino acid precursors are assembled to form **procollagen** molecules within fibrocytes. In subsequent steps, the extension peptides that assisted in procollagen assembly are removed, and **tropocollagen** molecules are secreted into the extracellular space for **polymerization** and eventual formation of **collagen fibers.**

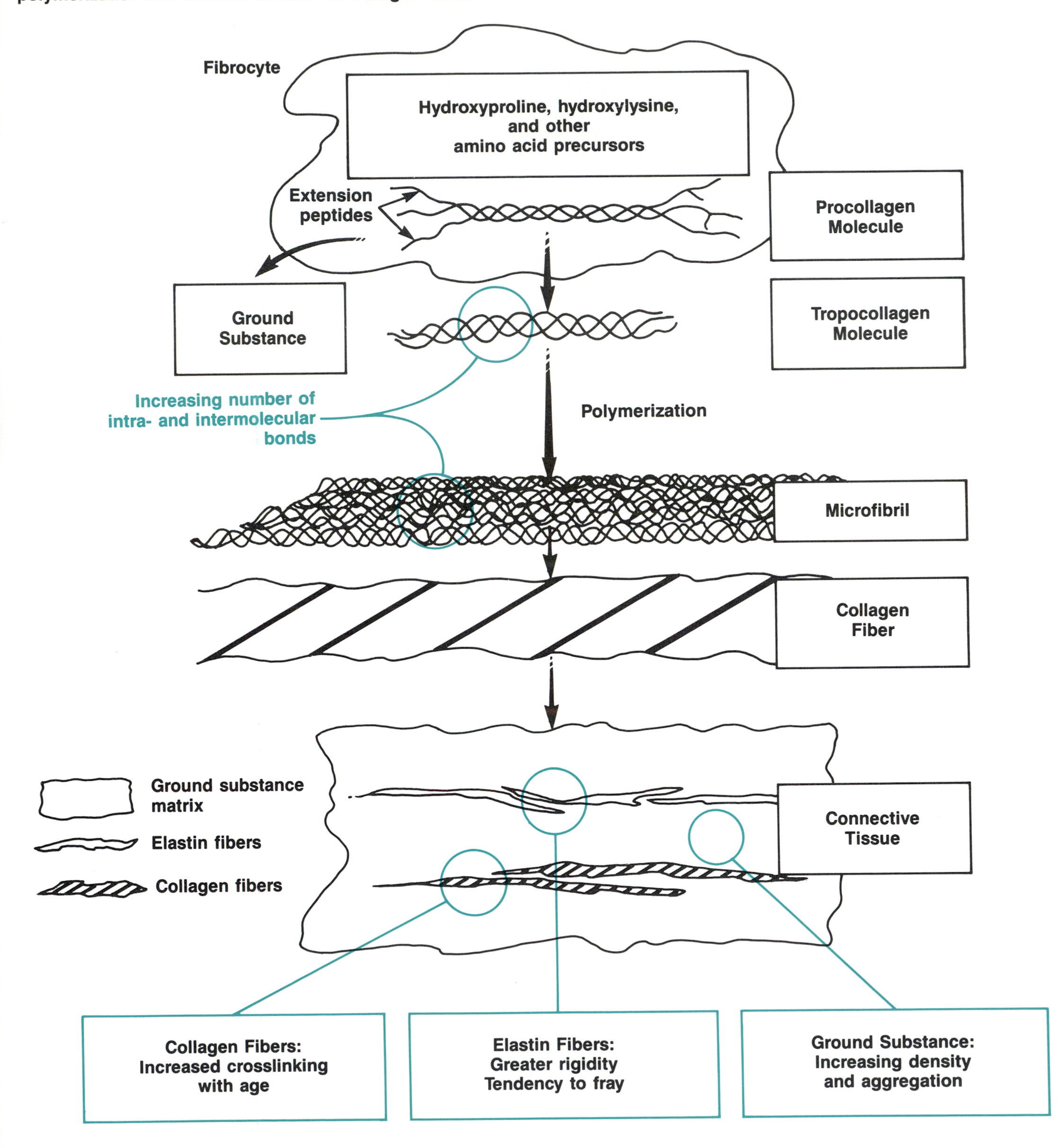

AGING AND THE GASTROINTESTINAL TRACT

Changes in the Mouth

Chewing and swallowing become more difficult with old age partly because of

- changes in the teeth,
- increasing fragility of the mucosal surfaces in the mouth, and
- weakening or atrophy of the muscles of mastication and swallowing.

Changes in the Esophagus

Changes in the esophagus include

- failure to initiate a peristaltic wave with each swallow;
- failure of the lower esophageal sphincter to relax on arrival of the peristaltic wave;
- appearance of localized, nonperistaltic contractions;
- progressive imprecision in the timing of segmental contractions that result in peristalsis;
- desynchronization of segmental contractions.

◆ **The net result of these changes is delay in the transport of food through the esophagus and a feeling of fullness in the substernal region.**

Changes in the Stomach and Duodenum

At least two factors contribute to the **decrease in basal gastric secretory activity (pepsin and HCl)** that is observed in those elderly who do not have ulcers. They are

1. atrophy in both secreting cells and endocrine cells that regulate secretion;
2. impaired sensitivity of secreting cells to hormonal and digestive stimuli (possibly caused by receptor changes).

Changes in the Liver

Most functional changes in human liver are moderate. A significant exception is the **marked decline in mixed-function oxidases,** causing reduced capacity for drug metabolism.

Changes in the Pancreas

The quantities of pancreatic secretions normally far exceed what is required for digestion. Hence, **the common age-related changes in the gland are not likely to compromise normal digestion.**

Changes in Jejunum and Ileum

Although the mucosal villi shrink in height, change their distribution pattern, and show reduced motility, at normal intake loads there is **no evidence of impaired absorption of carbohydrates, proteins, or amino acids.** There may be age-related decreases in the absorption of fat and of Ca^{++}, but they are caused by defects in the synthesis of specific and essential transport factors.

Changes in the Colon

The aging colon shows

- atrophy of muscular layers,
- loss of enteric nerve cells, and
- loss of coordination between the contractions of longitudinal and circular muscle.

AGING AND THE GASTROINTESTINAL TRACT

All regions of the GI system undergo changes with age.

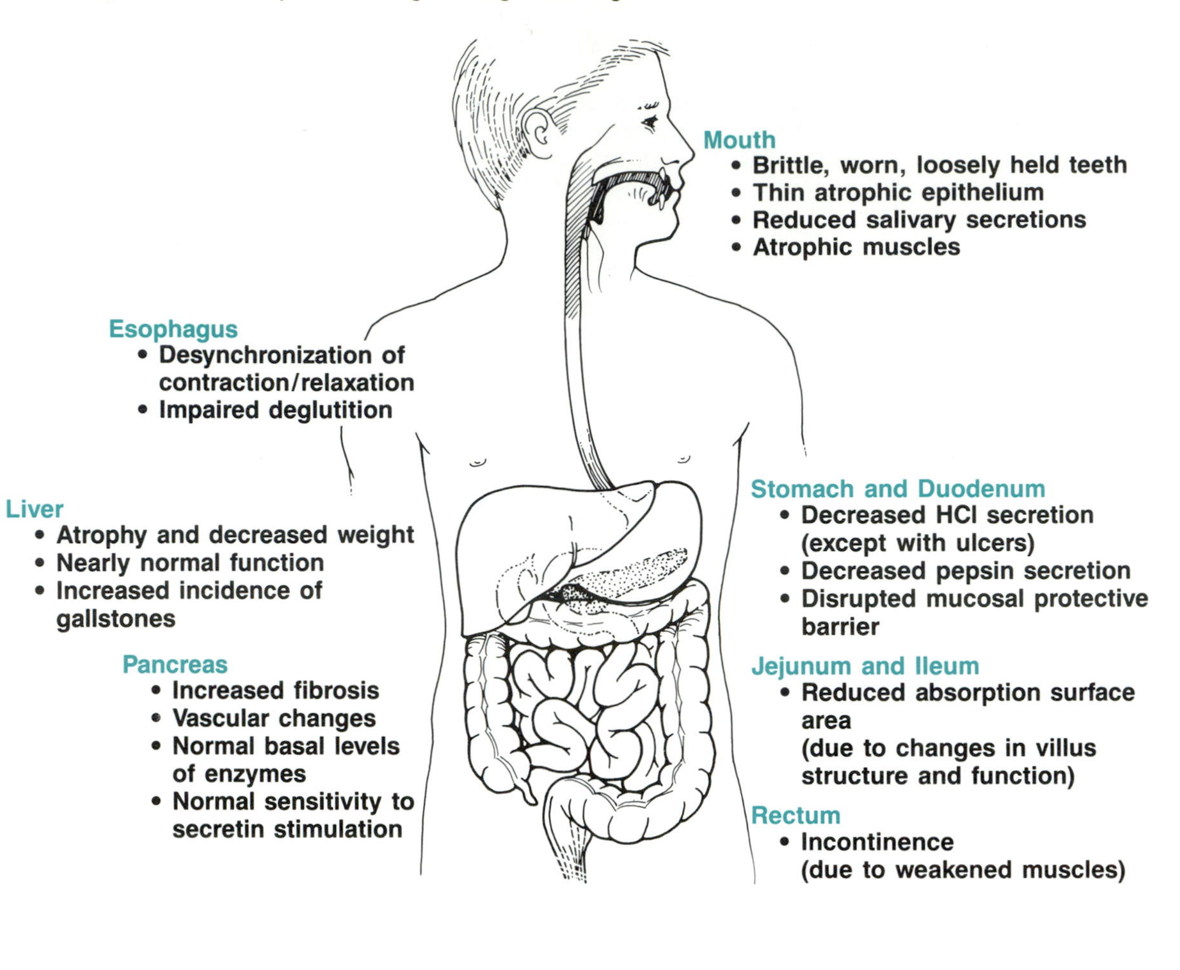

Mouth
• Brittle, worn, loosely held teeth
• Thin atrophic epithelium
• Reduced salivary secretions
• Atrophic muscles
Esophagus
• Desynchronization of contraction/relaxation
• Impaired deglutition
Liver
• Atrophy and decreased weight
• Nearly normal function
• Increased incidence of gallstones
Pancreas
• Increased fibrosis
• Vascular changes
• Normal basal levels of enzymes
• Normal sensitivity to secretin stimulation
Stomach and Duodenum
• Decreased HCl secretion (except with ulcers)
• Decreased pepsin secretion
• Disrupted mucosal protective barrier
Jejunum and Ileum
• Reduced absorption surface area (due to changes in villus structure and function)
Rectum
• Incontinence (due to weakened muscles)

AGING AND THE NERVOUS SYSTEM

Changes in Morphology

After the age of 60, normally aging human brain shows several morphologic changes:

- The **number of neurons decreases** in selected cortical regions.
- The **number of dendrites** and dendritic spines (site of synapses) **decreases.**
- Extracellular **senile plaques** appear occasionally, particularly in hippocampus, amygdala, and neocortex. They consist of an amyloid core surrounded by degenerating neuronal processes and astrocytes.
- **Intraneuronal tangles** (paired helical neurofilaments) appear, particularly frequently in individuals with Alzheimer's disease.
- There is an increase in the **number of glial cells.**
- **Axons** begin to be **demyelinated** and to swell.
- There is **less vigorous repair** of neuronal loss and damage and a **decline in neural plasticity.**
- Within neurons the number of **neurotubules decreases** and **lipofuscin accumulates.**
- Vascular changes are dominated by the appearance of **atherosclerotic lesions.**

Effects on Biochemistry

Global

Global changes that accompany normal aging are restricted to **decreased water content and decreased lipid synthesis.**

The metabolic rates for O_2 and glucose remain normal if there is no pathology.

Regional

During normal aging, many brain regions show patchy decreases in protein content. They represent **decreases in the levels and activities of several neurotransmitters and enzymes** at pre-, trans-, and postsynaptic sites.

Although the chemical changes may be subtle, the resulting **regional imbalance of neurotransmitters** may have profound functional consequences.

Effects on Function

Although the age-related changes in chemical composition and in the number of neurons are moderate and occur in regionally circumscribed patterns, their functional consequences affect **motor behavior, sleep patterns, and memory functions.** In addition, the changes that take place in peripheral nervous structures bring about age-dependent changes in **somatic sensation, special senses, and autonomic function.**

AGING AND THE NERVOUS SYSTEM

Changes in Central Neurotransmitter Synthesis (S) and Receptor Density (R)

	Acetylcholine	GABA	Dopamine	Norepinephrine	Serotonin
Basal ganglia	S: ↓ or ↔ R: ↓ or ↔	S: ↓ R: ↓	S: ↓ or ↔ R: ↓	S: ↔ R: ↔	
Cerebellum		S: ↓ R: ↑			
Hypothalamus	S: ↓	S: ↓	S: ↔	S: ↓	
Midbrain	S: ↓	S: ↓	S: ↓ or ↔	S: ↓	
Cortex	S: ↓ or ↔ R: ↔	S: ↓ R: ↔	S: ↔	R: ↔	R: ↔
Hippocampus	S: ↓	S: ↓	S: ↔		

Functional Consequences of Structural and Chemical Changes

Central

Loss of Neurons in:	Affects:
Basal ganglia Cerebellum Spinal motor neurons	Motor control
Locus ceruleus	Sleep patterns
Nucleus basalis of Meynert	Dementia?
Auditory cortex	Sound discrimination and sound perception

Peripheral

Change in:	Affects:
Connective tissue Peripheral nerve structures	Touch Pain perception Temperature perception
Optical properties of lens Loss of retinal cells	Vision
Cochlear hair cells	Hearing
Epithelium	Smell Taste
Autonomic receptors	Autonomic nervous system

Changes in Somatic Sensations with Age

Touch

Changes occur in mechanical and nervous aspects of touch sensitivity:

- **Increasing stiffness of connective tissue attenuates transmission of high-frequency vibrational stimuli** to cutaneous receptor organs (Meissner's and Pacinian corpuscles).
- **Changes in receptor organs,** their connecting fibers, or central nervous projections cause decreased sensitivity to all frequencies of periodically applied tactile stimuli.
- As a result of **central nervous changes,** there is a progressive loss in the ability to recognize the shapes of objects by touch only, and
- There is a decrease in the minimum distance by which two simultaneously applied point stimuli can be physically separated and still be perceived as two separate points.

Pain perception

The question of changes in pain perception with age is unresolved because of such complicating distractors as tolerance, emotion, and habituation.

Changes in Somatic Sensations With Age

Touch

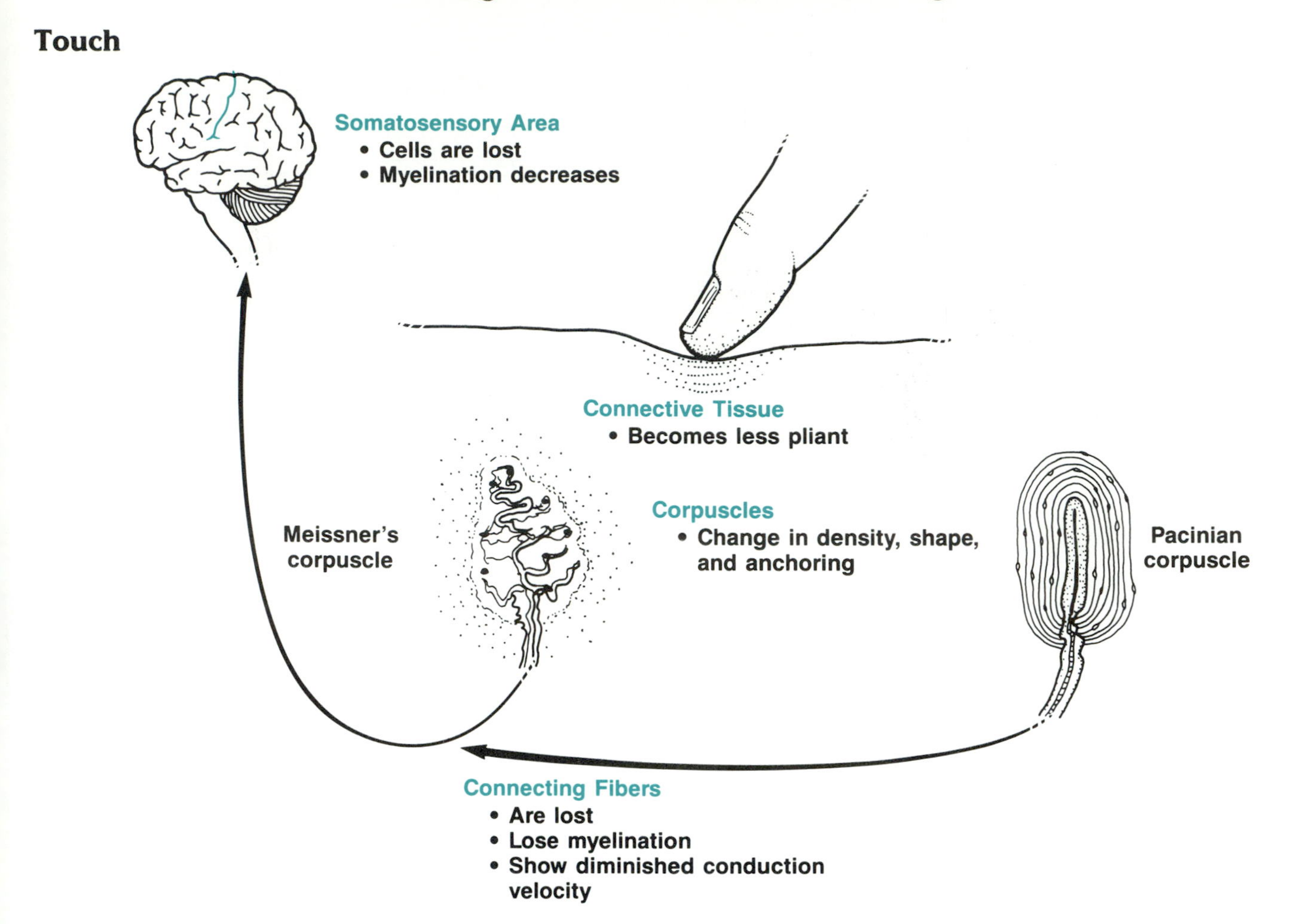

Changes in Special Senses With Age

Vision

Changes occur in optical aspects as well as in neural aspects, with five significant functional consequences:

1. As a result of changes in the refractive properties of the cornea and lens, there is greater likelihood of **astigmatism,** and the point of **near vision recedes** so that practically everyone over age 60 requires corrective lenses for near vision.
   (The point of near vision is the minimum object-to-eye distance required for formation of a clear image. It is 9 cm at age 10, but 84 cm at age 60.)
2. Thickening of the lens reduces the power of **accommodation.**
3. **Color perception is altered** because changes in the cornea and lens cause increased light scattering in the blue and yellow ranges.
4. **Visual threshold and acuity decrease** because of both pupillary constriction and changes in the retina, optic nerve, and visual cortex.
   (Visual threshold is the minimum amount of light necessary to perceive an object.)
5. Mechanical **obstruction of the Canal of Schlemm** by the thickening lens occurs.

Hearing

Age-related **hearing loss** (presbycusis) **is common** and may progress at different rates in the two ears. It involves **changes in hair cells and in central nervous structures.** As a result:

1. **The highest sound frequency that can be heard decreases** progressively with age. (It is 20 kHz at age 10 and normally declines to 4 kHz by age 80, but declines sooner in those who are often subjected to high sound levels.)
2. **Speech perception is altered because the elderly can hear vowels better than consonants.** (Vowels are produced by low-frequency sounds, whereas consonants are produced by high-frequency sounds.) Since consonants make speech intelligible and vowels make it audible, the elderly often complain that **they can hear but they cannot understand.**
3. **There is a decline in the ability to isolate some sounds and mask others** (for instance, when attempting to focus on one conversation at a party).

AGING AND THE NERVOUS SYSTEM—CONT'D.

Changes in Special Senses With Age

Vision

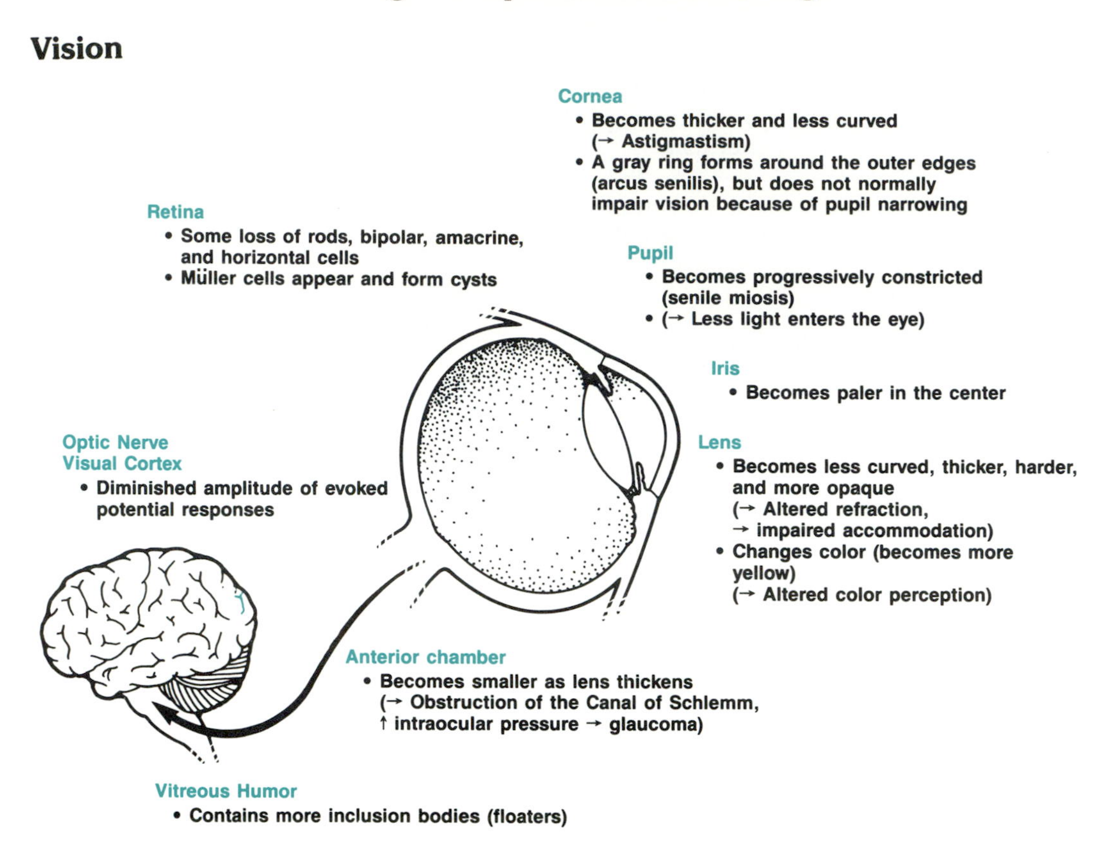

Cornea
• Becomes thicker and less curved (→ Astigmastism)
• A gray ring forms around the outer edges (arcus senilis), but does not normally impair vision because of pupil narrowing
Retina
• Some loss of rods, bipolar, amacrine, and horizontal cells
• Müller cells appear and form cysts
Pupil
• Becomes progressively constricted (senile miosis)
• (→ Less light enters the eye)
Iris
• Becomes paler in the center
Optic Nerve
Visual Cortex
• Diminished amplitude of evoked potential responses
Lens
• Becomes less curved, thicker, harder, and more opaque (→ Altered refraction, → impaired accommodation)
• Changes color (becomes more yellow) (→ Altered color perception)
Anterior chamber
• Becomes smaller as lens thickens (→ Obstruction of the Canal of Schlemm, ↑ intraocular pressure → glaucoma)
Vitreous Humor
• Contains more inclusion bodies (floaters)

Hearing

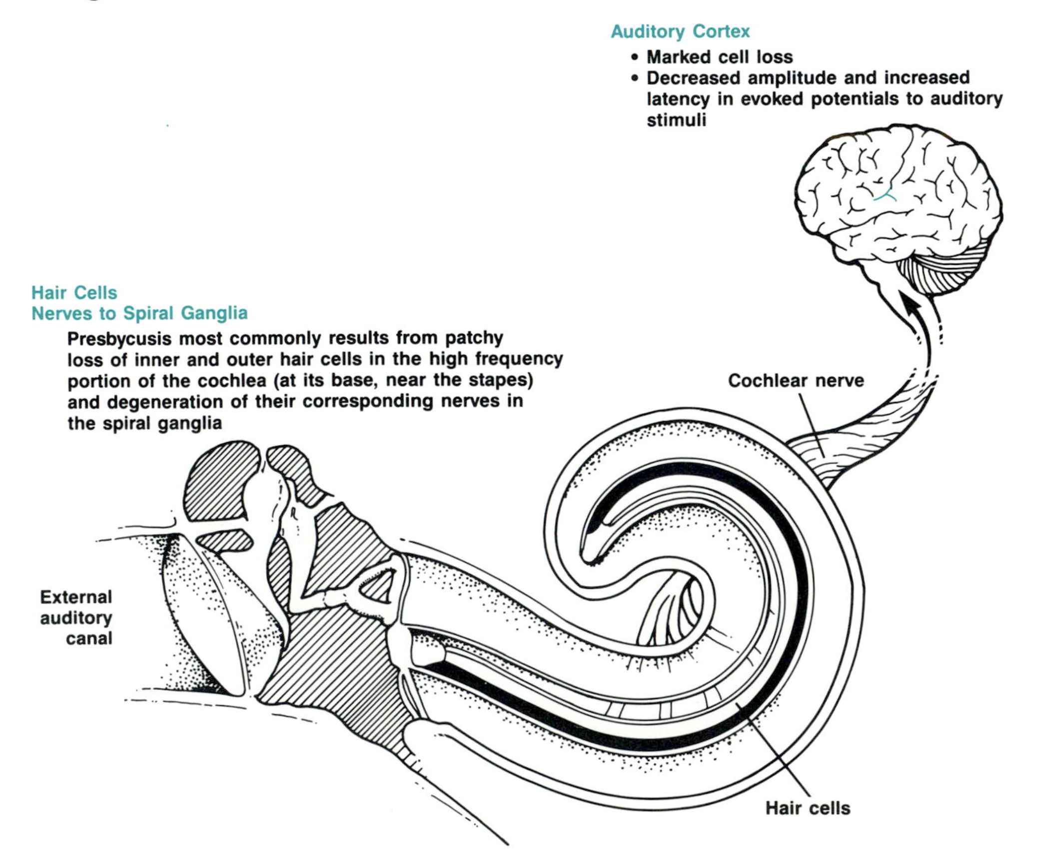

Auditory Cortex
• Marked cell loss
• Decreased amplitude and increased latency in evoked potentials to auditory stimuli
Hair Cells
Nerves to Spiral Ganglia
Presbycusis most commonly results from patchy loss of inner and outer hair cells in the high frequency portion of the cochlea (at its base, near the stapes) and degeneration of their corresponding nerves in the spiral ganglia
Cochlear nerve
External auditory canal
Hair cells

Changes in Special Senses With Age—Cont'd.

Olfaction

As a result of peripheral and central nervous changes, **olfactory discernment declines** with age. The rate of decline is not the same for all odors, and it is quite slow until the mid-sixties, but becomes marked after that:

- Higher concentrations of substances are required before they can be detected.
- It becomes more difficult to identify the constituent odors in a complex fragrance.

Taste

Sensitivity and discrimination decline with age in this chemical sense just as they do in the others. This is particularly true in the case of salty flavor.

AGING AND THE NERVOUS SYSTEM — CONT'D.

Changes in Special Senses With Age—Cont'd.

Olfaction

Taste

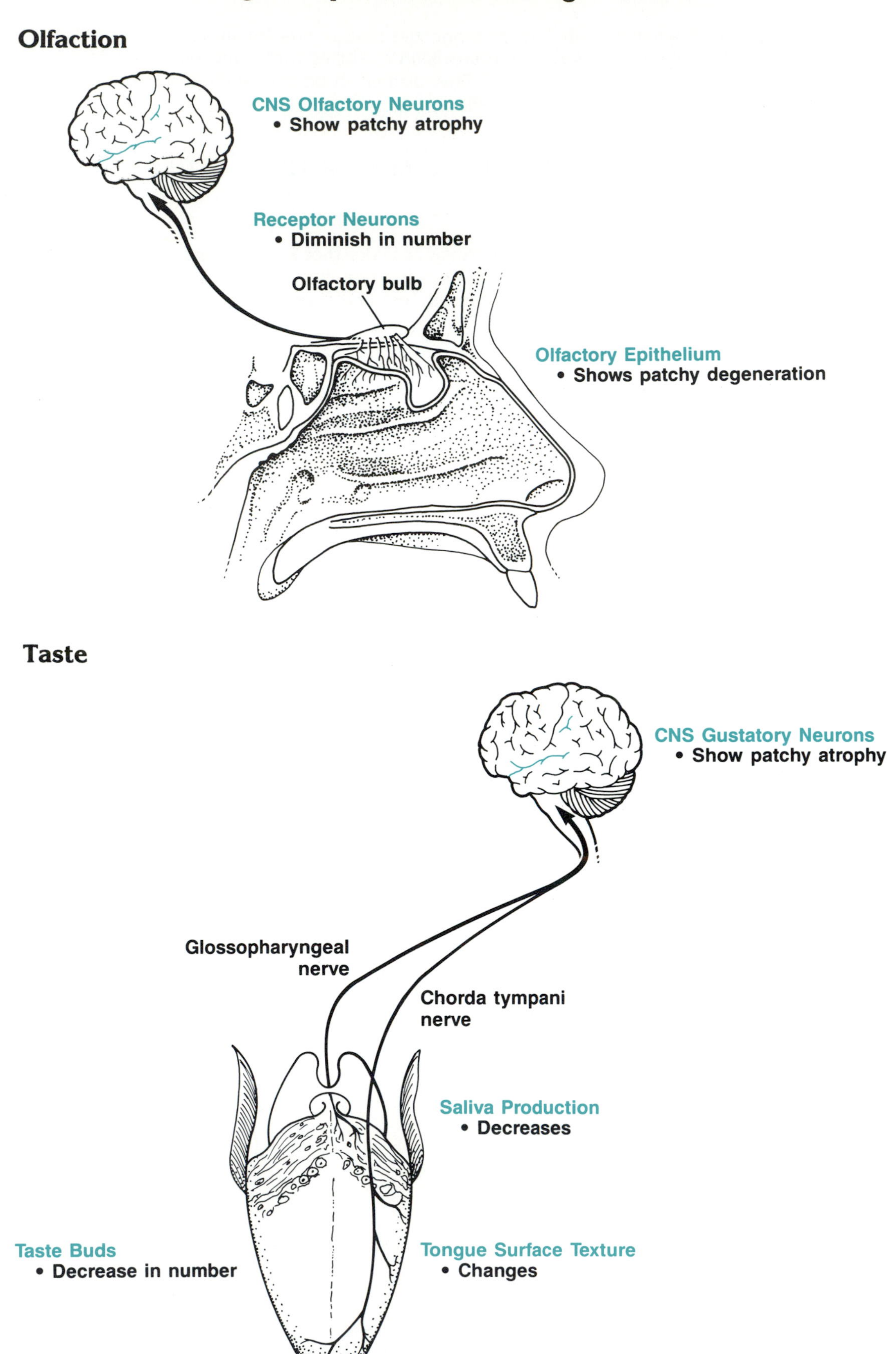

AGING AND THE NERVOUS SYSTEM—CONT'D.

Changes in Motor Functions With Age

Even in the absence of pathology in connective tissue, muscles, and joints, normal aging is accompanied by **diminished speed and precision of skilled motor movements.** Changes in the central nervous system are responsible. They also affect posture and gait, leading to **uncertain balance and increasing incidence of falls.**

CHANGES IN HIGHER NERVOUS FUNCTIONS WITH AGE

Memory

Aging brings a **decline in the performance of tasks that have a significant recent-memory component.** The central nervous cause of this memory loss is thought to be related to neurotransmitter physiology (particularly **vasopressin** and **cholecystokinin**).

Sleep and wakefulness

Sleep patterns change with normal aging:

- It takes longer to fall asleep.
- The number of wakenings per night increases.
- The time of deep sleep is markedly reduced.

All of these changes probably result from changes in **serotonergic neurons** of the Raphe nuclei.

Changes in Motor Functions With Age

Reflexes of posture and gait are altered

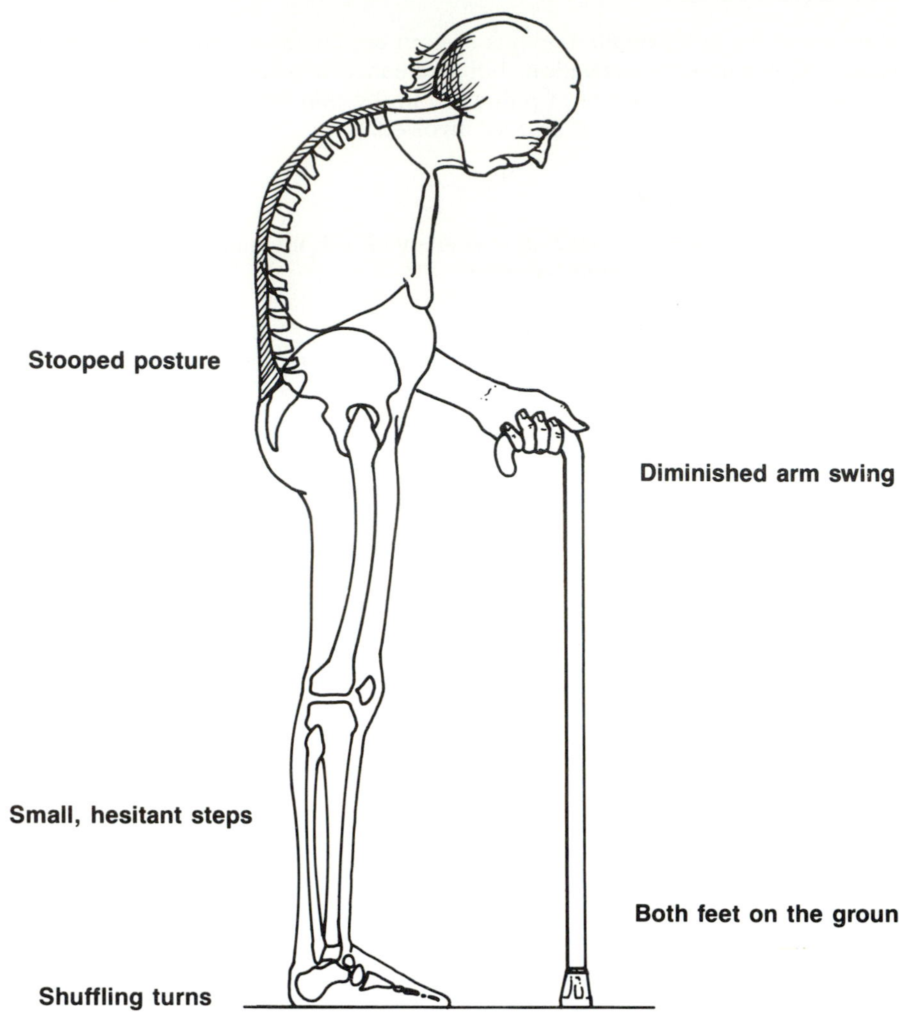

Changes in the Autonomic Nervous System With Age

Sympathetic nervous system

Age-related changes in the sympathetic nervous system are characterized by **augmented responses to situations that cause its activation,** but **decreased sensitivity of** its **target tissues.** The latter is most likely a consequence of **reduction in the number of adrenoreceptors.** Altered intracellular messenger dynamics may also be involved.

Parasympathetic nervous system

Vagal efferent activity in response to stimuli is preserved with increasing age. However, there is a **decline in sensitivity of some target tissues:**

- Cardiac muscarinic responses decline with age.
- Vascular and gastrointestinal muscarinic receptor responses remain unchanged.

AGING AND THE NERVOUS SYSTEM—CONT'D.

Changes in the Autonomic Nervous System With Age

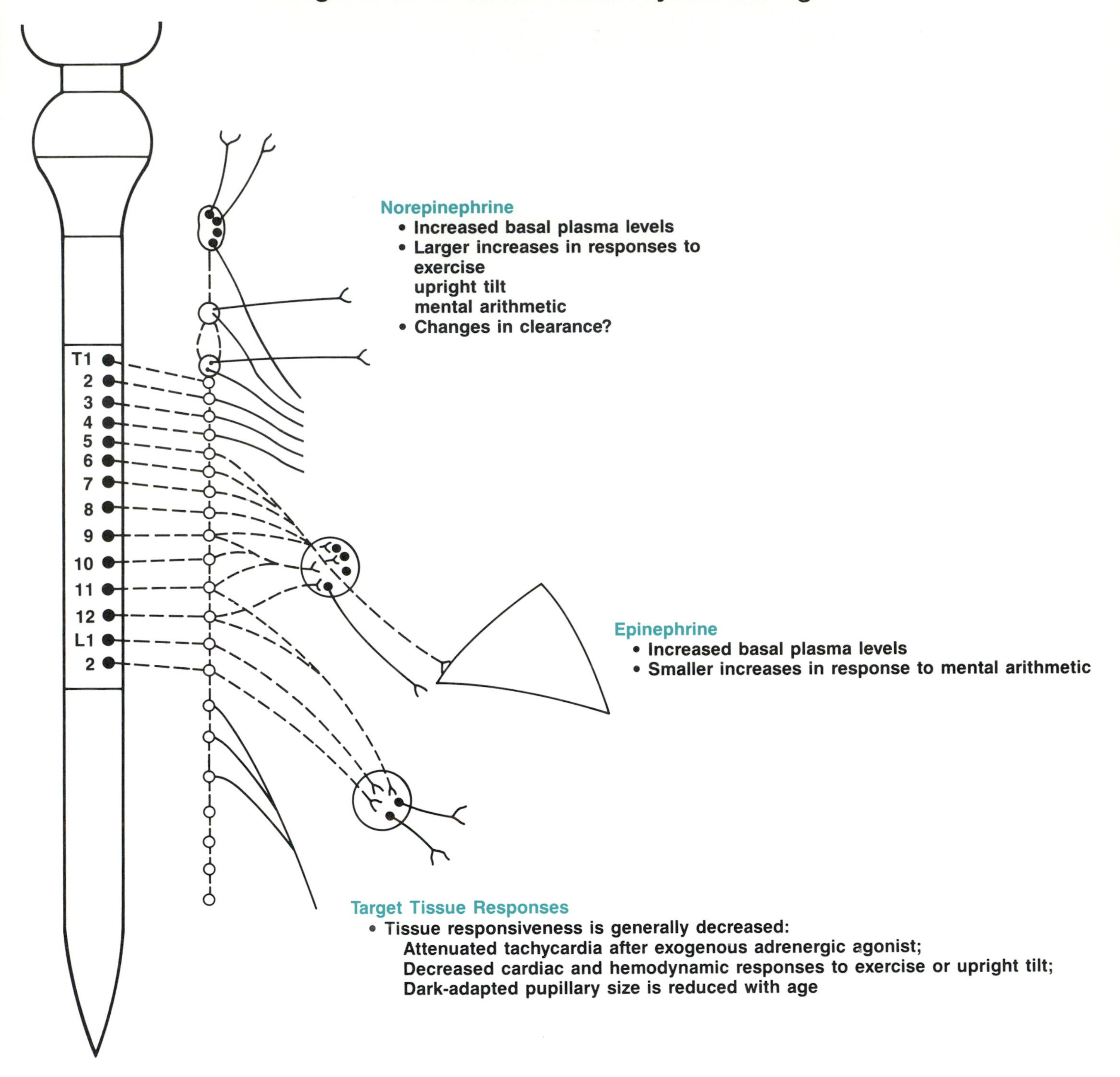

T1
2
3
4
5
6
7
8
9
10
11
12
L1
2
Norepinephrine
• Increased basal plasma levels
• Larger increases in responses to
exercise
upright tilt
mental arithmetic
• Changes in clearance?
Epinephrine
• Increased basal plasma levels
• Smaller increases in response to mental arithmetic
Target Tissue Responses
• Tissue responsiveness is generally decreased:
Attenuated tachycardia after exogenous adrenergic agonist;
Decreased cardiac and hemodynamic responses to exercise or upright tilt;
Dark-adapted pupillary size is reduced with age

INDEX

D

E

F

H

I

J

K

L

M

N

S

T

U

V

W

X

Y

Z